PHOTOSYNTHESIS,
PHOTORESPIRATION, AND
PLANT PRODUCTIVITY

PHOTOSYNTHESIS,
PHOTORESPIRATION, AND
PLANT PRODUCTIVITY

ISRAEL ZELITCH

Department of Biochemistry
The Connecticut Agricultural Experiment Station
New Haven, Connecticut

Academic Press   1971

New York and London

ACADEMIC PRESS, INC.
111 Fifth Avenue, New York, New York 10003

United Kingdom Edition published by
ACADEMIC PRESS, INC. (LONDON) LTD.
24/28 Oval Road, London NW1 7DD

LIBRARY OF CONGRESS CATALOG CARD NUMBER: 70-154386

PRINTED IN THE UNITED STATES OF AMERICA

Contents

SECTION I
Biochemical and Photochemical Aspects of Photosynthesis

SECTION II
Respiration Associated with Photosynthetic Tissues

Chapter 5. Dark Respiration and Photorespiration

Chapter 6. Glycolate Metabolism and the Mechanism of Photorespiration

210 ho

SECTION III
Photosynthesis and Plant Productivity in Single Leaves and in Stands

Chapter 7. Photosynthesis as a Diffusion Process in Single Leaves

Chapter 8. Environmental and Physiological Control of Net Photosynthesis in Single Leaves

Chapter 9. Relation of Photosynthesis, Total Respiration, and Other Factors to Control of Productivity in Stands

Preface

In this book I have attempted to provide advanced undergraduates, graduate students, teachers, and research workers in a number of disciplines with a perspective of photosynthesis and how it may be increased. Thus many processes other than photosynthesis that affect the productivity of plants have been examined from the standpoint of enzyme chemistry, chloroplasts, leaf cells, and single leaves. Finally, the function of all of these in a stand of plants growing outdoors and subject to the overriding influence of climate has been considered. I have had to deal with elements of physics, biochemistry, plant physiology, genetics, climatology, agronomy, and other sciences to varying degrees. Wherever possible I have tried to evaluate our more recent knowledge, to point out areas of uncertainty, and to indicate promising approaches for future investigation. Examples were taken from higher plants when they were available, although studies with algae and photosynthetic bacteria have also been examined.

Most of this book deals with the advances of the sixties, but sufficient historical background is provided so that the foundations on which this knowledge was based may be fully appreciated. The past decade may be remembered for the Green Revolution in which agricultural scientists demonstrated that with an input of suitable genetic material and increased nutrient supply the yield of certain crops which had only doubled in the previous one hundred years was at least doubled again. But scientists believe that the next large increases in productivity must come from the application of the newer findings about photosynthetic carbon assimilation. Fortunately, research during the last decade has created impressive advances in our understanding of these processes.

Some of the newer and often unexpected knowledge revealed during the sixties that I have considered includes the realization that rates of enzymatic reactions are often controlled by small regulator molecules; findings about the detailed structure and function of chloroplasts and other organelles and their variability; substantial evidence that photosynthesis requires at least two separate photoacts; the unraveling of many details about the mechanism of photosynthetic electron transport and photophosphorylation; the appreciation of the importance of diffusive resistances to carbon dioxide assimilation, especially the role of stomata; the grasping of the importance of dark respiration in diminishing productivity; the discovery that the great differences in net photosynthesis that occur between many species and varieties is largely caused by a newly recognized type of respiration—photorespiration turned on by light; and the coping with the complexities of illumination and climate within the real canopies made by the leaves of crops that feed us.

Readers may wonder how a biochemist became interested in these diverse subjects and even learned the language used by investigators in other disciplines. For this I must thank my colleagues at The Connecticut Agricultural Experiment Station with whom I have had the opportunity to discuss these important problems for many years. Thus by sometimes asking what might have appeared to be "foolish" questions, in due course I began to converse about these subjects relatively painlessly.

By sharing the fruits of these conversations with readers, it is my fondest hope that we may be able to work together more easily in order to alleviate the widespread suffering that will otherwise result from an inadequate world food supply.

ISRAEL ZELITCH

Acknowledgments

The orderly completion of this book involved the cooperation of a number of people. I wish to thank Dr. Graham Berlyn, Dr. Vernon E. Gracen, Jr., Mr. Fritz Goro, and Dr. Eldon H. Newcomb for providing illustrations, some of them before publication. Material in advance of publication was also received from Dr. William A. Jackson, Dr. Jack Preiss, and Dr. Peter Trip. Dr. Paul E. Waggoner kindly supplied unpublished data that was used extensively in Chapter 9. Miss M. Gregory, Jodrell Laboratory, Royal Botanic Garden, Kew, England, provided an analysis of earlier investigations about chloroplasts in the bundle-sheath cited in a footnote in Chapter 2.

Friendly critics in several departments of The Connecticut Agricultural Experiment Station (and nearby) offered suggestions about various chapters: Dr. Peter R. Day, Dr. Kenneth R. Hanson, Dr. Evelyn A. Havir, Dr. Jean-Yves Parlange (Engineering and Applied Science, Yale University), Dr. Raymond P. Poincelot, Jr., Dr. Ernest G. Uribe (Biology Department, Yale University), and Dr. Paul E. Waggoner. Two colleagues read the entire manuscript at one or more stages, and thus I am especially grateful for the counsel of Dr. H. B. Vickery (Biochemist Emeritus) and Dr. Gary Heichel (Department of Ecology and Climatology).

I have been greatly helped in many ways by Mrs. Ruth DiLeone, who drafted most of the figures and also typed the manuscript, and by Mr. John A. Marcucci, who assisted with library searches.

Conversion Units

LIGHT AND ENERGY

Illumination of 1 foot candle (ft-c) = 10.764 lumens m^{-2} (lux)
Total maximum incident sunlight = 10,000 ft-c = 108,000 lux
1 watt = 10^7 erg sec^{-1}
1 cal = 4.186 × 10^7 ergs
Maximum incident solar irradiation at earth's surface = 1.2 cal cm^{-2} min^{-1}
1 watt m^{-2} = 10^3 erg sec^{-1} cm^{-2} = 1.43 × 10^{-3} cal cm^{-2} min^{-1} (400 to 700 nm)
Maximum visible solar irradiation (400 to 700 nm) = 0.6 cal cm^{-2} min^{-1} = 42 × 10^4 erg sec^{-1} cm^{-2} = about 9 μeinsteins cm^{-2} min^{-1} absorbed by an average leaf

OTHER CONVERSIONS

1 nm = 10^{-9} m = 1 mμ = 10 Ångstrom (Å)
ppm of CO_2 = μl per liter by volume

300 ppm of CO_2 = about 9×10^{-6} M free CO_2 in solution in pure water
1 acre = 2.47 hectare (ha) or 2.47×10^4 m^2
1 pound per acre = 1.12 kg ha^{-1}
Wind speed of 100 cm sec^{-1} = 2 miles per hour

AVERAGE LEAF COMPOSITION

1 dm^2 projected area = 2.0 gm fresh weight of lamina = 3.0 mg chlorophyll = 6 mg protein N = 40 mg protein

Section I

Biochemical and Photochemical Aspects of Photosynthesis

1   *Morphology of Leaf Cells*

An aim of this monograph is to provide a basis for understanding the main factors concerned with regulating plant productivity in communities of plants. However, before one can evaluate how plants function in a stand, it is first necessary to examine the biochemical, physiological, and physical activities of single leaves. Hence it becomes necessary to study the nature of the kinds of cells in leaf tissues as well as the organelles they contain, and to examine especially the biochemical and physiological functions associated with these various parts.

Leaf cells possess the fundamental capabilities and carry out the processes common to all other living cells, but perhaps of greater importance in this discussion are the unique properties not encountered elsewhere. Some of these special attributes include the unusual cell wall structure surrounding the cell protoplasm; the effects of specific growth regulators; and the anatomic features specially adapted to the control of the uptake of inorganic salts and water and the movement of oxygen and carbon dioxide to and from the atmosphere. Higher plants are autotrophic organisms, obtaining most of their carbon compounds from atmospheric carbon dioxide; hence the photosynthetic apparatus is of special interest. The photosynthetic cells of higher plants are fairly autonomous and carry out large numbers of biochemical reactions, some of which result in the biosynthesis of unique natural products. Many such substances have still not been identified and many other compounds are known to be present for which nȯ metabolic role has yet been perceived. In this discussion, I shall consider only those biochemical and physiological aspects of higher plant metabolism that are believed to be closely related to productivity.

A. THE CUTICLE AND SURFACE WAXES

Certain anatomic features of leaves must first be considered. Examination of the cross section of typical leaf blades (Figs. 1.1, 1.2, 1.3) shows that a waxy layer covers the surface of the cuticle. These waxes, so-called because of their physical rather than their chemical characteristics, were studied intensively by A. C. Chibnall and his colleagues during the period between 1930 and 1950. They were able to isolate and identify many of the substances present by laborious fractional crystallization followed by X-ray analysis. Present-day techniques of separation and identification, especially thin-layer chromatography and gas chromatography, coupled with mass spectrometry, have made the task of identifying the compounds of leaf waxes much easier and more certain.

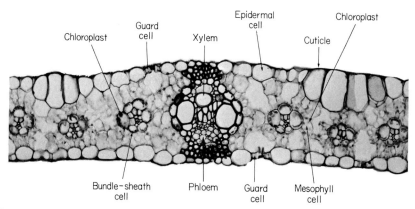

Fɪɢ. 1.1. Cross section of a maize leaf showing chloroplasts in the bundle-sheath as well as in the mesophyll (provided by G. Berlyn).

The cuticular waxes can be extracted by dipping a leaf into a solvent such as chloroform for 30 seconds. The surface wax of leaves may account for up to 4% of the fresh weight, and 50 mg/cm² may cover the leaf surface (Eglinton and Hamilton, 1967). An analysis of the composition of cabbage leaf wax obtained in this way is given in Table 1.1. The wax consists of a complex mixture of long-chain compounds, and 65% of the surface lipids in cabbage leaves consist of only seven different straight-chain compounds, all 29 carbon atoms in length (C_{29}) (Purdy and Truter, 1963). The rest are fatty acids, C_{12} to C_{24}; primary alcohols, C_{12} to C_{28}; and esters containing compounds of the same chain length as those in the free alcohols and acids. In the hydrocarbon fraction, one compound, nonacosane ($n\text{-}C_{29}$), accounts

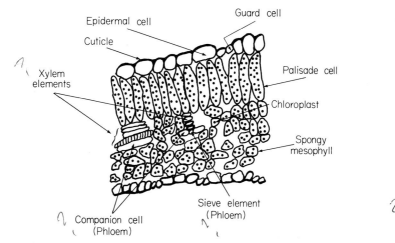

FIG. 1.2. Diagram of the cross section of a sugarbeet leaf showing the relation of the chloroplast-containing palisade and mesophyll cells (from material provided by P. Trip). A portion of the xylem and phloem is shown in longitudinal section.

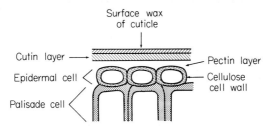

FIG. 1.3. Diagram of a leaf cuticle showing its relation to adjacent palisade cells (from Eglinton and Hamilton, 1967). (Copyright 1967 by The American Association for the Advancement of Science.)

TABLE 1.1   *Composition of Surface Lipids of Cabbage Leaf (Brassica oleracea)*[a]

Total lipids (%)		Distribution of hydrocarbon fraction (%)			
Hydrocarbons	36	n-C_{22}		n-C_{27}	1
Esters	13	n-C_{23}		n-C_{28}	1
Ketones	14	n-C_{24}	} 1	n-C_{29}	91
Secondary alcohols	11	n-C_{25}		n-C_{30}	1
Ketoalcohols	1	n-C_{26}		n-C_{31}	6
Primary alcohols	9	–		–	–
Acids	9	–		–	–

[a] From Purdy and Truter, 1963; Herbin and Robins, 1969.

for over 90% of the fraction. In most species the hydrocarbons in the surface waxes are saturated and consist predominantly of substances with an odd number of carbon atoms. However, in other species, branch-chained hydrocarbons are mostly present, but an odd-numbered chain from C_{25} to C_{35} is usually present in largest amount. Thus in the tobacco leaf, C_{31} predominates, and in ryegrass it is C_{29} and C_{31} hydrocarbons (J. D. Waldron *et al.,* 1961).

Studies on the mechanism of biosynthesis of components of the cuticular wax have been carried out by providing tissues with isotopically labeled precursors. The results of such experiments have recently been summarized (Kolattukudy, 1968, 1970). Originally, it was postulated that a C_{29} hydrocarbon could arise from the α-oxidation of palmitic acid (C_{16}) to produce pentanoic acid (C_{15}), followed by a condensation of two such identical fatty acids to produce a C_{29} ketone. The most probable ketone, 15-nonacosanone, is, in fact, the predominant ketone in cabbage wax, and it would then presumably be further reduced to the alcohol and finally to the corresponding paraffin as shown in the following sequence:

$$2\ C_{15}\text{—COOH} \xrightarrow{2\ CO_2} 2\ C_{14}\text{—COOH} \xrightarrow{CO_2} C_{14}\text{—CO—}C_{14}$$

Palmitic acid   Pentanoic acid   15-Nonacosanone

$$C_{14}\text{—}CH_2\text{—}C_{14} \longleftarrow C_{14}\text{—CHOH—}C_{14}$$

Nonacosane   15-Nonacosanol

However, recent evidence shows that when acetate-1-^{14}C or -2-^{14}C is supplied to cabbage leaves through the petiole or to leaf fragments, it is readily incorporated into the various fractions of the surface wax including the C_{29} hydrocarbon. The carbonyl carbon of the C_{29} ketone arose exclusively from the methyl carbon of acetate, suggesting that these long-chain compounds are produced by an elongation of preexisting fatty acids followed by a decarboxylation (Kolattukudy, 1965). Furthermore, palmitic acid labeled in the carboxyl carbon was incorporated into C_{29} hydrocarbon in cabbage leaves as effectively as was uniformly labeled palmitic acid. This suggested that palmitic acid is incorporated as a unit and subsequently elongated by successive C_2 units, finally being decarboxylated to yield the appropriate hydrocarbon (Kolattukudy, 1966):

$$CH_3\text{---}COOH \longrightarrow \longrightarrow C_{15}\text{---}COOH \xrightarrow{+(C_2)_7} [C_{28}\text{---}CH_2\text{---}COOH]$$

Acetic acid   Palmitic acid C_{30} acid

$$\downarrow\searrow \quad CO_2$$

$$CH_3\text{---}(CH_2)_{27}\text{---}CH_3$$

Nonacosane

Hydrocarbons are possibly synthesized in leaves by the condensation of nonidentical fatty acids, one serving as an acceptor and the other as a donor to produce a C_{30} acid equivalent, followed by a decarboxylation (Kolattukudy, 1968). Although much is known about the biosynthesis of the paraffins, a full description of the details of the mechanism of the biosynthesis of these long-chain compounds must await the availability of active cell-free systems that can carry out such reactions. The synthesis of these compounds likely occurs under the influence of a complex of enzymes in which carriers of intermediates are sequentially involved in such a manner that the intermediates do not dissociate from the complex until the end product is produced. More than one mechanism of elongation may also exist in leaves, since Kaneda (1969) and Herbin and Robins (1969) have shown that the hydrocarbons present in the internal tissues of the leaf have a very different composition from those on the surface. The range of compounds present inside the leaf is much broader, and in addition there is no preference for odd-numbered compounds as is observed in the surface wax of the same species.

The synthesis of waxy esters has been successfully investigated with partially purified enzyme preparations from leaves (Kolattukudy, 1967). Both C_{16} and C_{18} alcohols react by three distinct biochemical mechanisms with acyl compounds, mainly C_{16}, to produce the corresponding esters.

The paraffins in the cuticle are probably synthesized in the epidermis itself, and chloroplasts are likely not involved in these reactions except in the synthesis of C_{16} and C_{18} fatty acid precursors (Chapter 2.F) (Kolattukudy, 1968). For instance, the epidermis of *Senecio odoris* produced waxy esters as effectively as the entire leaf tissue, although the tissue with its epidermis removed synthesized esters at less than 10% of the rate of the whole tissue. There is uncertainty whether these long-chain compounds are extruded onto the leaf surface through pores in the cuticle, since some investigators claim to have evidence for the existence of pores while others state they are lacking. The waxy substances may move out through channels in the cuticle. These waxes are important in controlling the water balance of the leaf, in preventing mechanical damage, and in providing resistance to insects and invading fungi. Their presence makes the leaf interior largely impervious to diffusion by O_2 and CO_2.

Further reasons for the high diffusive resistance of leaf cuticles to O_2 and CO_2 are shown in Fig. 1.3. Just beneath the wax there is a layer of cutin, composed of cross-linked hydroxy fatty acids containing some waxy substances imbedded within it. This is then underlain by a layer of pectin, a complex polysaccharide consisting primarily of polygalacturonic acid residues, before one comes to the cell wall surrounding the epidermal cells.

Clearly this mixture of long-chain substances on the leaf surface, complex in structure and relatively uniform in composition for a given species, will continue to hold interest because of the importance of the chemical nature of the leaf surface in a number of physiological processes.

B. EPIDERMAL CELLS

These cells support the outer cuticle, and in dicotyledonous plants they are adjacent to the palisade cells on the upper surface. In all species they are located next to the mesophyll cells that occupy the leaf interior. Epidermal cells, with one important exception, generally do not contain chloroplasts. The exception is the guard cell in the epidermis which does possess chloroplasts, though they are not well developed, and this feature is undoubtedly concerned with the ability of the stomata on the outside of leaves to open in the light (Chapter 7.C). Because of the impervious character of the surface wax, the stomata play a vital role in the diffusion of carbon dioxide during photosynthesis as well as in the escape of water vapor from the leaf by transpiration. These diffusion processes take place through the stomatal pores which may number about 10,000/cm² of leaf surface. The pores may be 10 μ in diameter and 20 μ to 30 μ in length when open; their width is usually less than 1 μ when closed.

C. THE CELL WALL

Plant cell walls provide rigidity to the cells, and may consist of 20 to 60% of cellulose, an unbranched polysaccharide consisting of β-1,4-D-glucopyranose residues. Cellulose occurs in the form of microfibrils of indeterminate length and 50 to 250 Å in diameter. Plant cell walls also contain other polysaccharides (consisting of residues of D-mannose, D-galactose, D-xylose, and L-arabinose), lignin, and protein components that yield an unusually high proportion of hydroxyproline. The mechanism of synthesis of cellulose and other polysaccharides in plants has been studied extensively by Hassid (1967) and his colleagues. As is often the case for the synthesis of many complex carbohydrates, including glycolipids (Chapter 2.F), a nucleoside diphosphate sugar is the immediate precursor of cellulose:

$$\underset{\substack{\text{Guanosine} \\ \text{triphosphate}}}{\text{GTP}} + \alpha\text{-D-glucose-1-phosphate} \xrightarrow{\text{Pyrophosphorylase}} \underset{\substack{\text{Nucleoside diphosphate} \\ \text{sugar}}}{\text{GDP-D-glucose}} + \underset{\text{Pyrophosphate}}{\text{PPi}} \quad (1.1)$$

The nucleoside diphosphate sugar, guanosine diphosphate, then reacts with an acceptor in the presence of a second enzyme, the transferase, to produce cellulose:

$$n(\text{GDP-D-glucose}) + \text{acceptor} \xrightarrow{\text{Transferase}} \underset{\text{Cellulose}}{\text{acceptor-}(\beta\text{-1,4-D-glucose})_n} + n(\text{GDP}) \quad (1.2)$$

Soluble extracts of mung bean seedlings or leaves of spinach, buckwheat, mustard, and parsley contain a pyrophosphorylase capable of forming ^{14}C-labeled GDP-D-glucose from GTP and α-D-glucose-1-phosphate. The transferase reaction, in which cellulose is produced in the presence of an acceptor, has recently been studied in particulate preparations obtained from mung bean and lupine seedlings (Flowers *et al.,* 1969). However, there is still some doubt about whether the transferase produces only cellulose.

The biosynthesis of plant pectin has also been accomplished with particulate preparations isolated from mung bean seedlings (Villemez *et al.,* 1966). Pectic substances are linear polymers consisting primarily of D-galacturonic acid residues in α-1,4-glycosidic linkage with the carboxyl groups on C-6 frequently esterified with methyl alcohol. Again a pyrophosphorylase and a transferase mechanism participate in the synthesis of the α-1,4-linked polymers of D-galactopyranosyluronic acid. Although pectins are usually partially esterified with methyl groups, the enzyme preparations that produce pectin are unable to utilize a methyl-esterified galacturonic acid derivative of a nucleoside diphosphate as a substrate. Hence the mechanism of methylation is uncertain. Pectins are present in the middle lamella, where they function as a cementing substance between cells. Plant cell walls also contain other noncellulose polysaccharides, and the biosynthetic mechanism of substances such as mannans, galactans, xylans, and araban has not been studied.

Lignin, a polyphenolic substance, is also associated with plant cell walls. The synthesis of lignin-like substances has been achieved by the polymerization and co-polymerization of *p*-hydroxylated cinnamyl alcohols, especially coniferyl alcohol, in the presence of a polyphenol oxidase (laccase) obtained from mushrooms (S. A. Brown, 1966). The enzyme peroxidase is

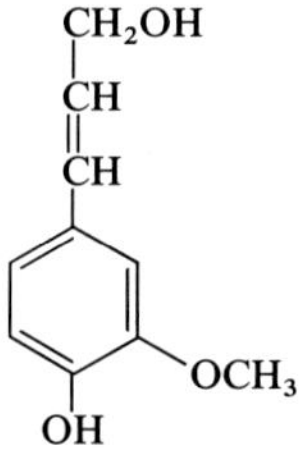

CH_2OH
CH
CH

OCH$_3$

OH

Coniferyl
alcohol

also known to catalyze the polymerization of a large number of phenols, and this reaction produces lignin-like products. In woody plants, according to studies with isotopically labeled precursors, all the enzymes necesssry to convert L-phenylalanine to lignin appear to be present, but so far the individual enzymes responsible for the postulated steps in the pathway have not been isolated except for the first step catalyzed by L-phenylalanine ammonia-lyase which has received considerable attention (Hanson and Havir, 1970):

$$\text{L-Phenylalanine} \xrightarrow{\overset{\text{Phenylalanine}}{\text{ammonia-lyase}}} \textit{trans-}\text{cinnamic acid} + NH_3 \qquad (1.3)$$

As shown in Fig. 1.4, the protoplast (the cytoplasm and nucleus of the cell) is surrounded by a cellulose cell wall, and the adjacent parenchyma cells are held together by the layer of pectic substances. Tobacco leaf tissue, as well as leaves from eighteen other species, have been treated with a crude polygalacturonase preparation to hydrolyze the pectin thereby freeing the mesophyll cells (Takebe *et al.,* 1968; Otsuki and Takebe, 1969). The isolated mesophyll cells were suspended in a cellulase solution to hydrolyze their cell walls, and intact protoplasts which were liberated retained their normal appearance for 2 to 3 days. Such preparations may be useful in biochemical studies where it is desirable to have a homogeneous population of higher plant cells in which only the plasma membrane is present to contain the cytoplasm.

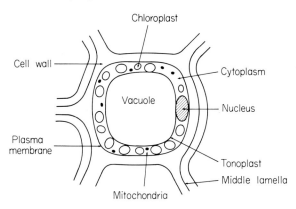

FIG. 1.4. Representation of a leaf parenchyma cell showing the relation of the cytoplasm and its organelles to the remainder of the cell.

D. MEMBRANE BARRIER SYSTEMS

The protoplasts of cells in the mesophyll are covered by a plasma membrane 100 to 150 Å in thickness. The lipoprotein nature of this membrane has been demonstrated chemically as well as by electron microscopy. The tonoplast, the membrane separating the cytoplasm from the vacuole, is presumably similar in composition. The more soluble a substance is in nonpolar solvents, the greater is the rate at which the substance can enter the cytoplasm and find its way into the vacuole. Small molecules tend to penetrate cells more rapidly when they are water soluble. Thus it can be inferred that the barrier system incorporates a lipid phase and a mosaic of pores filled with water through which water-soluble substances can also diffuse, since ions, water, and organic solutes must be transported across these membranes.

A number of models have been proposed to account for the solubility properties of solutes in these membranes. Clearly membrane permeability can be altered in lipoproteins either by changing the protein conformation or by effecting a separation of the hydrocarbon chains of the lipid layers.

The addition of lipids such as mono- and digalactosyl diglycerides (Chapter 2.C) to the root medium of bean plants strongly increased chloride transport as well as water transport to all plant parts (Kuiper, 1969). The diffusion of ions across the plasma membrane of the absorbing cells is the rate-limiting process in ion absorption in roots and not the rate of diffusion of ions to the plasma membrane (Epstein, 1969). Presumably the plasma membrane also plays some role in influencing the rate of CO_2 diffusion into the mesophyll cell during photosynthesis and in the translocation of organic solutes, but the manner in which these membranes function in these processes is not yet understood.

Membranes are present not only at the surface of the cytoplasm and next to the vacuole, but also at the surfaces of mitochondria, nuclei, chloroplasts, and peroxisomes and can also be seen inside mitochondria and chloroplasts (Figs. 1.5, 2.1) as an integral part of these organelles.

E. THE VACUOLE

The vacuole in plant cells is a large liquid inclusion that is separated from the cytoplasm by the tonoplast membrane. The tonoplast membrane has specific properties so that the inward and outward diffusion of solutes may occur at different rates. Thus many compounds are found at much higher concentrations in the vacuole than in the cytoplasm. For example, the vacuole of the tobacco leaf may contain malate at a concentration of 0.1 M, but one can add low concentrations of malate to preparations of leaf tissue and demonstrate that the externally supplied metabolite is readily oxidized in preference to the malate present in the vacuole. In this way, the vacuole provides the locus for a biochemical compartmentation of substrates that may be of considerable importance in regulating metabolic pathways (Chapter 4.M).

F. CYTOPLASMIC ORGANELLES

Studies of techniques for the isolation of cytoplasmic organelles from leaves have received considerable attention in recent years, and many metabolic processes can now be duplicated at reasonable rates in the laboratory with purified organelle preparations from leaves. Special precautions are necessary to obtain such biochemically active preparations, because difficulties may be encountered from substances present in the vacuole as well as the cytoplasm once the cells are broken. Leaf cells are usually first disrupted by grinding in a mortar, blending with a mechanical blender, or by cutting with razor blades. Once the cells have been broken in a suitable medium, the organelles are separated by differential centrifugation. Frequently further purification is carried out by density-gradient centrifugation with sucrose or another suitable solute. Some of the problems connected with devising a proper medium for disrupting leaf cells and for resuspending the organelles have been summarized by Öpik (1968).

The vacuole of leaf cells is usually acidic, and a pH of 4.0 or lower may be encountered in some species. Thus grinding media are generally adjusted to a high pH, up to 8 or 9, to neutralize the cell acidity. Reductants, such as cysteine, ascorbate, or citrate, are frequently added to inhibit oxidation of polyphenols. The oxidation of polyphenols can be troublesome because the resulting quinones react rapidly with amino groups on proteins and may cause considerable alteration of enzymatic activity. To avoid this difficulty, activated carbon (Zelitch and Zucker, 1958), polyvinylpyrrolidone (W. D. Loomis and Battaile, 1966), a substance which forms complexes with phenols, and Dowex-1 anion exchange resin (Lam and Shaw, 1970) have been added to grinding media. Bovine serum albumin has also been used to bind free fatty acids. Ethylenediaminetetraacetate (EDTA) is often added to chelate extracted metal ions, while specific metal cations such as magnesium are usually included. To maintain the osmotic properties of the organelles, solutes such as sucrose, NaCl, or mannitol, are always a part of the grinding and suspending media in concentrations ranging from 0.25 to 0.50 M.

1. RIBOSOMES

Ribosomes, the smallest particles that have been identified by electron microscopy, are about 200 to 300 Å in diameter. They generally contain roughly equal amounts of RNA and protein (J. Bonner, 1965). They are isolated from homogenates by first removing larger particles at lower centrifugal forces followed by centrifugation at 105,000 g for 1 to 2 hours. Magnesium ions are always added to the medium to prevent the dissociation of the ribosomes into smaller subunits.

Lyttleton (1962) showed that leaves have two kinds of ribosomes. He concentrated the ribosome fraction from an extract of spinach leaves and then separated two distinct ribosome components in the ultracentrifuge. These fractions had sedimentation coefficients respectively of 83 S and 66 S (S refers to the Svedberg unit which is a function of molecular weight). He found that the smaller ribosome was present exclusively in purified spinach chloroplasts, and that it contained between 42 and 45% of RNA.

Further evidence that photosynthetic tissues contain at least two kinds of ribosomes comes from work on *Euglena gracilis*. This organism can be grown in the dark on organic carbon sources, and under these conditions it produces no chloroplasts. However, ribosomes can be extracted with sodium deoxycholate from purified chloroplasts present in light-grown organisms. In this way, a comparison was made of the nucleotide composition of the RNA derived from ribosomes from the chloroplast with those present exclusively in the cytoplasm (Brawerman and Eisenstadt, 1967). The base composition of the two kinds of ribosomes differed greatly (Table 1.2), with the chloroplast ribosomes possessing a much higher content of adenylic acid and less cytidylic acid.

Ribosomes play an essential part in protein synthesis, since attachment of the ribosome to a strand of messenger RNA is an essential step in polypeptide formation. Further work on protein synthesis by chloroplasts will be described later (Chapter 2.F). It is interesting from an evolutionary standpoint that ribosomes with a sedimentation coefficient of about 70 S (leaf chloroplast ribosomes) also occur in bacteria and in blue-green algae,

TABLE 1.2   RNA Nucleotide Composition of Chloroplast and Cytoplasmic Ribosomes[a]

	Moles/100 moles	
Nucleotide	Ribosomes from chloroplasts	Cytoplasmic ribosomes
Adenylic acid	30.7	22.7
Guanylic acid	27.1	29.5
Cytidylic acid	17.1	27.1
Uridylic acid	25.1	19.6
Ratio A/C	1.80	0.84

[a] From Brawerman and Eisenstadt, 1967.

while those of 80 S (leaf cytoplasmic ribosomes) are found in yeasts, fungi, and mammalian tissues.

2. MITOCHONDRIA

Mitochondria vary in width from less than 1 μ to several μ. They have a smooth outer membrane, about 50 Å in width, and a membranous inner structure consisting of baffles, or cristae (Fig. 1.5). The cristae in leaves do not appear to be as ordered in structure as in other tissues. Because mito-chondria consist to such a great extent of membranes, their composition is chiefly lipoprotein in nature, together with small amounts of RNA and DNA.

Biochemically active mitochondria can be readily isolated by differential centrifugation at 10,000 g after grinding leaf tissue in suitable media. Such preparations are usually contaminated with broken chloroplasts, and purified mitochondria can be obtained by further centrifugation in sucrose gradients (Pierpoint, 1962, 1963). Isolated mitochondria from leaves carry out the reactions of the citric acid cycle, electron transport through the cytochrome system, and oxidative phosphorylation with good efficiencies as evidenced by their P:O ratios. Mitochondria appear to be largely responsible for the "dark" respiration of leaves, and their activity is doubtless also responsible for part of the CO_2 evolution in photosynthetic tissues in the light (Chapter 5.A).

3. PEROXISOMES

These cytoplasmic particles, which are 0.5 to 1.5 μ in size, have been known to exist in animal tissues for some time (de Duve and Baudhuin, 1966); their existence and possible biochemical importance in leaf metabolism has only recently been discovered, partly as a result of the highly developed skills in electron microscopy (Fig. 1.5). Peroxisomes were previously de-scribed in mammalian liver and kidney tissue, where they were known to contain a high concentration of catalase. A flavoprotein oxidase activity is also associated with them.

Breidenbach and Beevers (1967) discovered similar microbodies, which they called glyoxysomes, in extracts of the endosperm of germinating castor bean. Upon sucrose-gradient centrifugation, these particles were found to be denser than mitochondria and contained the key enzymes of the glyoxylate cycle, isocitrate lyase and malate synthetase, as well as glycolate oxidase (a flavoprotein) and catalase activity (Breidenbach *et al.,* 1968). The glyoxysomes from castor bean endosperm were recently shown to contain RNA, although the nature of this RNA is unknown and no

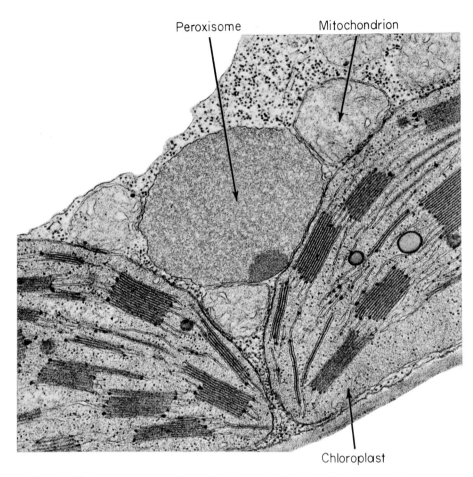

Fɪɢ. 1.5. Electron microscope picture of a tobacco leaf section showing a peroxisome containing a dense amorphous inclusion next to the limiting membrane (from Frederick and Newcomb, 1969a). The peroxisome is adjacent to three mitochondria (with poorly defined cristae) and two grana-containing chloroplasts ($\times$ 37,000). (Copyright 1969 by the American Association for the Advancement of Science.)

ribosomes have been detected in electron microscope pictures of peroxisomes (Gerhardt and Beevers, 1969); hence there is uncertainty whether this finding indicates contaminating fragments in the preparation.

Mollenhauer *et al.* (1966), in the course of electron microscopy studies, had observed the widespread occurrence of organelles that resembled animal peroxisomes in many tissues including species of algae, fungi, roots, and stems, as well as in leaves. The most characteristic morphological features of peroxisomes are their single limiting membrane and a finely granular matrix that occasionally shows dense regions of an amorphous or crystalline nature (Frederick *et al.,* 1968). The crystalline structure was shown by carefully controlled cytochemical tests to be associated with catalase activity (Frederick and Newcomb, 1969b). The crystalline inclusions are fewer in number or smaller in size in the particles from cells of the palisade layer, raising the question whether more than one kind of peroxisome is present in leaf tissue. They also appear to diminish in number as the cells mature. Peroxisomes were observed by electron microscopy to be a common component of the leaf mesophyll cells of bean, wheat, barley, tobacco, timothy, and other species (Frederick and Newcomb, 1969a). The organelle contains a single outer membrane and is frequently found pressed against chloroplasts (Fig. 1.5), and the structures are particularly large and numerous in leaf cells containing chloroplasts. The relative numbers of peroxisomes varies with plant species, and in some species there may be one-third as many peroxisomes as mitochondria.

Besides having been observed by electron microscopy, peroxisomes have also been isolated in reasonable yield from leaves of spinach, sunflower, tobacco, pea, and wheat (Tolbert *et al.,* 1968, 1969; Kisaki and Tolbert, 1969). A crude chloroplast fraction, obtained by centrifugation at 3000 *g* for 20 minutes contained the highest specific activity of glycolate oxidase, catalase, and NADH-glyoxylate reductase in comparison with other particulate fractions and the supernatant fluid. This chloroplast fraction was further fractionated on a sucrose density-gradient by layering the suspension on top of five layers of sucrose solution that varied in concentration from 1.3 *M* to 2.5 *M*. After centrifugation at from 45,000 to 107,000 *g* for 3 hours, the peroxisomes were found concentrated in a fraction that contained 1.9 *M* sucrose, and the three enzyme activities listed above were present in largest amount and with highest specific activity in this fraction (cf. Table 6.3). The leaf peroxisomes were somewhat fragile, and poor recoveries of intact organelles were obtained when NaCl or mannitol were substituted in the grinding media for sucrose. On the sucrose gradients, chlorophyll-containing particles were essentially completely separated from the peroxisomes, while enzymatic activities known to be associated with mitochondria, such as cytochrome c oxidase, were also completely absent from the peroxisome fraction.

These organelles may be important in the process of photorespiration, as well as in other biochemical pathways that originate from the oxidation of glycolate (Chapter 6.B). In endosperm tissue, the analogous particles, the glyoxysomes, are concerned with the biochemical conversion of storage lipids to sucrose through the glyoxylate cycle, which accomplishes a net synthesis of succinate from acetyl CoA units. The β-oxidation of fatty acids, which produces acetyl CoA, is also present in glyoxysomes (T. G. Cooper and Beevers, 1969). However, leaf tissues lack isocitrate lyase activity and thus cannot carry out the reactions of the glyoxylate cycle, which is restricted to fatty tissues (Carpenter and Beevers, 1959).

Before the discovery of plant peroxisomes these organelles were believed to function in "a special kind of respiration in which electrons are transferred to oxygen by way of hydrogen peroxide" and that they might be considered a "fossil organelle" remaining from the time when life originated under anaerobic conditions (de Duve and Baudhuin, 1966). Nevertheless, peroxisomes are present today, and their metabolic importance in many plant tissues is only beginning to be appreciated.

2 Chloroplasts and Their
Various Activities

The photosynthetic mechanism of green plant cells is contained in the chloroplasts, the chlorophyll-containing cytoplasmic organelles. Their importance is not restricted to photosynthesis, however, because chloroplasts are also the site of many biosynthetic reactions concerned with nucleic acid, protein, and lipid metabolism, and they serve as a storehouse of inorganic salts and of a portion of the end products of photosynthesis including carbohydrates. The functions of chloroplasts more directly related to photosynthesis are discussed later (Chapters 3 and 4). The structure and composition of chloroplasts, some of their activities *in vivo*, and related biochemical reactions associated with isolated chloroplasts are described here.

Higher plant chloroplasts are disk-shaped bodies from 5 to 10 μ in diameter and 3 to 4 μ in depth. Palisade cells in leaves of dicotyledonous plants contain the largest number of chloroplasts, as many as 200 per cell, while fewer are found in the cells of the spongy parenchyma of the meso- phyll. Chloroplasts are also found in the bundle-sheath parenchyma cells that surround the vascular bundles of small veins in certain species (Chapter 2.A) (Fig. 2.1 B), as well as in the guard cells that surround the stomata.

The membrane surrounding the chloroplast is about 100 Å thick, and by electron microscopy appears to be double. There are two main com- ponents within the membrane envelope: a lamellar system composed of membranous disks made up primarily of lipoprotein, and a matrix surround- ing it consisting of a granular stroma (Menke, 1966; von Wettstein, 1966). The lamellae are frequently stacked in groups called grana (or they may occur singly and are called stroma lamellae) which may contain from 15 to 60 lamellar disks, and the average chloroplast contains about 50 grana. The dimorphism in the structure of grana in chloroplasts of certain effi- cient photosynthetic species such as maize is not well correlated with efficient CO_2 fixation in all species (Chapter 6.E).

Various metabolic functions are distributed between the lamellae and the stroma of chloroplasts. The lamellae, for example, contain chlorophyll and carotenoids and are the site of the photochemical systems in photo- synthesis. The reactions concerned with oxygen evolution, photophos- phorylation, and the production of NADPH also take place in the lamellae. The stroma contains those enzymes associated with CO_2 uptake and the synthesis of amino acids, lipids, and proteins (Menke, 1962; von Wettstein, 1966). Plants grown in the dark contain proplastids, the precursors of chloroplasts, which completely lack the normal lamellar system. During exposure to light for about 5 hours the lamellar layers appear, although increases in certain enzymatic activities of maize plastids can be detected within minutes of exposure to illumination (Bogorad, 1967).

A. VARIATIONS IN CHLOROPLAST STRUCTURE

Chloroplasts in mesophyll cells in a given species are usually similar in appearance, although mutations are known in which the lamellar organiza- tion may be less well defined and where grana may even be lacking (Schmid and Gaffron, 1967). Certain species such as maize and sugarcane, however, have leaves that are anatomically unusual in that they possess a highly developed parenchyma bundle-sheath, containing specialized chloroplasts (Fig. 2.1 B), surrounding the xylem and phloem in the vascular bundles.

The vascular bundle-sheath cells are found only around the smaller or minor veins (Kiesselbach, 1949), and frequently contain starch-storing chloroplasts which lack grana. Adjacent mesophyll cells in these species contain smaller sized chloroplasts that sometimes store very little starch and possess large numbers of grana (Fig. 2.1 A).

The initial products of $^{14}CO_2$ fixation during photosynthesis appear to differ in the species with these two types of chloroplasts compared with species lacking bundle-sheath cells (Chapter 4.F), and there has been some speculation that a lack of photorespiration is associated with the presence of these bundle-sheath chloroplasts (Chapter 6.E).

The sheath around the vascular bundles is frequently arranged in layers consisting of an inner parenchyma sheath and an outer sheath of radially arranged palisade cells. The chloroplasts in the cells of the outer sheath are of average size, while in the inner sheath cells they are smaller and correspondingly more numerous, so that at low magnification the entire cytoplasm appears greenish. This type of layered-sheath has been observed in many species. The chlorophyll-containing sheath is believed to function as an assimilatory tissue and the efficiency of the sheath cells may be increased by their proximity to vascular tissue (Chapter 8.G), thus providing for the speedy translocation of the products of photosynthetic CO_2 assimilation. The bundle-sheath serves a taxonomic function (W. V. Brown, 1958), and it has been observed in many Gramineae and some dicotyledonous species.[1]

Chloroplasts of the bundle-sheath have been partly separated from mesophyll chloroplasts in homogenates prepared from sugarcane leaves. Since these bundle-sheath chloroplasts contain more starch, they move further down the tube during centrifugation in a dense sucrose solution (Baldry *et al.*, 1968). The mesophyll chloroplasts, containing grana, were thus located at the top of the centrifuge tube, while the bottom layer contained a mixture of both types, with the nongrana containing chloroplasts making up from 15 to 50% of this fraction. These partially purified bundle-

---

[1] *Saccharum officinarum* (sugarcane), *Zea mays* (maize), *Sorghum vulgare* (sorghum), *Spartina cynosuroides* (salt reed grass), *Andropogon foveolatus, A. hirtus, A. papilosus, A. ischaemon, Cynodon dactylon* (Bermuda grass), *Danthonia forskalii, Panicum turgidum, Pennisetum dichotomum, P. cenchyoides, Sieglingia* (heath grass), *Anthephora hochstetteri, Aristida sp.* (needlegrass), *Dactyloctenium aegypticum* (crowfoot grass), *Eragrostis pilosa, Microchloa setacea, Tragus oracemosus, Setaria sp.* (bristly foxtail), *Leersia oryzoides, Cyanastrum cordifolium,* and *Digitaria sanguinalis* (crabgrass).

The presence of bundle-sheath chloroplasts has also been detected in dicotyledons such as *Portulaca oleracea* (purslane), *Pectis humifusa, Bassia muricata, Atriplex halimus, A. leucocladum, A. roseum, A. sibiricum, A. nummularia, A. tataricum,* and *Amaranthus edulis* (pigweed).

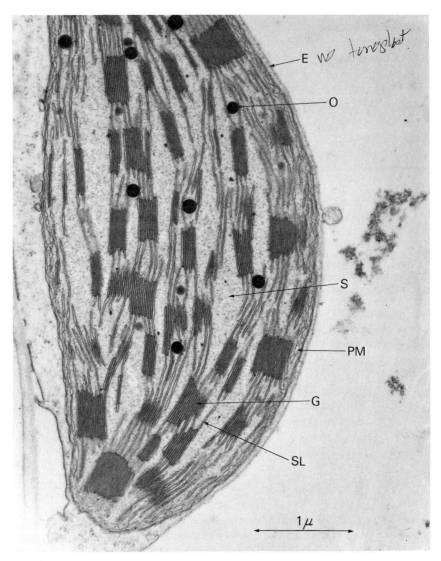

F‍IG. 2.1 A. Electron micrograph of a mesophyll chloroplast of maize (provided by Vernon E. Gracen, Jr.). The tissue was taken from fully expanded leaves and was fixed in 2% glutaraldehyde and post-fixed in 1% osmium tetroxide. The sections were post-stained with uranyl acetate followed by lead citrate. Abbreviations: E, chloroplast envelope, double membrane; G, lamellae stacked in grana; O, lipid-rich osmiophilic granules; S, stroma; SL, stroma lamellae; PM, peripheral reticulum or membrane system.

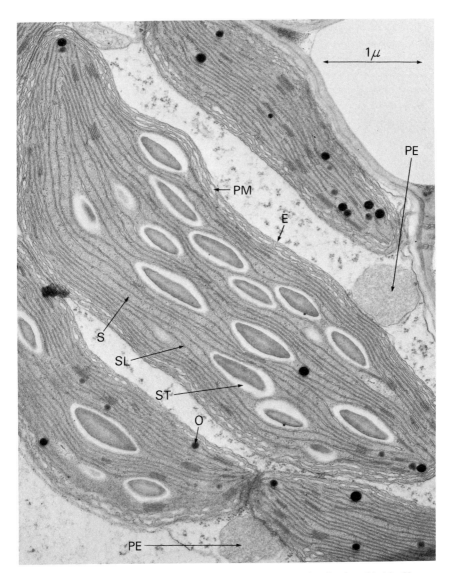

FIG. 2.1 B. Electron micrograph of bundle-sheath chloroplasts of maize (provided by Vernon E. Gracen, Jr.). Two peroxisomes (PE) are also shown. The section was prepared as in Fig. 2.1 A. E, chloroplast envelope, double membrane; ST, starch; O, lipid-rich osmiophilic granules; S, stroma; SL, stroma lamellae; PM, peripheral reticulum or membrane system.

sheath chloroplasts, however, reduced the dye dichlorophenolindophenol in the light at low rates of activity (Chapter 3.D), but rapid CO_2 fixation has not yet been described with such chloroplasts.

Guard cells contain chloroplasts without a well-developed lamellar system, with low concentrations of chlorophyll and generally no grana (W. V. Brown and Johnson, 1962; Zucker, 1963), and this is perhaps the most outstanding cytological feature which differentiates guard cells from other epidermal cells, which usually do not possess functional chloroplasts. Both chlorophyll a and b are present, and when chlorophyll cannot be detected in the plastids of guard cells, the stomata generally do not open in the light. The light-induced opening of stomata in etiolated leaves begins after 2 to 3 hours exposure to light, during which time chlorophyll synthesis is also initiated (Chapter 7.C).

B. ISOLATION OF CHLOROPLASTS

Biochemical studies with chloroplasts isolated from leaves have most often been carried out with plastids extracted in an aqueous medium. Density-gradient centrifugation is usually carried out at a later stage of purification in order to separate the chloroplasts completely from contaminating mitochondria and peroxisomes, as well as from broken chloroplasts. In aqueous media there is a tendency for a portion of the water-soluble compounds, including proteins, to be extracted and lost in the supernatant fluid. Nonaqueous solvents, such as mixtures of hexane–carbon tetrachloride, have therefore been used to isolate chloroplasts (Stocking, 1959; Stocking and Ongun, 1962). Ordinarily, one might not expect lipid components of isolated chloroplasts to be lost during isolation in aqueous media, but such loss can occur.

Purified chloroplasts commonly have a chlorophyll/protein N ratio close to unity (Sissakian, 1958; Orth and Cornwell, 1963; Bottrill and Possingham, 1969; Table 2.2), and this has sometimes been used as a criterion of purity.

Kahn and von Wettstein (1961) observed that when isolated spinach chloroplasts are suspended in weakly buffered 0.35 M NaCl, some chloroplasts appeared intact by electron microscopy and retained their normal boundary membranes, while others lacked their outside membranes and much of the stroma, but still retained their lamellar organization. Walker (1964, 1965) described a procedure for isolating chloroplasts from pea leaves so that a large proportion contained intact boundary membranes.

In the phase-contrast microscope intact chloroplasts appeared opaque with a halo around them, in contrast to chloroplasts which had lost their boundary membranes. These appeared dark and granulated and lacked a halo. The intact chloroplasts carried out CO_2 fixation in the light at much greater rates than the others, while the chloroplasts consisting of a naked lamellar system were superior in cyclic photophosphorylation (Chapters 3.E and 4.D). Walker's medium was modified somewhat by Jensen and Bassham (1966) who worked with spinach chloroplasts, and they achieved still faster rates of CO_2 fixation. Their procedure is discussed later (Chapter 4.D), but the innovations introduced by these laboratories in order to achieve the higher rates of CO_2 fixation with chloroplasts deserve enumeration: (*1*) sorbitol (0.33 *M*), buffered with dilute MES (Table 2.1) at pH 6.1 was used to maintain a suitable osmotic concentration in the grinding medium. This lower pH enhanced the stability of the chloroplasts, possibly by slowing lipase activity or altering what adheres to the surface. The use of these novel buffers, which are dipolar compounds that are derivatives of either *N*-substituted taurine or *N*-substituted glycine, was described by N. E. Good *et al.* (1966a). They bind metals negligibly, and produce faster rates of reaction than more conventional buffers in a number of biochemical reactions. (*2*) Leaves were macerated for only a very short time (5 seconds) in a Waring Blendor. This permits the isolation of a small number of intact chloroplasts in a short time. (*3*) The homogenate was filtered rapidly through muslin or layers of cheesecloth, rather than centrifuged, in order to remove the cell debris more rapidly. (*4*) The suspension was centrifuged at 0° at 2000 to 4000 *g* for no more than 50 to 100 seconds, and the centrifuge was rapidly stopped. The supernatant fluid was thus quickly separated from the chloroplasts and discarded.

Because of their superior photosynthetic CO_2 fixation, it is not surprising to find that enzymatic activities associated with photosynthetic reactions in intact chloroplasts are also greater than in the chloroplast preparations lacking boundary membranes and some of the stroma (Gibbs *et al.,* 1967) (Chapter 4.D). An examination of the protein and lipid concentration of these two types of chloroplast preparations by Leech (1966) also revealed some surprising chemical differences. The two classes of chloroplasts were separated from each other by sucrose density-gradient centrifugation. Intact chloroplasts had a higher protein concentration, and the total lipid of the chloroplast was also greater. These lipid components are probably at least partly responsible for the differences in the biochemical activities between the two classes of chloroplasts.

Intact chloroplasts that appear opaque by phase-contrast microscopy have been called Class I chloroplasts, while those which have lost their

TABLE 2.1   *Buffers Useful for the Isolation of Cytoplasmic Organelles and for Biochemical Reactions Associated with Chloroplasts[a]*

Abbreviation	Compound	Structural formula	pK at 20°
MES	2-(N-morpholino)ethanesulfonic acid	(morpholine ring) $\overset{+}{N}HCH_2CH_2SO_3^-$	6.15
TES	N-[tris (hydroxymethyl)methyl]aminoethanesulfonic acid	$(HOCH_2)_3C\overset{+}{N}H_2CH_2CH_2SO_3^-$	7.5
HEPES	N-2-hydroxyethylpiperazine-N'-2-ethanesulfonic acid	$HOCH_2CH_2\overset{+}{N}$ (piperazine ring) $\overset{+}{N}CH_2CH_2SO_3^-$	7.55
TRICINE	N-[tris(hydroxymethyl)methyl]glycine	$(HOCH_2)_3C\overset{+}{N}H_2CH_2COO^-$	8.15
BICINE	N,N-bis(2-hydroxyethyl)glycine	$(HOCH_2CH_2)_2\overset{+}{N}HCH_2COO^-$	8.35

[a] From N. E. Good *et al.*, 1966a.

outer membranes and much of the stroma are known as Class II. Karlstam and Albertsson (1969) have taken advantage of differences in the surface properties of these classes of chloroplasts to separate them by counter-current distribution between solutions of dextran and polyethylene glycol. The chloroplasts were partitioned between these two phases, and the proportion of the different classes of chloroplasts in spinach leaves and spinach cotyledons was determined. Intact chloroplasts (Class I) were separated into two fractions, Class IA and IB. Since both subclasses have an outer membrane, these membranes presumably differ in composition, and the subclasses will also in all likelihood have different biochemical activities. Even Class I chloroplast preparations are not intact in some respects and they may be less than fully active in certain of their biochemical reactions.

C. CHEMICAL COMPOSITION OF CHLOROPLASTS

Analyses of the composition of chloroplasts were suspect for many years because of uncertainty about the homogeneity of isolated chloroplast preparations and the difficulty in determining whether substances had been leached from the plastids during isolation. A review of the literature (Rabinowitch, 1945) thus concluded that chlorophyll a and chlorophyll b comprised 5 to 10% of the dry matter, while the carotenoids accounted for 2 to 4%. The volume occupied by the chloroplasts in an average cell was about twice as great as that of the remainder of the cytoplasm, and the plastids contained about 40 to 50% of the leaf protein. The lipid components were said to include fatty acids, aldehydes, phospholipids, and triglycerides. The chloroplast ash made up 8 to 10% of the dry weight, and iron, copper, calcium, magnesium, and phosphorous were present in the chloroplast in concentrations higher than that found in the whole leaf. Although the concentration of potassium in the chloroplast appeared to be less than that of the whole leaf, it was recognized that this element may be more easily extracted from chloroplasts during isolation. It was also acknowledged that several enzymes are associated with chloroplasts. In a later review by Rabinowitch (1956), it was judged possible that nucleic acids are present in chloroplasts, because a fractionation of tobacco leaf homogenates revealed that deoxyribonucleic acid (DNA) was generally distributed together with chlorophyll. By then, additional enzymes had also been found in chloroplast preparations, and newer evidence was presented that iron and copper are present in chloroplasts in concentrations higher than in the remainder of the cell.

TABLE 2.2   *Gross Chemical Composition of Isolated Leaf Chloroplasts, % Dry Weight*

Species	Chloroplast isolation method	Protein	Lipid	Carbohydrate, other	Inorganic elements	RNA	DNA	Reference
Spinach	Nonaqueous	48[a]	52[a]	–	–	–	–	Lichtenthaler and Park, 1963
Spinach	Nonaqueous	–	–	–	–	7.5	0.5	Biggins and Park, 1964
Bean (Phaseolus vulgaris)	Aqueous 4% formaldehyde	50–60[b]	30[b]	–	–	–	–	Mego and Jagendorf, 1961
Spinach, beet	Aqueous	–	–	–	–	–	–[c]	Chun et al., 1963
Broad bean (Vicia faba)	Aqueous	–	–	–	–	–	5.7×10^{-5d}	Kung and Williams, 1969
Snapdragon (Antirrhinum majus)	Aqueous and nonaqueous	40[e]	22	30[e]	–	1.0[e]	0.02[e]	Menke, 1966
Maize	Aqueous, density-gradient	60[f]	–	–	–	–	–	Orth and Cornwell, 1963
Tobacco, bean	Nonaqueous	76[g]	–	–	K, Mg, Ca, Na[h]	–	–	Stocking and Ongun, 1962
Spinach	Aqueous	–	–	–	Mn[i]	–	–	J. M. Anderson and Pyliotis, 1969
Evening primrose (Oenothera)	Aqueous	75[j]	–	–	–	–	–	Zucker and Stinson, 1962
Tobacco	Aqueous	26[j]	–	–	–	–	–	Ongun et al., 1968
Tobacco	Nonaqueous	63[j]	25[k]	–	–	–	–	Ongun et al., 1968

[a] Percent of total in the chloroplast present in the lamellae; chlorophyll entirely in the lamellae.

[b] Proplastids contained 40–50% protein and 18–20% lipid.

[c] Buoyant density on a CsCl density-gradient was 1.705 and 1.719 gm/cm^3 for chloroplast DNA and 1.695 for nuclear DNA.

[d] Grams DNA per chloroplast; about 0.33 gm DNA per 100 gm chlorophyll. DNA content of a nucleus was equivalent to 10^4 chloroplasts. 5-methylcytosine absent in chloroplast DNA in contrast to nuclear DNA.

[e] Dry weight of one chloroplast was 24.6×10^{-12} gm. Twenty-two percent of the chloroplast protein was lamellar and 18% was in the stroma. About 10% carbohydrate was associated with the lamellar proteins, which contained a relatively large amount of nonpolar amino acid residues and little cysteine. Fifty percent of the RNA was in the lamellae. The DNA contained no 5-methylcytosine.

[f] Chloroplasts purified by sucrose density-gradient centrifugation had a chlorophyll/protein N ratio (weight basis) of 1:1. In spinach chloroplasts prepared in 1% glutaraldehyde the ratio was 0.8 (Bottrill and Possingham, 1969).

[g] Percent of total leaf protein in chloroplasts; 40% of leaf protein in chloroplasts isolated in aqueous medium.

[h] Intact tobacco leaves on a dry weight basis contained 2.88% K, 2.00% Mg, 0.84% Ca, and 1.16% Na. Isolated chloroplasts contained (% of the total), 55% of the K, 72% of the Mg, 63% of the Ca, and 51% of the Na. Results for bean leaves were similar.

[i] One gram-atom Mn per 75 chlorophyll molecules.

[j] Percent of total leaf protein in chloroplasts.

[k] Loss of total lipid compared with amount in chloroplasts isolated by aqueous technique.

More recent work (Table 2.2) has confirmed some earlier observations, altered others, and provided additional insight about the detailed composition of chloroplasts and how these may relate to their cellular activities. For example, about 75% of the leaf protein, and not 50%, is present in the chloroplasts (Stocking and Ongun, 1962; Zucker and Stinson, 1962), and the lower figures previously found were undoubtedly a result of the extraction of proteins during the isolation procedures used. Chloroplasts also constitute a reservoir of many inorganic elements, including potassium, so that inorganic ions are not necessarily located largely in the vacuole. About 80% of the substances that make up the lipid fraction of chloroplasts have now been identified (Table 2.3). In addition, the presence of

TABLE 2.3   *Molecular Distribution of Lipid and Protein Components of Chloroplasts, Based on a Photosynthetic Unit of 230 Chlorophylls[a]*

Substance	Moles	Molecular wt.	% of Total	Formula
Phosphatides (Phospholipids)	116	90,800	4.7	
Phosphatidyl glycerol	52	–	–	p. 32
Phosphatidyl choline	42	–	–	p. 32
Others	22	–	–	
Digalactosyl diglyceride	144	134,000	7.0	p. 33
Monogalactosyl diglyceride	346	268,000	14.0	p. 33
Sulfodeoxyglucosyl diglyceride (Sulfolipid)	48	41,000	2.1	p. 33
Quinone Compounds	46	31,800	1.7	
Plastoquinone A	16	12,000	–	p. 34
Others	30	19,800	–	
Carotenoids	48	27,400	1.4	
β-Carotene	14	7,600	–	p. 35
Others	34	19,800	–	
Chlorophylls	230	206,000	10.8	
Chlorophyll a	160	143,000	–	p. 40
Chlorophyll b	70	63,400	–	
Unidentified		191,000	10.0	
Sum lipid		990,000	52.0	
Protein		929,000		
Mn, Fe, Cu		1,000		
Sum Protein		930,000	48.0	
Sum Lipid and Protein		1,920,000		

[a] From R. B. Park and Biggins, 1964. Copyright 1964 by the American Association for the Advancement of Science.

ribonucleic acid (RNA) and DNA have been clearly established (Table 2.2), and their role in the synthesis of polypeptide chains (Chapter 2.F) and in the genetic makeup of chloroplasts (Chapter 2.E) has been studied. Chloroplasts also frequently contain deposits of starch as well as soluble carbohydrates, primarily sucrose, and small amounts of other metabolites. The mechanism of synthesis of some of these photosynthetic products is described later (Chapter 4.K).

1. PROTEINS IN CHLOROPLASTS

Chloroplasts contain at least 50% of protein on a dry weight basis (Tables 2.2, 2.3), and this is about equally distributed between the lamellae and the stroma (Lichtenthaler and Park, 1963; Menke, 1966). The proteins of the stroma include enzymes concerned with CO_2 fixation (Chapter 4). The lamellar protein is in part the so-called structural protein (Criddle, 1966, 1969). This protein was isolated from spinach chloroplasts after they were disrupted sonically, and the lipids in the resulting fragments were removed by washing them with acetone and ethyl ether. The structural protein was still insoluble and was extracted with detergents. The protein was then precipitated with ammonium sulfate and the accompanying detergent was removed by solvent extraction. This procedure yielded a protein preparation that appeared homogeneous in the ultracentrifuge, accounted for about 40% of the total chloroplast protein, and had a molecular weight of about 25,000.

The structural protein of young bean and wheat leaves has also been isolated from chloroplasts by a similar procedure (Mani and Zalik, 1970). These proteins also appeared homogeneous in the ultracentrifuge and had a somewhat lower molecular weight than the structural protein from spinach chloroplasts. However, on electrophoresis the preparations could be separated into from two to five components. The amino acid composition was different from the spinach structural protein, but bean and wheat structural proteins were similar in composition and contained a relatively high proportion of amino acids with nonpolar side chains. These hydrophobic residues may provide centers that facilitate complexing with enzymes and control enzymatic activities, but this has yet to be established for chloroplast structural protein as it has been with structural protein isolated from mitochondria.

2. ACYL LIPIDS OF CHLOROPLASTS

Lipids, including chlorophylls and carotenoids, make up about 50% of the dry weight of the chloroplast (Tables 2.2, 2.3), and the orientation of these molecules must affect the photosynthetic activity of the plastids since chlorophyll carries out its photochemical activity largely in a lipid environment (Benson, 1963). Chloroplasts contain very small amounts of neutral triglycerides, and a summary of the known lipid constituents of chloroplasts shows that complex acyl lipids predominate (Table 2.3). An excellent survey of the subject of plant phospholipids and glycolipids has been prepared by Kates (1970).

In phosphatides, which are common lipid components, two of the hydroxyl groups of glycerol are esterified with fatty acids, and the third hydroxyl group is attached through an ester linkage with phosphoric acid which is then esterified with glycerol or a nitrogenous base such as choline:

$$
\begin{array}{l}
H_2C-O-PO_2-OR' \\
HC-O-COR \\
H_2C-O-COR
\end{array}
$$

R = Fatty acid residues, mostly unsaturated
R' = $-CH_2-CHOH-CH_2OH$ phosphatidyl glycerol or
    $-CH_2-CH_2-N^+(CH_3)_3$ phosphatidyl choline

The phospholipids of leaves usually have as their major fatty acids palmitic acid (hexadecanoic), oleic acid (9-octadecenoic acid), linoleic acid (9,12-octadienoic acid), and linolenic acid (9,12,15-octatrienoic acid). Generally the concentration of unsaturated fatty acids exceeds the saturated, and *trans*-3-hexadecenoic acid is found almost exclusively in phosphatidyl glycerol (Haverkate and van Deenen, 1965).

The lipids present in largest amounts in chloroplasts are the glycolipids, although they are rare in (anaerobic) photosynthetic bacteria (Constantopoulos and Bloch, 1967). These compounds were first isolated and identified in the lipid fraction of wheat flour. They, together with the phospholipids, serve to stabilize the membranes of chloroplasts. There is generally about twice as much monogalactosyl diglyceride present as digalactosyl diglyceride in chloroplasts. By comparison, the lipids of mitochondria are devoid of galactosyl glycerides (Benson, 1961). One of the most striking properties of the galactolipids is their high concentration of unsaturated fatty acids. In alfalfa, spinach, and bean, 67 to 95% of the fatty acids in the monogalactolipid and digalactolipid consist of linolenic acid, and over 90% of the fatty acids always contained three double bonds, the other fatty acid being hexadecatrienoic acid (Carter *et al.,* 1965). The water solubility of the galactosyl groups, together with the hydrophobic properties of the two tri-unsaturated fatty esters, gives these lipids the physical properties of a surfactant much like a nonionic detergent (Benson, 1963).

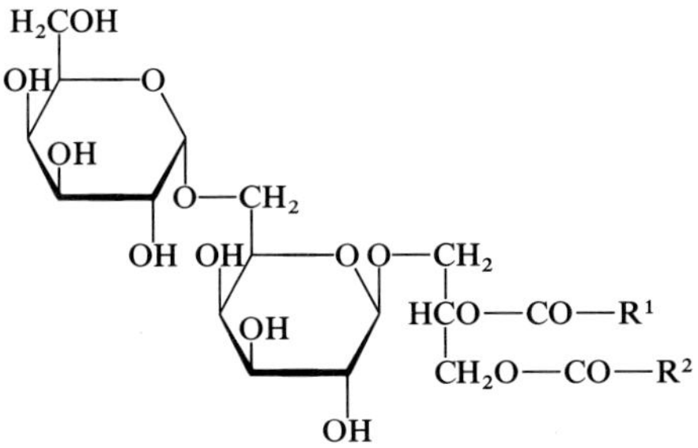

Digalactosyl diglyceride

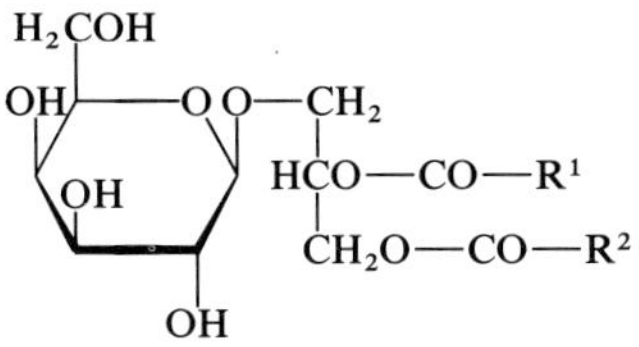

Monogalactosyl diglyceride

The plant sulfolipid is a sulfonic acid derivative of 6-deoxyglucosyl diglyceride. This compound was discovered by Benson and co-workers and appears to be universally present in photosynthetic organisms that produce oxygen, and is restricted to such organisms. Most of the leaf sulfolipid is found in chloroplasts, where it is concentrated in the lamellar membranes (Shibuya *et al.,* 1965). The sulfolipid, like the phospholipids, possesses saturated as well as unsaturated fatty acids, with 27 to 43% consisting of palmitic acid and the remainder comprising mostly linolenic and linoleic acids (Carter *et al.,* 1965).

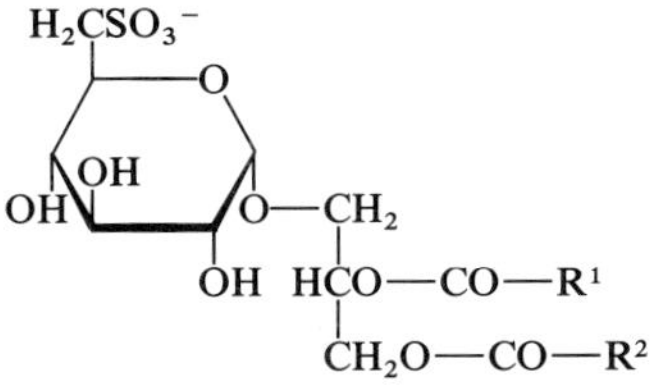

Sulfodeoxyglucosyl diglyceride (sulfolipid)

The concentrations of galactolipid and sulfolipid are greater in light-grown cells of *Euglena gracilis* than in dark-grown cells. A comparison of etiolated leaves with green ones also showed that only the phosphatidyl glycerol was absent from the etiolated tissues (A. T. James and Nichols,

1966). There was very little difference in fatty acid composition, except that when grown in light the fatty acid *trans*-3-hexadecenoic acid was specifically detected. No data were found to support the concept that the unsaturated fatty acids are chemically involved in any part of the photosynthetic process, but these investigators suggested that the function of lipids is expressed by a protein-enzyme-pigment-lipid complex that produces the appropriate spatial alignment of the lipid-soluble pigments and thus facilitates photosynthetic electron transport. However, these four acyl lipids, the mono- and digalactosyl diglycerides, sulfolipid, and phosphatides, are specifically associated with chloroplasts, and their association in the lamellae as well as in the boundary membranes in the correct configuration must also be important in regulating the CO_2 diffusion process during photosynthesis.

3. QUINONES

At least six quinones have been detected in significant quantities in the chloroplasts of higher plants (Crane *et al.,* 1966). Plastoquinone A is present in largest amount, and plastoquinones B, C, D, vitamin K, and α-tocopherylquinone are also present.

Plastoquinone A

Plastoquinone A has a side chain comprised of isoprene units. These may vary in length, but a C_{45} side chain is the most common. Although the structure of plastoquinone A has been determined, and plastoquinone B can be separated from it on thin-layer chromatography and is known to differ from it, the nature of the structural differences is still unknown. Plastoquinones C and D have been too unstable to permit a detailed structural analysis.

Quinones in chloroplasts have received considerable experimental attention because they are reduced in the light, as can be observed in the quinones extracted from chloroplasts or by direct spectroscopic examination of intact chloroplasts. Also when chloroplasts are extracted with *n*-heptane, several electron-transfer reactions demonstrable in isolated chloroplasts are abolished, and these can be restored by the addition of one or another of the naturally occurring quinones. None of the quinones, however, can restore the activity for all of the reactions abolished by solvent extraction. This suggests that there are multiple sites for quinone function and each shows considerable specificity for a particular quinone. The pathway of biosynthesis of chloroplast quinones is still uncertain, perhaps because their relatively low concentration makes such studies difficult. Naphthoquinone, however, a related substance, has been shown to be derived from shikimic acid in microorganisms and higher plants.

The chloroplast quinones are located in the osmiophilic granules or "plastoglobuli," the small (0.05 μ in diameter) lipid-rich inclusions that are observed in the stroma of chloroplasts under electron microscopy (Figs. 2.1A, 2.1B) in sections fixed with osmium tetroxide (Lichtenthaler and Sprey, 1966; Leech, 1968). The granules have been separated from several species after isolated chloroplasts were "broken" by placing them in a hypotonic medium, disrupting the chloroplasts sonically, and centrifuging at 100,000 g for 30 minutes. The osmiophilic granules were present in the supernatant fluid, and after further centrifugation at 145,000 g for 90 minutes, they were still found in the supernatant fluid. When they were extracted with acetone and acetone-petroleum ether mixtures, the globuli were found to be enriched in α-tocopherol, α-tocopherylquinone, plastoquinone, and vitamin K_1. They were essentially devoid of chlorophyll and carotenoids in comparison with the chloroplasts. These particles appear to provide a pool from which lipids can be utilized during membrane synthesis.

4. CAROTENOIDS

The carotenoids, like the quinones of the chloroplast, are also made up of isoprene units, and the formula of the most common carotenoid, β-carotene, shows that it is apparently made up of eight isoprenes by the head-to-head association of two units of four isoprenes (two diterpenes).

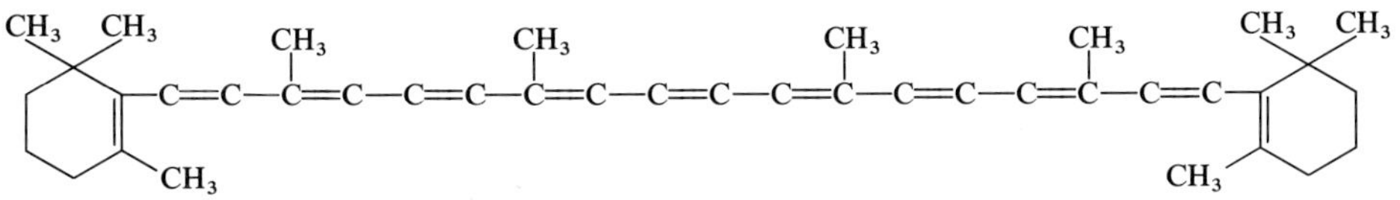

β-Carotene skeleton

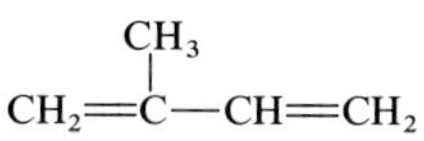

Isoprene

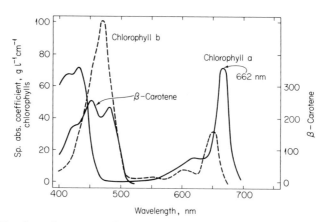

Fig. 2.2. The absorption spectra of chlorophylls a and b in methanol and of β-carotene in hexane (from French and Young, 1956). (Copyright 1956. Used with permission of McGraw-Hill Book Company.)

Carotenes have the elementary composition $C_{40}H_{56}$, and yellow oxygen-containing derivatives, $C_{40}H_{56}O_2$, are commonly called xanthophylls. β-Carotene, the predominant carotenoid of chloroplasts, absorbs light primarily in the blue region of the spectrum (Fig. 2.2).

Isoprenoid compounds, including carotenoids, are synthesized from acetate, which is converted to mevalonic acid by the sequence of reactions:

Acetate $\longrightarrow$ Acetyl CoA $\longrightarrow$ Acetoacetyl CoA $\longrightarrow$ Hydroxymethylglutaryl CoA

$$\underset{\text{Mevalonic acid}}{HO-CH_2-CH_2-\overset{\overset{\displaystyle CH_3}{|}}{\underset{\underset{\displaystyle OH}{|}}{C}}-CH_2-COOH} \quad (2.1)$$

All of the steps between acetate and mevalonate are not demonstrable with preparations isolated from plant tissues, but the conversion of mevalonate to isopentenyl pyrophosphate, the biological "isoprene unit," has been established (Goodwin, 1965; J. W. Porter and Anderson, 1967).

$$\text{Mevalonic acid} \xrightarrow[\text{+ ATP}]{\text{CO}_2} \underset{\text{Isopentenyl pyrophosphate}}{HO-\overset{\overset{\displaystyle O}{\|}}{P}-O-\overset{\overset{\displaystyle O}{\|}}{\underset{\underset{\displaystyle OH}{|}}{P}}-O-CH_2-CH_2-\overset{\overset{\displaystyle CH_3}{|}}{C}=CH_2} \quad (2.2)$$

Isopentenyl pyrophosphate is then converted to geranyl pyrophosphate (C_{10}), farnesyl pyrophosphate (C_{15}), and geranylgeranyl pyrophosphate (C_{20}). These C_{20} compounds then condense head-to-head to produce carotenoid structures. A sequential desaturation pathway probably produces the various carotenes, and oxygenated carotenoids arise late in the biosynthetic pathway from carotenes.

Although carotenoids do not usually absorb light directly in the photosynthetic process, carotenoids function to inhibit the photooxidation of chlorophyll. This was shown with mutant strains of photosynthetic bacteria that did not produce their normal complement of carotenoids. They were killed in the light in an aerobic environment, but survived in the dark under aerobic conditions or in the light if oxygen was excluded. In higher plants the absence of a colored carotenoid has also led to a photooxidative destruction of chlorophyll, but the destruction did not occur in an atmosphere of nitrogen. Such observations led Krinsky (1966) to suggest that the light-catalyzed epoxidation (addition of an oxygen atom across a double bond) of a xanthophyll may protect cells against lethal photosensitized oxidations. He demonstrated that such an epoxidation occurred, and that de-epoxidation could then be carried out enzymatically in the presence of NADPH. This cyclic process would result in the carotenoids drawing off oxygen in the neighborhood of the chlorophyll molecules, thus protecting them.

β-Carotene is probably present in spinach chloroplasts in the form of a complex with structural protein of the lamellae. In this complex, the absorption peak at 493 nm, observed with β-carotene in solution (Fig. 2.2), is shifted to 538 nm (Ji *et al.,* 1968). Extracted β-carotene could also be reassociated with pigment-free lamellar protein to form the complex. The carotenoid binding was specific for this chloroplast protein, and it combined in the ratio of 1 mole of β-carotene per mole of protein, thus providing evidence that carotenoids function in close association with chloroplast membranes.

D. CHLOROPHYLL

The chlorophylls are the primary light-absorbing pigments and constitute about 10% of the dry weight of the chloroplast (Table 2.3). They are found in leaves at a concentration of about 1.5 mg/gm of fresh weight and are extracted quantitatively with 80% acetone (Bruinsma, 1961). Chlorophyll a (Scheme 2.1) is generally present at twice the concentration of b. The tetra-

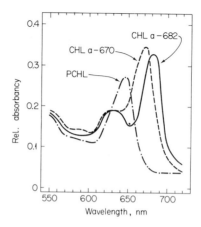

Fɪɢ. 2.3. Demonstration of two kinds of chlorophyll a in a bean leaf (from French, 1961). The protochlorophyll (PCHL) spectrum was obtained from dark-grown tissue. A few minutes after exposure to light the peak for chlorophyll a (CHL a-682) is observed, and after the same leaf is returned to darkness for 30 minutes the absorption spectrum for CHL a-670 is obtained.

pyrrole structure shown was worked out by Hans Fischer in 1940, the trans configuration of the "extra" hydrogens in ring D was proved by Ficken *et al.* (1956), and the total synthesis of chlorophyll a was accomplished by Woodward and co-workers (1960). Chlorophyll b differs in structure from a in that the methyl group in ring B is replaced by a formyl group (CHO). This slight difference changes the absorption spectrum in organic solvents so that the maximum in the red which occurs at 660 nm for chlorophyll a is now at 643 nm for chlorophyll b with a lower absorption coefficient (Fig. 2.2). The main red absorption band between 670 and 680 nm that is observed in all photosynthetic plants except the photosynthetic bacteria is caused by chlorophyll a. In absorption spectra of intact leaves the chlorophyll a bands can be recognized, but it has been known since 1870 that they are found at longer wavelengths than in acetone, petroleum ether, or ethyl ether solution, where the major bands occur at 662 nm, in the red, and at 430 nm (Fig. 2.2). In green plants chlorophyll a is present in several different forms; immediately after its formation in leaves it absorbs at 682 nm (Fig. 2.3), and after a few minutes in the light, or in the dark, the peak position shifts to 670 nm. These complexes, however, revert to ordinary chlorophyll a on extraction with organic solvents (French, 1961). *In vivo* chlorophyll b has an absorption maximum in the red at 643 nm and also absorbs at a longer wavelength, 650 nm (Holt, 1965).

In ring D, the propyl side chain of chlorophyll is esterified with phytol, a long-chain aliphatic alcohol, $C_{20}H_{39}OH$, and many of the physical properties of chlorophylls no doubt arise from the presence of this long hydrophobic molecule. The enzyme chlorophyllase may be responsible for the esterification reaction with phytol *in vivo*, but purified preparations of this enzyme isolated from leaves hydrolyze chlorophyll into chlorophyllide and phytol (A. Klein and Vishniac, 1961).

1. BIOSYNTHESIS OF CHLOROPHYLL

Many of the biochemical steps in the biosynthesis of chlorophyll also occur in the biosynthesis of porphyrins, including cytochromes, heme, and vitamin B_{12}, hence some of the biochemical reactions in the pathway were first discovered in tissues such as avian red blood cells. Most of the direct information about chlorophyll biosynthesis has come from studies with unicellular algae, where chlorophyll a is the predominant form, and with photosynthetic bacteria. Much of the progress in this area stems from the description by Granick in 1948 of mutants of *Chlorella* that could not synthesize chlorophyll and accumulated various intermediate products. There are several comprehensive reviews of this subject (Bogorad, 1965; Lascelles, 1965; Marks, 1966). A general outline of some of the more important reactions in chlorophyll biosynthesis is shown in Scheme 2.1.

The first reaction consists of the formation of δ-aminolevulinic acid by the condensation of succinyl CoA with glycine. This reaction, catalyzed by δ-aminolevulinic acid synthetase, has still not been detected in isolated chloroplasts (Porra and Irving, 1970); the reaction may occur outside the chloroplast or the enzyme may be lost during isolation of chloroplasts. In the next step, two molecules of δ-aminolevulinic acid are condensed, with loss of two molecules of water in a reaction catalyzed by δ-aminolevulinic acid dehydrase, to form the pyrrole, porphobilinogen. This enzyme is present in chloroplasts, and its increase in activity exactly parallels the increase in chlorophyll concentration with time after illumination of etiolated leaves (Steer and Gibbs, 1969). The enzyme has been detected in extracts of leaves of many species. It has been partially purified from tobacco leaves and porphobilinogen was identified as the product of the enzymatic reaction (Shetty and Miller, 1969).

The biosynthesis of uroporphyrinogen III, a tetrapyrrole, takes place through the condensation of four molecules of porphobilinogen and is catalyzed by the sequential action of at least two enzymes. The first of these, porphobilinogen deaminase, has been obtained in highly purified form from Swiss chard leaves and wheat germ (Frydman and Frydman, 1970). It transforms four molecules of porphobilinogen into the cyclic tetrapyrrole uroporphyrinogen I, which is different from III in that the

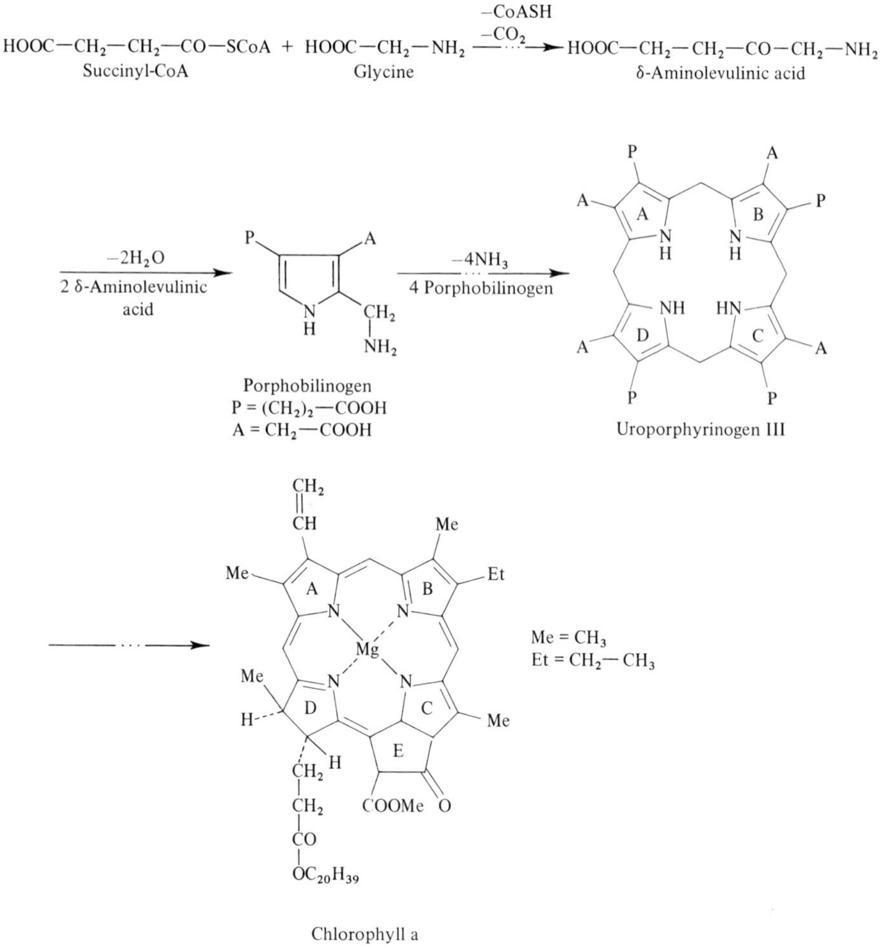

Scheme 2.1

positions of the carboxymethyl and carboxyethyl groups in ring D are in the same sequence as in the other rings. The activity responsible for polymerizing porphobilinogen and for cyclizing the polypyrrole appears to reside in the same enzyme molecule. The second enzyme, uroporphyrinogen III cosynthetase, isomerizes uroporphyrinogen I to give uroporphyrinogen III, in which the carboxymethyl and carboxyethyl groups in ring D are in a reverse order to the other three rings. This isomerization reaction is a common feature of cyclic tetrapyrrole biosynthesis, and its mechanism has been the subject of considerable speculation. Compound III is found exclusively when the cosynthetase is present in excess; I is formed when the enzyme is absent and only the deaminase is present; and mixtures of the two isomers are obtained at intermediate concentrations of the two enzymes. The synthetase must assemble porphobilinogen molecules, with the elimination of elements of NH_3, to form α-linked chains which cyclize spontaneously to give the expected isomer, compound I. Based on studies with molecular models, it was recently suggested that a cyclo-porphobilinogen composed of three pyrrole molecules is an intermediate. This would permit the ready formation of compound III when the fourth porphobilinogen molecule is introduced in the presence of the cosynthetase because formation of III may be sterically favored (Dalton and Dougherty, 1969).

The further steps in chlorophyll biosynthesis involve alteration of the side chains of the rings, the insertion of magnesium to produce magnesium protoporphyrin, the formation of ring E, and the esterification of phytol to the propionic acid side chain on ring D to produce protochlorophyll a. This intermediate is finally converted to chlorophyll a by the addition of the two "extra" hydrogen atoms in the trans configuration in ring D, a reaction in which there is a requirement for light (Holowinsky and Schiff, 1970).

The photoconversion of the protochlorophyll initially present in dark-grown leaves produces chlorophyll a but not chlorophyll b. The b is formed together with a as the chlorophyll concentration of the plant increases in the light. Most evidence thus favors the view that chlorophyll b is synthesized from a (Marks, 1966).

Leaves grown in the dark contain only a small amount of the chlorophyll precursor protochlorophyllide (chlorophyll lacking the phytol group and the two "extra" hydrogens in ring D), while an equal weight of fully green leaves may contain from 100 to 300 times more chlorophyll than the amount of protochlorophyllide present in the yellow tissue (Bogorad, 1967). After brief illumination, the protochlorophyllide in the leaf (maximum at 650 nm) is converted to chlorophyll a (maximum at 682 nm, Fig. 2.3). Light is apparently also necessary for the production of the chlorophyll precursor δ-aminolevulinic acid, perhaps because its precursor, glycine, is synthesized primarily in the light from glycolate (Chapter 6.D). When δ-aminolevulinic acid was supplied to etiolated barley leaves, they produced up to ten times more protochlorophyllide than control leaves in the dark, indicating that the enzymes needed for the production of protochlorophyllide are present in yellowed leaves but that the intermediate δ-aminolevulinic acid is lacking (Bogorad, 1967; Nadler and Granick, 1970).

2. MOLECULAR ARRANGEMENT OF CHLOROPHYLL

A complex probably exists between chlorophyll and protein in green cells although the chemical nature of such a linkage is unknown (Wassink, 1963). The phycobilins provide a well-documented analogous example since these linear tetrapyrroles present in algae have been characterized as chromoproteins. In addition, chlorophyll-containing extracts prepared from cells in the presence of various detergents show absorption spectra similar to that of chlorophyll *in vivo*. The use of detergents to overcome the insolubility of the chlorophyll–protein complex was first described by E. L. Smith (1941). Practically all of the pigments associated with photosynthesis undergo absorption shifts to lower wavelengths, sometimes as great as 20 to 100 nm, when the pigments are extracted with organic solvents (compare Fig. 2.2 and 2.4). Recent studies of the properties of two isolated chlorophyll–protein complexes have been described by Thornber *et al.* (1967). Chloroplasts of several species were extracted with detergents, and the resulting pigment complexes were separated by electrophoresis on polyacrylamide gels. Two fractions were obtained, complexes I and II, which contained over 75% of the proteins of the intact lamellae. These complexes varied greatly in their ratios of chlorophyll a and b, and evidence was provided that complex I (with a high concentration of chlorophyll a) was derived from particles associated with Photosystem I, and complex II (rich in chlorophyll b) originated from chloroplast particles concerned in Photosystem II (Chapter 3.D). More recently, it was found that complex I con-

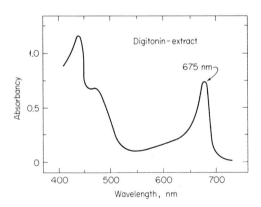

FIG. 2.4. The absorption spectrum of an extract of spinach leaf prepared with 1 to 5% digitonin (from E. L. Smith, 1941). The absorption maximum of chlorophyll a occurs at 675 nm compared with 660-665 nm (Fig. 2.2) in organic solvents because the detergent extracts a chlorophyll–protein complex which shifts the spectrum to higher wavelengths.

tains the reaction center (P700) of Photosystem I, and that it is completely oxidized by light, showing that it still has some photochemical activity even in the isolated state (Thornber, 1969).

Some of the differences between the absorption spectra of chlorophyll in organic solvents compared with those obtained *in vivo* may also be caused in part by the tendency of chlorophyll molecules to form aggregates, since dimers form at concentrations of about 10^{-2} M (Brody and Brody, 1963). The different absorption peaks of chlorophyll *in vivo* may therefore represent different states of aggregation of chlorophyll bound with proteins and lipids of the chloroplast membranes (Krasnovsky, 1969).

The degree of aggregation and orientation of chlorophyll molecules *in vivo* will determine the efficiency of the pigment to carry out its photochemistry. The lipid and protein components of the chloroplast are also important in the functioning of chlorophyll, and a number of models have been proposed to account for these interactions. The older view is that chlorophyll is deployed as a two-dimensional aggregate sandwiched between protein and lipid layers. The porphyrin "head" of the molecule is bound to protein and the phytol chain "tail" extends into the lipid layer.

E. CHLOROPLAST GENETICS

The structure and function of chloroplasts are sometimes under strict control by nuclear genes since many chromosomal mutants are known which control chloroplast development, including the lamellar structure, the electron transport system, CO_2-fixing enzymes, enzymes concerned with photophosphorylation, and enzymes related to the synthesis of pigments (von Wettstein, 1966, 1967). On the other hand, there is considerable evidence that plastids of higher plants are also autonomous structures capable of self-replication (Kirk and Tilney-Bassett, 1967).

The most striking evidence of cytoplasmic inheritance is the failure to observe nuclear segregation among the progeny from a cross between two parents which differ because of plastid determinants. In such cases the progeny resemble the maternal parent. Maternal inheritance occurs where the male gamete fails to transmit the cytoplasmic determinant at fertilization. The bulk of the cytoplasm of the zygote is contributed by the egg cell and because the amount of cytoplasm introduced through the pollen tube with the fertilizing nucleus is so small, cytoplasmic determinants are only rarely transmitted by the male parent. Plants resulting from such a cross have identical nuclear genes, but the phenotypes are those of the maternal parent. Examples are also known in which flowers on green branches of a plant will produce green offspring, white (pale green or yellow) branches produce progeny that are white, and variegated branches produce offspring with green, white, and variegated individuals in widely varying ratios. In all of these examples the pollen has no effect on the phenotype of the progeny (Rhoades, 1946), thus meiosis could not have been involved in such inherited characters.

In the nucleus of higher plants, there are two sets of genes present in each cell, one provided by the female germ cell and the other by the male. Each gene is present as a pair, and only when one member of a pair becomes defective is it possible to study the function associated with these genes. Genetic studies of the cytoplasm may be more difficult, because there may be many different kinds of cytoplasmic genes in a single cell, since each mesophyll cell, for example, may have as many as 100 plastids each of which may vary in their genetic makeup (Granick, 1965).

1. CHLOROPLAST DNA

At one time there was considerable debate about whether cytoplasmic inheritance in fact occurs or whether it is an altered form of nuclear inheritance, but there is now considerable evidence to show that plastids, as well as other cytoplasmic organelles such as mitochondria, exhibit extranuclear inheritance. The reality of cytoplasmic inheritance has become even more convincing since the clear demonstration that chloroplasts contain their own DNA, RNA, and the enzymes necessary for the synthesis of nucleic acids and proteins (Granick, 1965).

Evidence for the presence of a unique DNA in chloroplasts has been summarized (Gibor, 1965; Kirk, 1966). Observations under the electron microscope showed that cells treated with uranyl acetate, which combines with nucleic acid, contain chloroplast fibers 25 to 30 Å thick which disappeared when the cells were treated with RNase. The incorporation of tritiated thymidine into chloroplasts has been demonstrated by autoradiographic techniques. There is also chemical evidence for the existence of DNA, but the problem of contamination by elements derived from the nucleus were of concern until it was shown that enucleated cells of the giant alga *Acetabularia* contain DNA.

The base ratio of chloroplast DNA in higher plants is different from that of nuclear DNA and it has a different buoyant density (Table 2.2). Chloroplast DNA contains no 5-methylcytosine, and has the usual bases adenine, thymine, cytosine, and guanine. The moles of adenine equal those of thymine, and cytosine equals guanine. It has a molecular weight of 20 to 40 million, about the same as nuclear DNA, and a chloroplast contains at least 1×10^{-16} gm of DNA. Kirk (1966) has estimated that this is sufficient DNA to code for about 160 proteins containing an average of 200 amino acid residues, and this calculation is based on the lower limit of the DNA concentration.

The synthesis of DNA takes place in the chloroplast, as has been shown in some ingenious experiments conducted with *Euglena gracilis*. The proplastids lost their ability to develop into chloroplasts when dark-grown cells were treated with ultraviolet light, at 260 nm, and blue light photoreactivated the irradiated cells, suggesting that nucleic acids are the targets for the ultraviolet action (Lyman *et al.,* 1961). Gibor and Granick (1962) extended these experiments by shielding the nucleus of the cells and irradiating only the cytoplasm with a microbeam of ultraviolet light. Under these conditions proplastids do not develop into chloroplasts, whereas irradiation of the nucleus alone had no such effect. Thus the DNA of the chloroplast is not derived from the nucleus and genetic information concerned with chloroplast development resides in the nucleic acids of the proplastids.

If chloroplasts have at least a partial genetic autonomy, the DNA within the chloroplasts must be synthesized within the plastid. The enzyme DNA polymerase has now been found in isolated chloroplasts of tobacco and spinach leaves (Spencer and Whitfeld, 1969). Homogenates were treated with a detergent to disperse chloroplast fragments, a treatment that apparently leaves nuclei intact so that they could be removed by centrifugation. Such purified chloroplast preparations incorporated the tritiated triphosphate of deoxythymidine into a material which was insoluble in cold 5% trichloroacetic acid (DNA) only when the triphosphates of deoxyguanosine, deoxyadenosine, and deoxycytidine were present. Neither DNA nor DNA polymerase activity was lost from the chloroplast preparations during two washes in a hypotonic medium, hence DNA and the enzyme necessary for its synthesis are probably closely associated with the chloroplast lamellar structure. Moreover, these workers showed that the ^{3}H-labeled DNA produced by the chloroplast preparation hybridized readily with chloroplast DNA while nuclear DNA combined only very slightly under identical conditions, again demonstrating the distinct properties of this DNA.

A DNA-dependent RNA polymerase activity has also been demonstrated in broad bean chloroplasts (Kirk, 1966). The enzymatic incorporation of ^{14}C-ATP (adenosine triphosphate) into an acid-insoluble material was carried out in the presence of cytidine, guanosine, and uridine triphosphates

(CTP, GTP, UTP). It was necessary for the other three nucleoside triphosphates to be present for the reaction to proceed, it was inhibited by DNase, and the product had properties of a polyribonucleotide by a number of criteria. Experiments indicated the properties of this RNA polymerase are different from that in the nucleus, and the enzyme was also detected in enucleated *Acetabularia* cells. Hence the chloroplast DNA is similar in many respects to DNA found in other genetic material.

2. SIGNIFICANCE OF CHLOROPLAST GENETICS

The genetic autonomy of chloroplasts is of great potential importance in plant metabolism. Granick (1965) has pointed out that cytoplasmic inheritance provides a buffer system to the cell because the large numbers of plastids and their multigenic nature make it unlikely that an unfavorable mutation in chloroplast DNA will wipe them all out. The cytoplasmic organelles also provide an organism with versatility if the environment should change, because they represent a reservoir of mutations that can produce useful phenotypes on demand providing the mutations occur in meristematic or egg cells. Rhoades (1946) emphasized that chloroplasts are not entirely autonomous and that there are numerous examples of nuclear-controlled plastid characters. In fact, over a hundred genes are known in maize alone that produce phenotypes such as colorless, yellow, and striped leaves. Certainly if a nuclear gene controls an important intermediate that is needed in the synthesis of a chloroplast component, this would provide evidence of the interaction between both kinds of genes. In the synthesis of chloroplast membrane proteins, there is evidence that the synthesis is directed from within the chloroplast as well as from elsewhere (Chapter 2.F). Chloroplast mutations exist in barley which, under nuclear gene control, fail to develop a normal lamellar system unless they are supplied with the amino acid leucine (von Wettstein, 1966). An interesting speculation would be the possibility of finding comparable mutations which can be repaired in a similar way but which are cytoplasmically inherited.

Other cytoplasmic organelles (such as mitochondria and peroxisomes) will probably also influence the phenotype of chloroplasts just as nuclear genes do. The transplantation of mitochondria in fungi has been accomplished from one cell type to another by surgical means (Fincham and Day, 1971). Therefore, the possibility of transforming useful chloroplast phenotypes, perhaps even from one species to another, seems an exciting one in view of the likely diversity of the capabilities of chloroplasts.

F. ENZYMES ASSOCIATED WITH PROTEIN AND LIPID METABOLISM

Chloroplasts must have evolved primarily, although not exclusively, for their function in photosynthesis. Since they are at least partly autonomous structures, the plastids carry out the synthesis of enzymes, pigments, and factors directly concerned with photosynthesis. Chloroplasts are also able to carry out the synthesis of proteins and lipid components that make up the membrane structures and stroma just as they are able to synthesize the enzymes and nucleic acid that are important in their self-replication.

1. PROTEIN SYNTHESIS

Since at least 50% of the leaf protein is located within the chloroplast (Table 2.2), it is not surprising that the plastids are the primary site of protein synthesis in the leaf. Studies on the chemistry of leaf proteins were probably initiated by Osborne and Wakeman (1920), who presented a paper in which chemical analyses of the proteins precipitated from extracts of spinach leaves were given. Spinach was chosen "because a fresh supply can be obtained throughout the greater part of the year." This publication demonstrates how to determine when one is at the frontier of a new subject. The paper is completely devoid of any reference citations to previous work since there had been none for at least a century.

Much of our newer knowledge about protein synthesis results from work with bacteria, where it has been shown that deoxyribonucleic acid (DNA) directs the synthesis of ribonucleic acid (RNA), and this in turn directs protein synthesis. A number of similarities have been found in protein synthesis between bacterial and chloroplast systems. DNA is the genetic material which self-replicates. Current theory suggests two complementary strands of the molecule unwind and act as templates for the synthesis of new complementary strands whose composition is determined by the rules of base pairing (J. D. Watson, 1965). Genes are made up of stretches of one of the two strands of a DNA molecule, and they serve in the synthesis of a single-stranded molecule of messenger RNA (mRNA) complementary to the base sequence in one of the DNA strands of the gene (transcription). The genetic information encoded in the sequence of bases in the DNA and the unstable mRNA determines the sequences of amino acids in proteins (translation). The translation mechanism consists of ribosomes, the soluble low molecular weight polyribonucleotide called transfer RNA (tRNA), mRNA, and the appropriate enzymes.

Amino acids are first "activated" to yield aminoacyl-tRNA's. These molecules diffuse to the ribosome. The template for protein synthesis is mRNA, a polyribonucleotide which determines the sequence of amino acids in polypeptide chains. The mRNA attaches to ribosomes and moves across ribosomes to bring successive nucleotide triplets (codons) into position to select the correct aminoacyl-tRNA precursors. A given mRNA molecule generally works simultaneously on many ribosomes (polyribosome), hence many codons function at a given moment. As the amino acids are polymerized into proteins on ribosomes, there is a concomitant release of tRNA.

Initiation of protein synthesis occurs in ribosome subunits, which can be detected by their lower molecular weights on sedimentation in the ultracentrifuge. Presumably each of the ribosome subunits has a different function. Since amino acids have an amino (NH_2) group and a carboxyl (COOH) group, and peptide bonds are formed between the carboxyl group of the first amino acid and the amino group of the second, and so on, there has to be some mechanism to insure that the proteins are synthesized in this direction, that is from amino terminal to carboxyl terminal. In bacteria, and presumably in plants (as discussed below), a complex between a ribosome subunit, a mRNA molecule, and a molecule of formylmethionine-tRNA serves the function of initiating the synthesis of all proteins. The formyl group blocks the amino terminal end, hence the polymerization of the amino acid must occur with the free carboxyl group, and so chain elongation proceeds only in one direction. When the protein has been synthesized, it is liberated from the ribosome, and the subunits dissociate and are ready to repeat the cycle.

Many of the reactions associated with protein synthesis have been demonstrated in photosynthetic tissues. The first step involves an activation of an amino acid in the presence of ATP and tRNA catalyzed by aminoacyl-tRNA synthetase:

$$tRNA + amino\ acid + ATP \longrightarrow aminoacyl\text{-}tRNA + AMP + PP_i \qquad (2.3)$$

There appears to be a specific tRNA for each amino acid, and the tRNA is different from other kinds of RNA in that it has a low molecular weight, about 25,000. The first phase in amino acid synthesis may be subdivided into two steps:

$$Enzyme + ATP + amino\ acid \longrightarrow enzyme\text{-}AMP\text{-}amino\ acid + PP_i \qquad (2.4)$$

$$Enzyme\text{-}AMP\text{-}amino\ acid + tRNA \longrightarrow aminoacyl\text{-}tRNA + AMP + enzyme \qquad (2.5)$$

The activation of amino acids [Reaction (2.3)] was observed in spinach chloroplasts as measured by the amino acid-dependent exchange between pyrophosphate and ATP (Marcus, 1959; J. Bové and Raacke, 1959). The sequence by which the amino acid-tRNA is converted to protein may be summarized in the following fashion:

$$\text{Aminoacyl-tRNA} \xrightarrow[\substack{\text{ribosomes, enzymes,}\\ \text{cofactors}}]{\text{Polynucleotide message,}} \text{polypeptide} + \text{tRNA} \qquad (2.6)$$

Experiments showing incorporation of ^{14}C-labeled amino acids into chloroplasts provided evidence of protein synthesis in chloroplasts as early as 1956, but many of the first experiments are suspect because inadvertent bacterial contamination may have caused the results observed. Spencer (1965) avoided this pitfall by using an aseptic technique and showed that extracts and chloroplast preparations from young spinach leaves incorporated ^{14}C from a mixture of radioactive amino acids into protein in a reaction that was dependent on ATP and was inhibited by RNase. The rate of ^{14}C incorporation was 20-fold greater with the chloroplast fraction than with the cell-free leaf extract. Similar results on the localization of protein-synthesizing activity in chloroplasts have been obtained with preparations from wheat (Bamji and Jagendorf, 1966) and bean leaves (Parenti and Margulies, 1967).

Leaf ribosomes incorporate ^{14}C-leucine in the presence of ATP and an ATP-generating system (Williams and Novelli, 1964), and the incorporation is greatly enhanced in illuminated leaves compared with dark-grown plants, suggesting that chloroplast ribosomes are primarily responsible for the enhanced rate of protein synthesis. Chloroplasts had previously been found to have a unique type of ribosome compared with other cytoplasmic ribosomes (Lyttleton, 1962). A detailed study of the incorporation of ^{14}C-valine by chloroplast (70 S) and cytoplasmic (80 S) ribosomes from spinach leaves was made by Boardman *et al.* (1966). Maximal rates of incorporation required the presence of a mixture of nucleoside triphosphates, an ATP-generating system, a mixture of amino acids, and the supernatant fraction from the preparation of the ribosomes. The reaction was initiated by the addition of ^{14}C-valine, and in such a system chloroplast ribosomes were far superior in their ability to incorporate amino acids than were the cytoplasmic ribosomes.

Active systems for protein synthesis by ribosomes from chloroplasts have now been obtained using more highly purified fractions (Hadziyev and Zalik, 1970). These investigators obtained chloroplasts from 4-day-old wheat seedlings, because the RNase activity was low at this age. They isolated polyribosomes and ribosomes, tRNA, and a partially purified preparation of aminoacyl-tRNA synthetase [Reaction (2.3)] from these chloroplasts and demonstrated aminoacylation of tRNA and amino acid incorporation by polyribosomes. Amino acid incorporation by intact chloroplasts, like polyribosomes, required magnesium ions, ATP and a regenerating system, and GTP for optimum activity (Table 2.4). Incorporation of phenylalanine by

TABLE 2.4   *Amino Acid Incorporation by Intact Chloroplasts and Polyribosomes Isolated from Chloroplasts of Wheat Seedlings[a,b]*

	Amino acid incorporation	%
	cpm/mg chlorophyll	
Complete (intact chloroplasts,		
10 μmoles Mg^{2+})	3080	100
1 μmole Mg^{2+}	1300	42
NO Mg^{2+}	300	10
NO ATP and ATP-generating system	308	10
NO GIP	1050	34
	cpm/mg ribosomal protein	
Complete (polyribosomes)	35,500	100
NO ATP and ATP-generating system	2740	8
NO GTP	21,300	60
NO aminoacyl-tRNA synthetase	151	1
Plus 50 μg polyU	40,800	115
Plus 100 μg polyU	48,300	136
Plus 5 μg RNase	4450	12
Plus 50 μg RNase	1280	4

[a] From Hadziyev and Zalik, 1970.

[b] The reaction mixture for amino acid incorporation by intact chloroplasts contained ATP, GTP, creatine phosphate and creatine phosphokinase (the ATP-generating system), chloroplast suspension, and a mixture of 13 U-^{14}C-amino acids (265,000 cpm) in a final volume of 1.0 ml. The reaction was carried out at 30° for 30 minutes, and the incorporation represents material that was insoluble in hot trichloroacetic acid.

Polyribosomes were obtained by disrupting intact chloroplasts with a detergent (Triton X-100) in the presence of bentonite. The polyribosomes were obtained by centrifugation at 144,000 g. Amino acid incorporation was measured in a reaction mixture containing MgCl$_2$ (10 μmoles), ATP, GTP, 19 amino acids without phenylalanine, creatine phosphate and creatine phosphokinase (the ATP-generating system), wheat chloroplast tRNA, chloroplast aminoacyl-tRNA synthetase, polyribosomes, and U-^{14}C- phenylalanine in a final volume of 1.0 ml. Controls were run to correct for possible bacterial contamination. After 30 minutes, the reaction was stopped by addition of trichloroacetic acid and unlabeled phenylalanine, and the precipitate was collected in a filter and washed. Polyuridylic acid (polyU) is known to serve as a synthetic mRNA that specifically stimulates phenylalanine incorporation in bacterial systems.

polyribosomes required aminoacyl-tRNA [Reaction (2.3)], was stimulated by the synthetic mRNA polyuridylic acid (polyU), and was inhibited by DNase. The bulk of the ribosomes were probably bound to the chloroplast lamellae, since it was necessary to use a detergent to obtain them in suspension. Somewhat similar requirements to these for amino acid incorporation by chloroplast fractions from *Euglena* have also been described (Brawerman and Eisenstadt, 1967).

As indicated above, *N*-formylmethionyl-tRNA effects the initiation and direction of protein synthesis in bacterial systems. Apparently the same type of tRNA has also been found in bean seedlings by Burkard *et al.* (1969) who extracted the tRNA's from chloroplasts, removed contaminating ribosomal RNA's and DNA, and then further purified the tRNA's by chromatographic procedures. In the presence of chloroplast enzymes, the chloroplast tRNA reacted with added ^{35}S-methionine and formyltetrahydrofolate so that upon hydrolysis a formylmethionyl residue was detected. Thus the initiation of protein biosynthesis in chloroplasts involved a similar mechanism to that found in bacteria, which also possess 70 S ribosomes.

Experiments with the inhibitors chloramphenicol (which inhibits chloroplast ribosomes) and cycloheximide (which acts on cytoplasmic ribosomes) on *Chlamydomonas* cells indicated that both antibiotics are required to completely inhibit the synthesis of lamellar protein *in vivo*. Each inhibitor alone only diminished the total amino acid incorporation about 50% (Hoober *et al.,* 1969; Smillie, 1969). Thus some of the membrane components of the chloroplast are probably synthesized in the cytoplasm, although apparently most enzymes associated with photosynthetic CO_2 fixation and electron transport are synthesized in the chloroplast.

2. LIPID SYNTHESIS AND BREAKDOWN

Long-chain fatty acids are synthesized by chloroplasts starting with acetyl CoA and malonyl CoA by a sequence of seven enzymatic reactions like those worked out for *Escherichia coli* by P. R. Vagelos and his coworkers. This subject has been reviewed by Mudd (1967), Nichols and James (1968), and Stumpf (1969), and the reactions are shown in Scheme 2.2.

All of the reactions of fatty acid synthesis in this pathway occur with the substrate bound in a thioester linkage to acyl carrier protein (ACP). These proteins contain about 80 amino acid residues and have a molecular weight of about 10,000. ACP from bacterial and plant sources (including spinach leaf) have been obtained in highly purified form and their properties have been compared (Simoni *et al.,* 1967). The prosthetic group of ACP contains 1 mole of phosphate, 1 mole of pantoic acid, a β-alanine residue,

$$\text{Acetyl-S-CoA} + \text{HS-ACP} \underset{\text{transacylase}}{\overset{\text{Acetyl}}{\rightleftharpoons}} \text{acetyl-S-ACP} + \text{CoA-SH} \tag{2.7}$$

$$\text{Acetyl-S-CoA} + \text{HCO}_3^- + \text{ATP} \underset{\text{Acetyl CoA carboxylase}}{\rightleftharpoons} \text{malonyl-S-CoA} + \text{ADP} + \text{P}_i \tag{2.8}$$

$$\text{Malonyl-S-CoA} + \text{ACP-SH} \underset{\text{transacylase}}{\overset{\text{Malonyl}}{\rightleftharpoons}} \text{malonyl-S-ACP} + \text{CoA-SH} \tag{2.9}$$

$$\text{Acetyl-S-ACP} + \text{malonyl-S-ACP} \underset{\text{synthetase}}{\overset{\beta\text{-Ketoacyl-ACP}}{\rightleftharpoons}} \text{acetoacetyl-S-ACP} + \text{CO}_2 + \text{ACP-SH} \tag{2.10}$$

$$\text{Acetoacetyl-S-ACP} + \text{NADPH} + \text{H}^+ \underset{\text{reductase}}{\overset{\beta\text{-Ketoacyl-ACP}}{\rightleftharpoons}} \text{D}(-)\text{-}\beta\text{-hydroxybutyryl-S-ACP} + \text{NADP}^+ \tag{2.11}$$

$$\text{D}(-)\text{-}\beta\text{-hydroxybutyryl-S-ACP} \underset{\text{dehydrase}}{\overset{\beta\text{-Hydroxyacyl-ACP}}{\rightleftharpoons}} \text{crotonyl-S-ACP} + \text{H}_2\text{O} \tag{2.12}$$

$$\text{Crotonyl-S-ACP} + \text{NADPH} + \text{H}^+ \underset{\text{reductase}}{\overset{\text{Enoyl-ACP}}{\rightleftharpoons}} \text{butyryl-S-ACP} + \text{NADP}^+, \text{etc.} \tag{2.13}$$

SCHEME 2.2

and 2-mercaptoethylamine combined in the following manner (Majerus *et al.,* 1965; Pugh and Wakil, 1965):

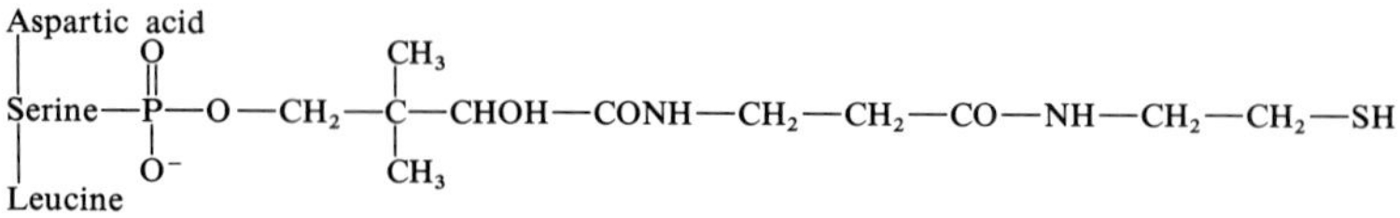

Aspartic acid
$$\text{Serine}\underset{\underset{\text{Leucine}}{|}}{\overset{\overset{\text{O}}{\|}}{-}}\text{P}\overset{|}{\underset{\text{O}^-}{-}}\text{O}-\text{CH}_2-\overset{\overset{\text{CH}_3}{|}}{\underset{\text{CH}_3}{\text{C}}}-\text{CHOH}-\text{CONH}-\text{CH}_2-\text{CH}_2-\text{CO}-\text{NH}-\text{CH}_2-\text{CH}_2-\text{SH}$$

Prosthetic group of acyl carrier protein

In spite of the identity of the prosthetic group of ACP with a portion of the coenzyme A (CoA) molecule, ACP is specifically required for fatty acid elongation, and its concentration and that of acetyl CoA carboxylase [Reaction (2.8)] may limit fatty acid synthesis in chloroplasts (Stumpf, 1969).

Chloroplasts from spinach (Mudd and McManus, 1962) and lettuce leaves (Stumpf and James, 1963) incorporate ^{14}C-acetate into long-chain saturated fatty acids in the presence of added ATP, CoA, NADPH, and magnesium ions. ATP is necessary to "activate" acetate in the presence of CoA in order to synthesize acetyl CoA. Fatty acid synthesis was stimulated by bicarbonate, which is known to be involved in the production of malonyl CoA from acetyl CoA [Reaction (2.8)], and it was stimulated by light presumably because the NADPH production was increased and it is needed for the reductive steps [Reaction (2.11) and (2.13)]. More recently, the conversion of acetate-2-^{14}C, acetyl CoA-1-^{14}C, and malonyl CoA-2-^{14}C into fatty acids was compared in a soluble system obtained by high speed centrifugation of chloroplast fragments from spinach and lettuce (Brooks and Stumpf, 1965, 1966). In this system there was an absolute requirement for NADPH and ACP while sulfhydryl compounds and NADH also stimulated the incorporation (Table 2.5). Malonyl CoA is the better precursor in this soluble system, which produced palmitic and stearic acids, because it is first converted to malonyl-S-ACP [Reaction (2.9)] before fatty acid elongation takes place. Fractions obtained from spinach chloroplasts bind acetate or acetyl CoA and this binding is inhibited by reagents that react with sulfhydryl groups, suggesting that the chloroplast lamellar fraction is concerned with fatty acid synthesis when acetyl CoA is the substrate (Devor and Mudd, 1968).

TABLE 2.5 *Incorporation of ^{14}C-Precursors into Long-chain Fatty Acids by Soluble Preparations Obtained from Chloroplasts[a,b]*

Species	^{14}C-Substrate		Substrate incorporated into long-chain fatty acids, nmoles
Lettuce	Malonyl CoA-2-^{14}C	Complete	13.2
		-ACP	2.8
		-NADP$^+$, G 6-P	1.6
		-GSH	3.0
		-NADH	8.4
		-MgCl$_2$	13.4
		-acetyl CoA	14.6
		-all	0.1
Spinach	Acetate-2-^{14}C	Complete	1.3
		-ACP	0.7
	Acetyl CoA-1-^{14}C	Complete	1.9
		-ACP	0.8
	Malonyl CoA-2-^{14}C	Complete	19.6
		-ACP	0.9
Lettuce	Acetate-2-^{14}C	Complete	0.7
		-ACP	0.3
	Acetyl CoA-1-^{14}C	Complete	3.4
		-ACP	1.7
	Malonyl CoA-2-^{14}C	Complete	8.4
		-ACP	2.2

[a] From Brooks and Stumpf, 1965, 1966.

[b] Isolated chloroplasts were disrupted by extrusion in a French press. The fragments were centrifuged at 105,000 g for 25 min and the supernatant fraction was used. The reaction mixture contained glycyl-glycine buffer (pH 8.2), glutathione (GSH), glucose 6-phosphate (G 6-P) and its dehydrogenase (a NADPH-generating system), NADP$^+$, NADH, purified acyl carrier protein (ACP), ATP, potassium bicarbonate, acetyl CoA, and ^{14}C-substrate (0.1 μmole). Palmitic and stearic acids were the major fatty acids synthesized with either acetyl CoA or malonyl CoA.

The mechanism of the biosynthesis of unsaturated fatty acids, which may account for 60% of the chloroplast fatty acids, has still not been fully resolved (Harris *et al.*, 1967), although the synthesis of oleic acid probably occurs by direct desaturation of the stearic acid-ACP thiol ester. Whole leaf tissue can convert saturated fatty acids from C_2 to C_{14} to unsaturated fatty acids, and oleic acid (one double bond) is converted to linoleic acid (two double bonds) when atmospheric oxygen is present. Stearyl-ACP (but not stearyl CoA) is converted to oleate by spinach chloroplasts in a reaction requiring ferredoxin (Chapter 3.D) and NADPH (Nagai and Bloch, 1966). These workers also demonstrated the same reaction with a soluble system obtained from *Euglena* that was fractionated into three components containing a NADPH oxidase, ferredoxin, and a desaturase enzyme requiring the other two fractions. In *Chlorella vulgaris* a desaturase system exists in the chloroplast fraction while an oleic acid activation system is located in the cytoplasm. This system converts oleate to linoleate and linolenate (three double bonds) (Harris and James, 1965). In *Euglena gracilis* the degree of polyunsaturation in the fatty acids of monogalactosyl diglyceride increased greatly as the light intensity under which the cells were grown was increased from 120 to 610 ft-c (Constantopoulos and Bloch, 1967). An increase in the rate of dye reduction in the Hill reaction (Chapter 3.C) by these cells was also strongly correlated with the increased amounts of linolenate and C_{16} polyenoic acids, suggesting these lipids play a role in electron transport (as shown below).

The synthesis of phosphatidic acid and glycerides has been studied in fractions from spinach leaves supplied with ^{14}C-L-glycerol-3-phosphate (Cheniae, 1965). These reactions required ATP, CoA, $MgCl_2$, and a fatty acid such as palmitic acid, and the activity was greatest in the microsomal (very small particle) fraction and was low in the chloroplast fraction. Sastry and Kates (1966) also found incorporation of DL-glycerol-3-phosphate-^{32}P into phosphatides to be greatest in the microsomal fraction, and only 10% of the spinach homogenate activity was in the chloroplasts. Some phosphatidyl glycerol and phosphatidyl inositol was also produced in their system.

Presumably the reactions involved are analogous to those described by E. P. Kennedy and coworkers for the synthesis of phosphatides in animal tissues whereby fatty acyl CoA's react with glycerol phosphate to produce phosphatidic acid.

$$\text{Phosphatidic acid} + \text{cytidine triphosphate (CTP)} \longrightarrow \text{CDP-diglyceride} + \text{PP}_i \quad (2.14)$$

$$\text{CDP-diglyceride} + \text{inositol} \longrightarrow \text{phosphatidyl inositol} \quad (2.15)$$

The biosynthesis of the highly unsaturated galactolipids was first carried out using ^{14}C-UDP-D-galactose as the glycosyl donor with spinach chloroplasts supplying the necessary enzymes as well as the endogenous acceptor (Neufeld and Hall, 1964). A more active system from spinach chloroplasts has recently been described by Ongun and Mudd (1968) in which over 90% of the total activity in the lipids was in monogalactosyl diglyceride. Both mono- and digalactosyl diglyceride were produced, as well as some trigalactosyl diglyceride, but the lamellar fraction synthesized proportionately more monogalactosyl diglyceride than the stroma fraction. For the synthesis of galactolipids by an acetone powder preparation obtained from spinach chloroplasts with UDP-D-galactose as the donor, highly unsaturated diglycerides (isolated from spinach phosphatides) must be added in contrast to chloroplast preparations which contained endogenous acceptor (Mudd *et al.,* 1969). It is still not known whether mono- can be converted to digalactosyl diglyceride or whether they are synthesized separately, although it has been noted earlier that the fatty acid composition of these galactolipids also has some interesting differences.

The enzymatic hydrolysis of galactolipids by leaf extracts has also received some attention. Sastry and Kates (1964) observed that homogenates of Scarlet runner bean (*Phaseolus multiflorus*) hydrolyzed galactolipids and liberated free fatty acids. The activity was greatest in the chloroplast fraction with monogalactosyl and digalactosyl dilinolenin as substrates and the pH optimum was 7.0 and 5.6 respectively. High activity was also found in preparations from leaves of mung bean (*Phaseolus aureus*), and kidney bean (*Phaseolus vulgaris*). Galactolipase activity was low in spinach leaves, but not detectable in soybean, cabbage, sugar beet, squash, and carrot leaves. This enzyme has been purified 50-fold from runner bean leaves, and both galactolipids still functioned as substrates as in crude extracts (Helmsing, 1969). Cysteine, at 10^{-3} M, but not dithioerythritol, inhibited the enzyme 100%, and this fact may be useful in slowing this reaction during the isolation of chloroplasts. In this connection, a study has been made of galactolipid transformation and fatty acid increase in isolated spinach chloroplasts stored for 5 to 22 hours at room temperature in various media (Wintermans *et al.,* 1969). In 0.5 M mannitol or sucrose (pH 7.5 or 6.0) there was little difference in rate of galactolipid breakdown compared with a medium containing 0.35 M NaCl, although in NaCl solution mainly the monogalactolipid was hydrolyzed and some extra free fatty acids were produced. Of course, in such studies considerable changes could have occurred in the chloroplasts during isolation and before the galactolipid composition was examined at zero time.

After illumination of dark-grown *Euglena* cells, the formation of chloroplasts is characterized by the appearance of chlorophyll and galactosyl diglycerides in a relatively fixed ratio (A. Rosenberg, 1967). This has suggested some special role of galactolipids in photosynthesis. The addition of mono- or digalactosyl diglyceride to spinach chloroplasts greatly increases the rate of the photoinduced reduction of cytochrome c (Chang and Lundin, 1965), thus providing direct evidence of some function for galactolipids in facilitating electron transport.

Detailed experiments on the sulfolipid biosynthetic pathway do not seem to have been carried out, although there is evidence that the sulfolipid is concentrated in the lamellar membranes in *Lemna* and spinach leaves (Shibuya *et al.*, 1965). The presence of sulfolipase activity that produces free fatty acids has been described in crude extracts of *Scenedesmus* and alfalfa leaves (Yagi and Benson, 1962).

3 Photochemistry and Photosynthetic Electron Transport

Photosynthesis can be described by the equation:

$$CO_2 + 2H_2O + light \rightarrow (CH_2O) + H_2O + O_2 (+ 112,000 \text{ cal/gm-atom of C}) \quad (3.1)$$

in which (CH_2O) represents the reduced product of CO_2 at the carbohydrate level. Photochemical energy derived from the absorption of light by chlorophyll is used in the first stages to remove electrons from water and produce O_2. Electrons are then transported to $NADP^+$, and the reducing power thus available in NADPH together with the ATP produced by photophosphorylation during electron transport is finally used to reduce CO_2 to the level of carbohydrate.

Photosynthesis in green plants requires that electrons are pumped uphill against a free energy gradient. Since the oxidation–reduction potential of the water–oxygen system is about $+0.8$ volts and that of the hydrogen–hydrogen ion is -0.4 volts, the transfer of a single electron requires an input of 1.2 electron volts of energy. The overall equation requires that four electrons or hydrogen equivalents be moved for the reduction of one mole of CO_2. Thus a total of 4.8 electron volts are needed to transfer the four electrons for each mole of CO_2 reduced and O_2 liberated. Since one electron volt equals 23 kcal per equivalent, this is about 112,000 calories per mole.

Electrons are actually transferred in at least two separate photoacts so that eight electron activations are required for a net transfer of four electrons. Photosystem II produces O_2 from water and forms a weak reductant, while Photosystem I yields a strong reductant with sufficient energy to reduce CO_2. Only the first stages in photosynthesis are discussed in this chapter, while problems related to CO_2 fixation are examined in succeeding chapters.

A. PHOTOCHEMISTRY

1. PRIMARY EVENTS OF LIGHT ABSORPTION

Certain basic features of the photochemistry of photosynthesis should first be appreciated. In order to transmit its energy and be useful in photosynthesis, light must be absorbed by the pigments in the chloroplasts, primarily by the chlorophylls that have absorption bands in the red and in the blue (Figs. 2.2 and 2.3). In intact green leaves, the maximum absorption band of chlorophyll in the red is broader than it is in organic solvents or in extracts prepared with detergents (Fig. 3.1). Although pigments other than the chlorophylls may absorb light which is useful in photosynthesis, the fluorescence spectrum of illuminated plant cells is always that of chlorophyll a, indicating that, if there are other initial absorbers of light, they can transfer the energy of excitation to the chlorophyll a molecules which then transmit the energy.

This book is primarily concerned with aspects of photosynthesis related to carbon metabolism, and the physical events of the photosynthetic process are already over before any chemical products of photosynthesis can be detected. During the interval between 10^{-15} seconds and 10^{-9} seconds, the quantum absorption and conversion processes that involve radiation and solid state physics take place. Between 10^{-9} seconds and 10^{-4} seconds, the accumulation of the stabilized products of photochemistry such as the oxidants and reductants is completed, and only between 10^{-4} seconds and 10 seconds after the initial photoact does sufficient CO_2 assimilation and initiation of the biosynthetic reactions occur to be detected (Kamen, 1963).

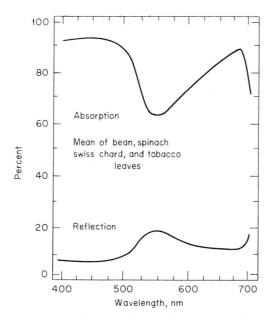

Fɪɢ. 3.1. The average absorption and reflection of light by leaves of six species (from R. A. Moss and Loomis, 1952). The absorption spectra were obtained by placing leaf sections in a specially constructed spectrophotometer, and the reflection normal to the leaf surface was measured with a phototube. There was considerable uniformity in the absorption and reflection spectra of the six species, with a maximum about 680 nm and a sharp drop beyond this wavelength.

Light is transmitted in individual packets called quanta. The energy (E) of a quantum of electromagnetic radiation is given by the formula,

$$E = hc/\lambda \tag{3.2}$$

Planck's constant (h) is 6.55×10^{-27} erg sec, c is the velocity of light (3×10^{10} cm/sec), and λ is the wavelength (cm). Thus the energy available from a quantum of light is inversely proportional to the wavelength and directly proportional to the frequency. An einstein (mole quantum, or 6.03×10^{23} quanta) of blue light represents about 60 kcal and one of red about 40 kcal (Fig. 3.2). Einstein predicted that the energy of a single light quantum (a photon) is transferred in photochemical reactions to a single electron. Since the energy available from chlorophyll excited by an einstein of light is about 30 kcal in the long-lived triplet state, it would require a minimum of 4 einsteins to provide sufficient energy to reduce one mole of CO_2 (Equation 3.1), and experimentally it is usually found that 8 quanta are needed

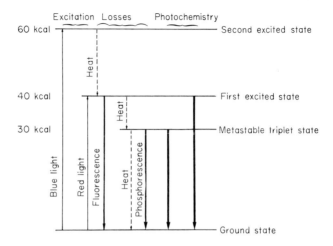

FIG. 3.2. Various energy states in the chlorophyll molecule and their role in photochemical reactions. Each horizontal line represents a different energy state. Those states above the ground state are unstable in varying degrees and lose energy by heat dissipation, fluorescence, phosphorescence, or photosensitization. Only the last process provides chemical energy useful in photosynthesis.

(the quantum requirement), or 2 quanta for each electron transported from water to NADPH. Thus the maximum quantum yield (moles O_2 released/einstein absorbed) equals $\frac{1}{8}$ or about 12% efficiency.

2. FLUORESCENCE

Pigment molecules can occupy a number of energy states when a quantum of light is absorbed (Fig. 3.2). The transitions between the different energy states of a pigment molecule can occur by the absorption or emission of radiation. Energy states above the ground state are known as excited states. When a molecule absorbs energy, changes occur in the positions of the electrons within that molecule. There are many possible excited states for a given molecule, and the particular state attained depends on the wavelength of light absorbed. Photosynthesis is concerned with the manner in which the excited molecule returns to the ground state, since chlorophyll acts catalytically, and all of the changes must therefore be reversible. It is important that as little of the energy of the excited molecule as possible escape as fluorescence, heat dissipation, or phosphorescence if the energy is to be used in photosynthesis.

Electrons move only in certain paths, or orbitals, around the atomic nuclei of the atoms of a molecule, and the energy of an electron depends on the orbital that it occupies. When a molecule is excited, one of its electrons is driven to an orbital of higher energy. This results in a molecule that is essentially novel, and it is usually more chemically reactive than when it is in the ground state. In general, two electrons occupying the same orbit must be "paired," that is, they must have opposite spins. After a transition of one electron of an original pair to a higher energy state, the spins may be either opposite or alike. When the spins are opposite, it is known as a singlet state, and when the spins are alike the triplet state is produced. The triplet state is generally achieved by decay from the first excited singlet state which usually has a higher energy level (Fig. 3.2). If a pigment in the triplet state survives sufficiently long in solution to collide with a different molecule to which the excess energy may be transferred, the second "acceptor" molecule may then undergo photochemical change in a process which is said to have been photosensitized. The pigment is called the sensitizer.

Blue light excites the second singlet state at a level of about 60 kcal in chlorophyll. Part of this energy is lost as heat within 10^{-11} seconds, so that 40 kcal are retained in the first excited singlet state. This energy state is also reached directly by absorption of red light, and its lifetime is approximately 10^{-9} seconds (Gaffron, 1960). In the living cell, the chances for photochemical use of the 40 kcal of the singlet state are small because of its short lifetime. With the sacrifice of at least 10 kcal by internal conversion into the metastable triplet state, the chances for photochemical reaction are increased a million times, because this excited state lasts about 10^{-2} seconds.

The ability to emit light while being irradiated results in the decay of an excited singlet state to the ground state, and is known as fluorescence. This process is temperature-independent and is rapid, with a lifetime of less than 10^{-8} seconds. The wavelengths of a fluorescence spectrum are always longer than the corresponding absorption spectrum; thus chlorophyll a absorbs light with maxima at 430 and 662 nm and emits it as fluorescence at 668 nm.

If the excited singlet state undergoes a transition to the triplet state, and photochemical action is excluded, decays occur to the ground state by heat dissipation or by the process of phosphorescence. This process is different from fluorescence in several important respects. It is temperature-dependent and is characterized by a very much slower afterglow (10^{-4} to 2 seconds). In phosphorescence, the spectrum is shifted to very much longer wavelengths, in the infrared. The slowness of phosphorescence is an advantage in permitting photochemical reactions to occur.

A molecule in the triplet state in solution may survive long enough to collide with a different solute molecule to which the excess energy may be transferred. As indicated, photosynthesis probably makes use only of the metastable triplets which live long enough to allow for all kinds of photochemical energy transfers to take place. Readers who wish further information about the primary events of light absorption related to the photochemistry of photosynthesis may find the following references useful. They are listed in approximate order of increasing detail from a physical standpoint: J. L. Rosenberg (1965), Glass (1961), Kamen (1963), Kok (1969), Franck (1951), Gaffron (1960), Clayton (1965), and Hoch and Knox (1968).

3. ENERGY TRANSFER AND TRAPPING CENTERS

The energy in an excited molecule can be transferred to an acceptor molecule, and the light emitted corresponds to the fluorescence spectrum of the acceptor molecule. The acceptor will then fluoresce as if it had originally absorbed the light. By such an energy transfer, chlorophyll b is shown to be the donor for chlorophyll a, but the reverse is not true. Chlorophyll molecules in the chloroplasts are arranged so that in the excited state the energy transfer can occur through several hundred light-harvesting chlorophyll molecules. The other pigments of the chloroplasts may also be involved in the gathering of light energy and its transfer to chlorophyll a.

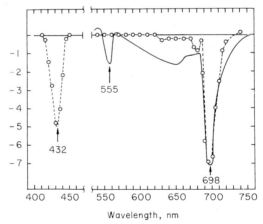

Fɪɢ. 3.3. Oxidized minus reduced difference spectrum (solid line, absorbancy × 100) and light minus dark spectrum (dashed line, relative units) of isolated chloroplast particles extracted with acetone (from Kok, 1961). In two parallel samples one was treated with ferricyanide and the other with ferrocyanide (oxidized minus reduced) to obtain a peak near 700 nm closely identical to the one induced by flashes of light (15 per second) and measurement of the absorbancy before and after each flash (light minus dark).

These necessary pigments, lipids, and proteins that function with chlorophyll are known collectively as the "photosynthetic unit," and for each of the two photoacts the unit appears to consist of about 200 chlorophyll molecules.

Located within the photosynthetic units are special chlorophyll molecules towards which the energy derived from light absorption is funneled, and in which it is trapped at the endpoint of the energy transfer. These reaction centers act much like light-harvesting antennae. The photochemistry is initiated in these trapping molecules. When leaves or isolated chloroplast suspensions obtained from several species are given a flash of intense white light, differences in the absorption spectrum can be observed in a comparison between light and dark (Kok, 1956, 1961; Kok and Hoch, 1961). The most striking shift in spectrum occurs in chloroplasts at about 700 nm (Fig. 3.3). This change in absorbancy characterizes a special chlorophyll molecule, P700. It is bleached (oxidized) by the light flash and although some regeneration of the pigment occurs during the dark period, a sizable fraction remains in the bleached form as long as the illumination is applied continuously.

The bleaching of P700 is sensitized by chlorophyll a and carotenoids, and although the velocity of bleaching is essentially instantaneous, the return reaction occurs in the dark with a half-time of about 5 milliseconds. The P700 can be oxidized either temporarily by light or permanently by ferricyanide (Fig. 3.3). About one P700 molecule is found for every 400 molecules of light-harvesting chlorophyll a, and when a photon arrives at this molecule while it is photochemically active (reduced), the excitation results in a photochemical reaction in which an electron is transferred from the P700 to some unidentified acceptor, and thus the P700 is seen in its oxidized (colorless) form. The oxidation–reduction potential of P700 is $+0.43$ V, indicating that it serves as a mild oxidant. Photosynthetic bacteria, which do not evolve oxygen in the light, also possess analogous trapping centers that have been purified with respect to the bulk chlorophyll. These have an absorption maximum in the far red at 800 nm or longer wavelengths. Although it is difficult to purify P700 because light-harvesting chlorophyll a must be present for its assay, the reaction center chlorophyll of green plants has been purified to an extent such that the particles in which it is present contain only 30 light-harvesting chlorophyll molecules for each P700 present (Vernon et al., 1969).

4. FLASHING LIGHT AND QUANTUM YIELD

In the 1920's, O. Warburg found that when *Chlorella* cells were supplied with high light intensities and high CO_2 concentrations a given quantity of light reduced more CO_2 when it was supplied in flashes rather than as continuous light. He thus concluded that after a dark period, a short flash of light would find a higher concentration of a CO_2 acceptor produced in a dark reaction than in continuous light. Emerson and Arnold (1932) obtained increases in the light yield of O_2 evolution of three- to fourfold by using about 50 flashes per second and .making the light flashes much shorter than the dark periods. The minimum dark time needed between individual flashes for each to have a maximum effect at 25° was about 0.03 seconds. Their experiments indicated that photosynthesis required a light reaction which was not affected by temperature, that was capable of proceeding at great speed, and a dark reaction which was temperature-dependent, and required a longer time. The maximum yield obtained was 1 mole of O_2 evolved for approximately 3000 moles of chlorophyll. Since there was not sufficient time for a chlorophyll molecule to become excited twice during a single flash, the size of the photosynthetic unit could be estimated. Assuming that each chlorophyll molecule absorbs one quantum and that 8 quanta are required per molecule of oxygen, the photosynthetic unit would be present to the extent of 8/3000 or about 1 for each 400 molecules of chlorophyll, a figure similar to that previously discussed for the concentration of P700 as used to estimate the size of the photosynthetic unit (both photoacts). The O_2 yield per flash differs depending on whether the chloroplasts are excited by periodic flashes (5×10^{-5} seconds) or by flash groups (two flashes per group and a long constant dark time between the groups, flash 5×10^{-5} seconds). Because of this discrepancy, doubt has been raised about the interpretation of the length of the dark time on the O_2 yield in periodic flashes as being equivalent to a half-life of the rate-determining step (Bültemann *et al.*, 1964).

Work on the quantum yield of photosynthesis unexpectedly provided some of the evidence that there are two successive photoacts involved in the production of the photochemical reductant for CO_2. In green plants, illuminated at wavelengths longer than 685 nm, the observed quantum yield was appreciably less than was expected from the absorption at this wavelength ("red drop"), while at shorter wavelengths the quantum efficiency was approximately constant throughout the visible region as expected (Emerson and Lewis, 1943). This decrease in the photosynthetic efficiency in the far red occurred in the region where chlorophyll a was the only known pigment absorbing light. Emerson *et al.* (1957) showed that the decrease in quantum yield observed with saturation far-red light is overcome when the cells are simultaneously provided with a supplementary light at high intensities of shorter wavelength (Fig. 3.4), the "enhancement effect." Thus Emerson (1958) concluded that a quantum requirement of

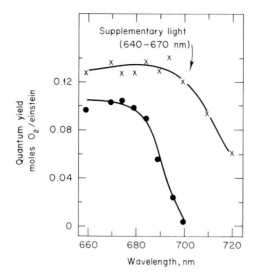

Fig. 3.4. Effect of supplementary light on the quantum yield in photosynthesis (from Emerson *et al.*, 1957). *Chlorella pyrenoidosa* cells were supplied with light in band widths not exceeding 10 nm. Control experiments demonstrated that the effect of supplementary light on the quantum yield in the far-red region cannot be attributed only to the higher rate of photosynthesis in the supplementary light. It was suggested that the maintenance of maximum efficiency requires the excitation of another pigment besides chlorophyll a.

eight was most likely and photosynthesis involved two photochemical reaction steps sensitized by two different pigments, both of which must be excited for efficient rates of photosynthesis to occur. The decrease in efficiency observed in the far red occurred because in these regions only one of these pigments (chlorophyll a) was absorbing radiation, and hence only one of the two steps is activated. The photosystem activated at wavelengths longer than 685 nm is known as Photosystem I. The excitation that occurs at shorter wavelengths, with a maximum at about 670 nm, has its first excited state at an energy level higher than that of the first excited state of chlorophyll a and is known as Photosystem II.

B. CYTOCHROMES OF CHLOROPLASTS

The hydrogen equivalent generated by the photochemical act in the chloroplast is transferred to CO_2 by the electron transport process involving a series of intermediate carriers. During dark respiration, which occurs

TABLE 3.1   Properties of chloroplast cytochromes[a]

Cytochrome	α-Band in reduced state (nm)	E_0' (volts)	Concentration in chloroplast moles chlorophyll/ mole cytochrome	Associated with Photosystem I or II	Autoxidizable	Other properties
f	554	+0.365	400	I	No	Oxidized by light at $-150°C$; reduced by light (Photosystem II), ascorbate, NADPH; oxidized by ferricyanide and Photosystem I.
b_6	563	−0.060	200	I	Yes	Not reduced by ascorbate; does not combine with CO; oxidized prior to illumination.
b_{559}	559	+0.055	200	II	No	Oxidized by light at $-150°C$; does not combine with CO; tightly bound to chloroplast lamellae.

[a] Sources: Bendall, 1968; Bendall and Hill, 1968; Boardman, 1968; Boardman and Anderson, 1966; Duysens and Amesz, 1962; Hill, 1954; Hill and Bonner, 1961; Knaff and Arnon, 1969; Hind and Nakatani, 1970.

primarily in the mitochondrion, electron transport is largely conducted by the reduction and oxidation of the heme–iron prosthetic group of the cytochrome components in a one-electron process. The oxidation–reduction properties of these proteins result from the presence of the iron atom which is alternately oxidized (ferric) or reduced (ferrous). Cytochromes have an absorption maximum near 400 nm characteristic of heme compounds (the Soret band), and in the reduced state have three absorption peaks in the visible spectrum designated as the α-, β-, and β'-bands. The strong α-band absorbs in the region between 550 and 610 nm. Although the process of photosynthesis involves the movement of electrons against a free energy gradient, in contrast to dark respiration where the electrons flow downhill, certain cytochromes have been found to be specifically associated with the electron transport in the chloroplasts of green plants and in chromatophores of photosynthetic bacteria (Table 3.1).

Investigations by R. Hill and his co-workers provided evidence for the existence of cytochrome f and b_6 in intact leaf tissue as well as in isolated chloroplasts. Changes in the oxidation and reduction of the α-band of cytochromes can easily be followed spectrophotometrically, but there is usually a strong absorption by chlorophyll in this region of the spectrum, and hence studies on cytochrome oxidation–reduction have been made using etiolated plant material, mutants deficient in chlorophyll, or with chloroplasts from which some of the chlorophyll has been extracted.

Cytochrome f is a c-type cytochrome that can be extracted from green leaves with alkaline ethyl alcohol. The purified cytochrome f (Forti *et al.*, 1965) has a molecular weight of about 245,000 and contains four hemes per molecule. Fractionation of chloroplasts with nonionic detergents (Boardman and Anderson, 1966) and spectrophotometric studies with chloroplasts (Duysens and Amesz, 1962) have shown that cytochrome f is associated with so-called Photosystem I and that light of these wavelengths causes its oxidation. Moreover, this cytochrome within the chloroplast is reduced by light of "Photosystem II." The reduction, but not its oxidation, is inhibited by low concentrations of DCMU [3-(3,4-dichlorophenyl)-1,1-dimethylurea], indicating that it is a carrier in photosynthetic electron transport that operates between Photosystem II and I (Fig. 3.5). The oxidation–reduction potential of $+0.365$ V is appropriate for such a position in the electron transport chain.

Cytochrome b_6, with an α-band at 563 nm (Table 3.1), is found in chloroplasts at twice the molar concentration of cytochrome f. It is autoxidizable and thus is usually found in the oxidized state. Its reduction in chloroplasts can be demonstrated by addition of sodium dithionite ($Na_2S_2O_4$). This cytochrome has not been purified from leaves, and although its spectrum is readily observed in etiolated mung bean tissue, no light-induced optical

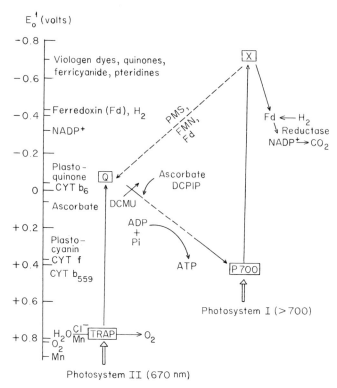

FIG. 3.5. The role of Photosystems II and I in photosynthetic energy transport and photophosphorylation shown as a "Z" diagram. The oxidation–reduction potentials (E_0') of various cofactors and possible intermediates are given to illustrate the region where they might function as carriers in the transport chain that begins with the photoxidation of water and ends in the reduction of CO_2. Recent evidence places the oxidation-reduction potential of cytochrome b_{559} at $+0.055$ V.

responses corresponding to this cytochrome were observed in green tissue. Nevertheless the reduction of cytochrome b_6 in the light has been observed with broken chloroplast preparations (W. Bonner and Hill, 1963).

A second b-type cytochrome, that has an α-band at 559 nm, is also found in chloroplasts. In pea chloroplasts the so-called cytochrome b_{559} apparently had a redox potential of $+0.37$ V (Bendall, 1968), but recent work with fractionated spinach chloroplasts gave a value of $+0.055$ V (Hind and Nakatani, 1970). Thus chloroplasts have at least two cytochromes of the b-type with similar α-bands in the reduced state but differing in oxidation–reduction potential by about 115 mV. Fractionation of spinach chloroplasts with digitonin revealed that cytochrome b_{559} is associated with Photosystem II particles (Boardman and Anderson, 1966), and this cytochrome component appears to be tightly attached to the grana lamellae.

Further evidence of the possible importance of this cytochrome in Photosystem II comes from recent experiments by Knaff and Arnon (1969). They observed the photooxidation of b_{559} in broken spinach chloroplasts only at low temperatures (below $-150°$ C), where enzymatic rates would be negligible, and showed that light at 664 nm was more effective than 714 nm, thus providing further evidence that this cytochrome is linked with Photosystem II. Because of the light-induced oxidation of this cytochrome below $-150°$ C, these workers argue that cytochrome b_{559} acts primarily as a participant in one of the primary light reactions of photosynthesis. They have proposed that there are in fact two separate Photosystem II reactions, not merely one, and that in the first light-driven step, a newly found constituent of chloroplasts, c_{550}, captures the electrons and passes them on to cytochrome b_{559} by a series of light-independent steps. However, based on present knowledge, the rate of turnover of cytochrome b_{559} appears to be too slow to account for such a primary role in photosynthesis (Hind, 1968; Ben Hayyim and Avron, 1970). Perhaps the oxidation of this cytochrome in the light affects photosynthesis by some indirect means such as by controlling the conformation of intermediates in electron transport rather than as serving as an intermediate in this process itself.

C. THE HILL REACTION

In 1941, C. B. Van Niel summarized his work with the green sulfur bacteria in which CO_2 is fixed by the reaction.

$$CO_2 + 2 H_2\overset{*}{S} \longrightarrow (CHOH) + H_2O + 2 \overset{*}{S} \tag{3.3}$$

By analogy, this suggested that all of the O_2 produced by green plants is derived from water and none from CO_2,

$$CO_2 + 2 H_2\overset{*}{O} \longrightarrow (CHOH) + H_2O + \overset{*}{O}_2 \tag{3.4}$$

A test of this hypothesis was made by Ruben and colleagues (1941) who supplied *Chlorella* with bicarbonate in the light under conditions where either the water or the bicarbonate was enriched with ^{18}O. The evolved oxygen, measured in a mass spectrometer, showed isotopic properties characteristic of the water and not the bicarbonate, as suggested by the above equation. Control experiments were carried out to correct for the rate of isotopic exchange between bicarbonate and water, but some workers believe that these rates were sufficiently high so that it was not conclusively established that water is the source of all of the O_2 evolution.

In 1937 R. Hill made the important discovery that when isolated chloroplasts are supplied with ferric salts in the light, the ferric salts are reduced to the ferrous form and O_2 is evolved. Illuminated isolated chloroplasts were able to produce O_2, in an apparent reversal of dark respiration, in the presence of a number of suitable hydrogen acceptors. These include two-electron acceptors, such as quinones and dyes, and one-electron acceptors such as ferricyanide. The stoichiometry of the "Hill reaction" with ferricyanide as hydrogen acceptor shows that four hydrogen equivalents are transferred for each mole of O_2 produced,

$$4 \text{ Fe}^{3+} \text{ (cyanide)} + 2 \text{ H}_2\text{O} \xrightarrow[\text{chloroplasts}]{\text{light}} 4 \text{ Fe}^{2+} \text{ (cyanide)} + 4 \text{ H}^+ + \text{O}_2 \qquad (3.5)$$

The direction of the flow of electrons during photosynthesis can be shown to reverse in the dark, when respiration predominates, with a crude mixture containing chloroplasts, mitochondria, and some leaf extract maintained in an aerobic environment (Hill and Bonner, 1961). In such a mixture, by observing changes in the spectrum of cytochrome c (present in the mitochondria), one finds that the cytochrome is oxidized in the dark as a result of the respiration of endogenous substrates in the crude leaf preparation. In the light, however, the cytochrome becomes reduced as electrons are transported against a free energy gradient in the chloroplasts.

During photosynthesis if X and Y are the hypothetical acceptors of the products of the photolysis of water, then photosynthetic electron transport may be conveniently expressed as,

$$\text{H}_2\text{O} + \text{X} + \text{Y} \xrightarrow{\text{light}} \text{XH} + \text{YOH}$$
$$\text{XH} \longrightarrow \text{X} + \text{(H)}$$
$$\text{YOH} \longrightarrow \text{Y} + \tfrac{1}{2}\text{H}_2\text{O} + \tfrac{1}{4}\text{O}_2$$
$$\text{Net:} \quad \tfrac{1}{2}\text{H}_2\text{O} \xrightarrow[\text{chloroplasts}]{\text{light}} \text{(H)} + \tfrac{1}{4}\text{O}_2$$

In such a formulation, X represents hydrogen acceptors of the Hill reaction such as ferricyanide, and XH is the reductant that provides electrons to eventually reduce CO_2.

Such experiments and equations led Hill and Bendall (1960) to postulate that cytochrome b_6 and f function to connect the two light-driven photosystems that transfer electrons against the free energy gradient. In this scheme, sometimes called a "Z" diagram (Fig. 3.5), oxidized cytochrome b_6 is reduced by Y to produce YOH, and cytochrome f is oxidized by X to give XH. The reactions between the cytochromes would be enzymatic, and if they could be inhibited, one would expect that in the light cytochrome f would become oxidized and b_6 reduced. Such results were in fact obtained on illumination of pale yellow-green leaves. Moreover, the suggestion by Hill and Bendall was consistent with the findings of Emerson (1958) on the enhancement of photosynthesis dependent on the absorption of the light by two different pigments (Chapter 3.A.4). As expected from the scheme in Fig. 3.5, cytochrome f was oxidized by light of Photosystem I and its reduction was obtained by excitation produced by Photosystem II (Duysens and Amesz, 1962). Thus Photosystem II reduces cytochrome f and produces O_2 from H_2O, while Photosystem I reduces $NADP^+$ and oxidizes cytochrome f.

These views greatly stimulated research progress on photosynthetic electron transport although they are not universally accepted (Chapter 4.J). Warburg has suggested that CO_2 is necessary for the production of O_2 in photosynthesis, yet according to Fig. 3.5, CO_2 should not be necessary for O_2 evolution in the Hill reaction. The reduction of the dye 2,6-dichloro-phenolindophenol by pea chloroplasts was, however, greatly accelerated by low concentrations of CO_2 (N. E. Good, 1963). Such effects of CO_2 have been confirmed by Hill (1965). The stimulation by CO_2 differs somewhat depending on the electron acceptor used in the Hill reaction. For example, with ferricyanide there was no CO_2 effect at low light intensities, while with dichlorophenolindophenol the addition of CO_2 increased the rate of dye reduction at all light intensities (J. West and Hill, 1967). The results with ferricyanide as acceptor indicate that the CO_2 stimulation is exerted upon a dark reaction that becomes rate-limiting at higher light intensities, and also suggests that the CO_2 effect on the Hill reaction may be an indirect one rather than representing the behavior of a CO_2-compound which reacts directly with the photochemical system as suggested by Warburg.

D. PHOTOSYNTHETIC ELECTRON TRANSPORT

1. PHOTOSYSTEM II

Present evidence indicates that photosynthetic electron transport takes place through two separate photoacts, and that the Hill reaction by which O_2 is evolved is the result of the action of the short wave-length Photosystem II. Differential centrifugation of digitonin-treated chloroplasts permits a partial separation of the two Photosystems (Boardman, 1968). The heavier fragments are active in the Hill reaction with water as the electron donor and dyes as oxidants, while the smaller particles photoreduce $NADP^+$ when provided with ascorbate or reduced dichlorophenolindophenol, ferredoxin, and ferredoxin-$NADP^+$ reductase (Fig. 3.5). The smaller particles are thus enriched in Photosystem I. Fractionation of the two Photosystems has also been achieved in the presence of other agents such as the anionic detergent Triton X-100 (Vernon *et al.*, 1967, 1969).

In isolated chloroplasts, Photosystem II is less stable to aging and to the presence of detergents. This photoact is strongly inhibited by low concentrations of a number of herbicides including DCMU (Chapter 3.E), CMU [3-(*p*-chlorophenyl)-1,1-dimethylurea], and *s*-triazines such as simazine [2-chloro-4,6-bis(ethylamino)-*s*-triazine], and detergents (Wessels, 1963). As shown in Fig. 3.5, DCMU inhibits electron transport on the reducing side of Photosystem II rather than on the oxidizing side (N. I. Bishop, 1958), and in the presence of DCMU, electron transport can be made to bypass Photosystem II and thus circumvent the need for water as an electron donor if substitute electron donors such as ascorbate, or better still a reduced dye such as dichlorophenolindophenol are provided.

The washing of isolated spinach chloroplasts in 0.8 M tris(hydroxymethyl)aminomethane buffer (Tris) at pH 8 for 10 minutes also inactivates Photosystem II (T. Yamashita and Butler, 1968). Chloroplasts treated in this way show a marked inhibition of dye reduction in the light, as well as in the photoreduction of $NADP^+$ in the Hill reaction. These chloroplasts contain about two-thirds as much manganese, and these various photoreactions are restored 50 to 80% by the addition of 0.1 mM Mn^{2+} (Itoh *et al.*, 1969; K. Yamashita *et al.*, 1969). When spinach chloroplasts (Mn/chlorophyll ratio of 1:80) were heated at 40°, they also lost most of their manganese. Further heating, however, did not remove the residual manganese (Mn/chlorophyll ratio of 1:400) (Jenkins and Griffiths, 1970). The portion of the manganese that was more difficult to extract is probably associated with chloroplast membranes, and there are thus at least two classes of manganese that are bound in different ways in chloroplasts. Isolated tomato chloroplasts also contain a manganese fraction that is not easily extracted by repeated washing, lysis, or during purification by density-gradient centrifugation (Possingham and Spencer, 1962).

The role of manganese in O_2 evolution (Photosystem II) had previously been suggested, but it seemed uncertain because many early experiments were carried out with manganese-deficient plants. This technique raised doubts because manganese deficiency also has an effect on the structural organization of the chloroplast lamellae as well as on the Hill reaction.

Cheniae and Martin (1967), however, found a restoration of the Hill reaction within 15 to 20 minutes, independent of *de novo* chlorophyll or protein synthesis, when $MnCl_2$ was added to manganese-deficient algae. This restoration of activity required light absorbed by Photosystem II, since DCMU, in concentrations that completely inhibited the Hill activity, also fully inhibited the photoreactivation of the inhibited cells. Since this restoration of activity, unlike earlier studies, occurred in a very short time, these experiments ruled out the involvement of significant structural changes in the chloroplast lamellae as being related to the manganese requirement in Photosystem II.

O. Warburg showed in 1948 that chloride is also necessary for oxygen evolution by isolated chloroplasts in the presence of artificial electron acceptors. This conclusion was confirmed more recently because broken spinach chloroplasts require the addition of chloride ion for several different reconstructed photochemical reactions in which oxygen is produced (J. H. Bové *et al.*, 1963). During the intervening years a number of investigators believed that chloride ion is needed merely for the protection of chloroplast components rather than serving a role as a cofactor. Whole pea and spinach chloroplasts do not show a chloride requirement for the Hill reaction until their internal chloride ion is released by some treatment that damages the outer membrane and promotes swelling (Hind *et al.*, 1969), but reactions associated with Photosystem II are dependent on added chloride ions. These workers have concluded that both the manganese and chloride ion-requiring sites are located close together and on the oxidizing (water) side of Photosystem II, as shown in Fig. 3.5.

Much less is known about the intermediates between water and Photosystem II than in other regions of the electron transport chain because the intermediates involved do not seem to show measurable changes in light absorption as do other portions of the chain and thus cannot easily be measured. For the most part, studies on intermediates in O_2 evolution have been restricted to kinetic analyses in attempts to measure the pool sizes of carriers involved in oxygen evolution (Kok and Cheniae, 1966). Such experiments suggest that two manganese enzymes are involved in oxygen evolution, both acting on the oxidant side of Photosystem II (Kok and Cheniae, 1969).

2. PHOTOSYSTEM I

Electron transport mediated by the long wavelength photoact produces the strong reductant that eventually reduces CO_2 (Fig. 3.5). Photosystem I is bound to a small particle, is associated with membranes, and is stable during isolation. It is associated with the trapping center P700 (Chapter 3.A), and small particles from chloroplasts of a number of species have recently been obtained that contain one P700 for about 30 chlorophyll molecules (Vernon *et al.*, 1969). These particles showed a light-induced oxidation of P700 as did preparations (one P700 per 16 chlorophylls and a high ratio of chlorophyll a to chlorophyll b) obtained from a stroma lamellae fraction of spinach chloroplasts (Sane and Park, 1970).

Plastocyanin is a blue copper-containing protein that appears to function as an intermediate in the reducing system of Photosystem I. The reduced form is not autoxidizable, and it has an E_0' (pH 7) of $+0.40$ V, slightly higher than that of cytochrome f. In its oxidized form plastocyanin has a maximum absorbancy at 597 nm and it is reduced in the light by chloroplast preparations and shows no absorbancy in the reduced form. Katoh and Takamiya (1965) observed that $NADP^+$ reduction by sonically disrupted chloroplasts is almost completely inhibited, and this inhibition was largely reversed on addition of about 2×10^{-5} M plastocyanin. The role of plastocyanin in electron transport was also demonstrated with a mutant of *Chlamydomonas* which did not synthesize plastocyanin, but could be grown on acetate (Gorman and Levine, 1966). No photoreduction of $NADP^+$ was found unless plastocyanin was added when the rate of photoreduction by chloroplast fragments obtained from this mutant was measured. In the absence of plastocyanin, cytochrome f was not oxidized but was reduced. Hence it followed that electron transport in these cells must have occurred by the sequence (Fig. 3.5):

Photosystem II ⟶ cytochrome f ⟶ plastocyanin ⟶ Photosystem I

3. THE STRONG REDUCTANT AND FERREDOXIN

The chemical identity of the strong reductant ("X" in Fig. 3.5) produced by Photosystem I is still uncertain. Such a substance, however, probably occurs in bacterial photosynthesis, where O_2 is not evolved and Photosystem II is lacking, as well as in photosynthesis carried out by higher plants. Photochemical systems involving Photosystem I can reduce viologen dyes and bipyridinium compounds having oxidation–reduction potentials of less than -0.7 V (Zweig and Avron, 1965), thus suggesting that the naturally occurring strong reductant must have a potential at least as low as this.

Certain pteridine derivatives have a redox potential of about -0.7 V and these can also be reduced in light in the presence of a bacterial chromatophore fraction, thus suggesting that compounds of this type are related to "X" in Fig. 3.5 (Fuller and Nugent, 1969). C. F. Yocum and San Pietro (1969, 1970) have partially purified a factor called the ferredoxin reducing substance (FRS) which stimulates methylviologen, ferredoxin, and NADP$^+$ photoreduction by spinach chloroplast fragments. No detailed information is yet available about the chemical nature of FRS, but it has some properties in common with electron transport factors found by other workers. A system containing chloroplast fragments, saturating concentrations of ferredoxin, ferredoxin-NADP$^+$ reductase, plastocyanin, and NADP$^+$, with ascorbate and dichlorophenolindophenol (DCPIP) as the donor system, produced little reduction of NADP$^+$ unless FRS in the reduced form was provided (Fig. 3.5). Illumination of the chloroplast fragments and the other components of the donor system resulted in the formation of sufficient reducing capacity in FRS so that it could then reduce NADP$^+$ in a subsequent dark period. The identification of FRS, which appears to have properties required of the naturally occurring strong reductant including a redox potential less than -0.5 V, will be awaited with great interest.

The strong reductant produced in Photosystem I transfers its electrons to ferredoxin, an iron-sulfur protein with a redox potential (-0.43 V) equal to or less than that of the hydrogen electrode. This protein was first extracted from chloroplasts and partly purified as "photosynthetic pyridine nucleotide reductase" (PPNR) and shown to be important in NADP$^+$ reduction by San Pietro and Lang (1956). The purified protein from spinach was later called ferredoxin (Tagawa and Arnon, 1962) because of its similar properties to the "iron-redox protein" or "ferredoxin" isolated from nitrogen-fixing bacteria. Ferredoxins have molecular weights of 6,000 to 12,000. They contain 2 moles of nonheme iron per mole of enzyme and inorganic sulfur in equivalent amounts, and the sulfur is acid labile, yielding hydrogen sulfide on acidification. The chemical nature of the acid-labile sulfur is uncertain, although there is some evidence that it arises from sources other than the cysteine residues of the polypeptide (Hong *et al.,* 1969). The isolated plant and algal proteins are red in color, and have absorption maxima at 277, 330, and 420 nm in the oxidized form. Reduced ferredoxin is strongly autoxidizable. During the catalytic activity of ferredoxin, its nonheme iron undergoes a change in valence state so that when the absorbancy at 420 nm is completely bleached, about 50% of the iron content is transformed to the ferrous state (San Pietro, 1967). Since there are two iron atoms per molecule of ferredoxin, only one functions as an electron carrier, and one electron is thus transferred per molecule of ferredoxin.

Ferredoxin is reduced in chloroplasts on illumination by Photosystem I, and it can also be reduced in isolated systems by the action of added hydrogenase in the presence of H_2 (Fig. 3.5). Reduced ferredoxin in chloroplasts might be oxidized by O_2 except that this leakage is believed to be prevented by the strong affinity of the reduced ferredoxin for the flavoprotein $NADP^+$-reducing system called ferredoxin-$NADP^+$ reductase. The reaction catalyzed by this enzyme is,

$$2 \text{ Ferredoxin} + NADP^+ \xrightarrow{\text{Reductase}} 2 \text{ ferredoxin (oxidized)} + NADPH + H^+ \quad (3.6)$$

Such a role in transferring electrons from ferredoxin to $NADP^+$ was established in several laboratories, and the reductase from spinach leaves has been obtained in crystalline form (Shin *et al.*, 1963). In addition to its reductase activity, this flavoprotein also catalyzes NADPH-diaphorase and pyridine nucleotide transhydrogenase reactions. The NADPH thus produced in the light is subsequently used to reduce CO_2 to the level of carbohydrate by enzymatic reactions (Chapter 4.C) which do not generally require light for their activities.

E. PHOTOSYNTHETIC PHOSPHORYLATION

The scheme for photosynthetic electron transport (Fig. 3.5) illustrates how NADPH results from the sequential action of the two photoacts. Besides providing NADPH, photosynthetic electron transport results in the production of adenosine triphosphate (ATP), whose energy when released on hydrolysis of the pyrophosphate bond may be used to promote endergonic reactions in many biological systems. Thus the ATP produced by photophosphorylation is essential in the process of photosynthetic CO_2 assimilation (Chapter 4.C) and in ion and solute movement in plant cells. Two types of photophosphorylation, designated as cyclic and noncyclic, are recognized.

Isolated chloroplasts carry out photophosphorylation when provided with catalytic amounts of certain redox cofactors such as phenazine methosulfate (PMS), riboflavin phosphate (FMN), ferredoxin, and the factors needed for phosphorylation ($ADP + P_i + Mg^{2+}$). In this type of photophosphorylation there is no net change in any exogenous electron acceptor and the ATP formation is thus said to be coupled to a *cyclic* flow of electrons. *Noncyclic* photophosphorylation occurs in the presence of oxidants of the Hill reaction such as ferricyanide, benzoquinone, dichlorophenolindophenol, or $NADP^+$. Noncyclic photophosphorylation with $NADP^+$ differs from cyclic by its dependence on a continuous production and uptake of O_2.

1. CYCLIC PHOTOPHOSPHORYLATION

Cyclic photophosphorylation occurs in organisms that are independent of the presence of O_2 in the external atmosphere but require light, as in photosynthetic bacteria and hydrogen-adapted algae. Since Photosystem II is missing in such organisms, it is not surprising that cyclic photophosphorylation takes place in their chromatophores or chloroplasts when they are supplied with suitable cofactors in the presence of inhibitors of O_2 evolution such as DCMU. In the presence of high concentrations of ferredoxin, chloroplast preparations also carry out cyclic photophosphorylation, and do it best in light of Photosystem I (Fig. 3.6). In systems consisting of broken chloroplasts from higher plants, rates of ATP synthesis greater than 1000 μmoles mg chlorophyll^{-1} hr^{-1} have been obtained (Avron, 1960), far in excess of rates observable in noncyclic photophosphorylation (about 300 μmoles mg chlorophyll^{-1} hr^{-1}). No O_2 uptake in cyclic photophosphorylation is observed by manometry or by measurement in a mass spectrometer, thus clearly distinguishing it from oxidative phosphorylation that occurs in mitochondria (Chapter 5.A), where ATP formation occurs at the expense of energy liberated by the oxidation of a substrate. Cyclic photophosphorylation may be represented as follows:

$$\text{Light (long wavelength)} + 2H_2O \rightarrow 2(H) + 2(OH)$$
$$\text{Carrier} + 2(H) + ADP + P_i \rightarrow \text{Carrier-}H_2 + ATP + H_2O$$
$$\underline{\text{Carrier-}H_2 + 2(OH) \rightarrow \text{Carrier} + 2H_2O}$$
$$\text{Net:} \quad \text{Light (long wavelength)} + ADP + P_i \rightarrow ATP + H_2O$$

Thus the cofactor (carrier) becomes alternately reduced and reoxidized by a component of the electron transport chain and not by O_2. Fig. 3.5 shows only a single region of photophosphorylation, and indicates that this site is the same as for noncyclic photophosphorylation. However, as discussed below, it is uncertain whether cyclic and noncyclic photophosphorylation take place at the same site or also in regions other than shown in Fig. 3.5.

It is still not known whether cyclic photophosphorylation, which is essential for anaerobic photosynthetic organisms, accounts for much ATP synthesis in the chloroplasts of higher plants. If it is important in higher plants, the identity of the natural cofactor is also not rigidly established. In the alga *Nitella* the active uptake of chloride ion requires the participation of Photosystem II, but in contrast the active uptake of potassium ions may be supported by cyclic electron transport since it occurs in far-red light in the presence of DCMU (Mac Robbie, 1965). Additional evidence for a role of cyclic photophosphorylation in algae has been obtained. In *Chlorella* a light-dependent uptake of glucose under anaerobic conditions was completely inhibited by a concentration of the inhibitor salicylaldoxime

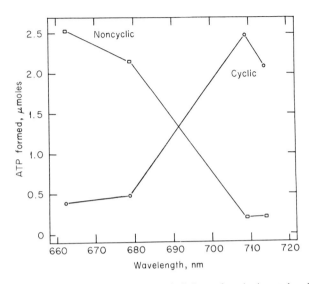

FIG. 3.6. Effect of red and far-red monochromatic light on ferredoxin-catalyzed photophosphorylation by spinach chloroplast preparations [from Arnon *et al.*, 1967. Copyright MacMillan (Journals) Ltd.] In addition to the chloroplasts, the cyclic system contained 140 nmoles of ferredoxin and the noncyclic system contained 14 nmoles of ferredoxin and 5 μmoles of NADP$^+$ in a final volume of 1.5 ml. Both systems contained 5 μmoles each of ADP and K$_2$H^{32}PO$_4$. Time of illumination, 10 minutes; radiation 1.56 einsteins min^{-1} cm^{-2}; gas phase, argon; temperature, 20°.

which did not affect O$_2$ evolution or CO$_2$ uptake (W. Tanner *et al.*, 1969), suggesting cyclic photophosphorylation was functioning under these conditions. Sufficient ATP was apparently produced by the cyclic process in *Chlorella* to provide energy for the synthesis of an adaptive enzyme, but CO$_2$ fixation required light of both photosystems and thus presumably noncyclic photophosphorylation was essential for CO$_2$ assimilation (Syrett, 1966). With isolated spinach chloroplasts it has been claimed that CO$_2$ assimilation depends on cyclic photophosphorylation because the inhibitor desaspidin blocked CO$_2$ uptake under conditions where noncyclic photophosphorylation was unaffected (Ramirez *et al.*, 1968). However, such reactions with reconstructed systems for cyclic photophosphorylation contained large quantities of added ferredoxin, and interpretation of these experiments is uncertain because the results are given as "percent of control" and no data are given for the absolute rates of CO$_2$ assimilation or photophosphorylation by the isolated chloroplasts. The role of cyclic photophosphorylation in CO$_2$ assimilation by higher plants is therefore not yet established, although it may have a role in the movement of solutes, especially in algae. Evidence suggesting that cyclic photophosphorylation does not function significantly in higher plants is presented later.

2. NONCYCLIC PHOTOPHOSPHORYLATION

This type of photophosphorylation is essential in O_2-evolving plants. It depends on the joint contribution of the two photochemical steps to produce ATP during photosynthetic electron transport and provides the bulk of the ATP that is necessary for the assimilation of CO_2.

The exact stoichiometry of noncyclic photophosphorylation is still a matter of controversy, but this form of photophosphorylation does occur as a stoichiometric reduction process,

$$2H_2O + light \rightarrow 2(H) + 2(OH)$$
$$2(OH) \rightarrow H_2O + \tfrac{1}{2}O_2$$
$$\underline{NADP^+ + 2(H) + ADP + P_i \rightarrow NADPH + H^+ + ATP + H_2O}$$
$$Net: \quad Light + NADP^+ + ADP + P_i \rightarrow NADPH + H^+ + \tfrac{1}{2}O_2 + ATP$$

Noncyclic photophosphorylation is inhibited by substances that block oxygen evolution, such as DCMU, but it can occur with isolated chloroplasts in the presence of DCMU if artificial electron donors are supplied (Fig. 3.5). In the presence of ADP and inorganic phosphate, with $NADP^+$ as the oxidant, the minimum stoichiometry is $NADP^+$ reduced: phosphate esterified: atoms O_2 equals $1:1:1$ (Arnon, 1958). Thus the P/2e ratio would equal at least unity.

$$ADP + P_i + NADP^+ + 2H_2O \rightarrow ATP + \tfrac{1}{2}O_2 + NADPH + H^+ + H_2O \qquad (3.7)$$

Ferredoxin may serve as an electron acceptor in noncyclic photophosphorylation, where light of Photosystem II is best (Fig. 3.6), whether present in catalytic amounts as a carrier to $NADP^+$ or in stoichiometric concentrations as the final acceptor (Arnon *et al.*, 1967).

Arnon has shown that ferricyanide, an oxidant in the Hill reaction, substitutes for $NADP^+$ in noncyclic photophosphorylation by chloroplast preparations. Then the ferricyanide reduced: phosphate esterified: atoms O_2 equals $2:1:1$. Since four hydrogen equivalents are needed for the reduction of one mole of CO_2 to the level of carbohydrate, this requirement is met as follows:

$$4Fe^{3+} (cyanide) + 2H_2O + 2P_i + 2ADP \rightarrow 2ATP + O_2 + 4Fe^{2+} (cyanide) + 4H^+ \quad (3.8)$$

With broken chloroplasts supplied with ferricyanide the addition of ADP and inorganic phosphate greatly stimulated the rate and extent of O_2 evolution in the light. Thus the Hill reaction as normally carried out, without phosphorylation, represents an uncoupled phosphorylation reaction or a "basal" rate of photosynthetic electron transport whose rate is enhanced in the presence of a phosphorylating system. Presumably this results because an intermediate is produced that slows down electron transfer unless its energy is utilized toward the production of ATP or unless it is "uncoupled" by substances that break down the high energy state. The considerable "basal" rate of electron transport may change during noncyclic photophosphorylation, and this point has a considerable bearing on the stoichiometry.

The stoichiometry for noncyclic photophosphorylation as described would provide one molecule of ATP for each molecule of NADPH produced, or a P/2e ratio of 1.0. However, one of the primary reactions for fixation of CO_2 in photosynthesis by the photosynthetic carbon reduction cycle (Chapter 4.C) requires 3 moles of ATP and 2 of NADPH for each mole of CO_2 fixed, a P/2e ratio of 1.5. Thus if noncyclic photophosphorylation were the only means of producing ATP in the light, the yield would be insufficient for CO_2 fixation. Winget *et al.* (1965) obtained P/2e ratios of 1.2 to 1.3 consistently in noncyclic photophosphorylation with ferricyanide as acceptor, and suggested that the true stoichiometry is in fact 2.0. Similar results were obtained in other laboratories (Table 3.2; Horton and Hall, 1968; K. R. West and Wiskich, 1968; Forti, 1968).

When this point was reinvestigated in Arnon's laboratory, they still concluded that the P/2e ratio is 1.0, although higher ratios were obtained under certain conditions which were attributed to analytic errors in the ATP determination (Del Campo *et al.*, 1968). The analytic methods used to determine both ferricyanide reduction (absorbancy at 420 nm) and ATP synthesis appeared to be straightforward and are not likely the major cause of such discrepancies. The main difficulty in interpretation probably arises because there are two kinds of electron transport in isolated chloroplasts, a nonphosphorylating or "basal" electron transport and a phosphorylating type. Isolated chloroplasts show a considerable electron transport without phosphorylation (Table 3.2), and should this continue during phosphorylation, the overall observed stoichiometry would then represent the average of the two processes only one of which serves to produce ATP. If the rate of the "basal" nonphosphorylating transport is unchanged in the presence and the absence of phosphorylation, then a P/2e ratio of 2.0 is regularly observed. Izawa and Good (1968) showed that these two pathways of electron transport in isolated chloroplasts are parallel and do not compete with each other. Providing this interpretation holds there would

TABLE 3.2   *Inhibition (by Dinitrophenol) and Uncoupling (by Ammonium Sulfate) of Noncyclic Photophosphorylation by Isolated Spinach Chloroplasts[a,b]*

M Inhibitor	Ferricyanide reduced (μmoles)		ATP Formed μmoles	P/2e Observed
	$-$ADP	$+$ADP		
2,4-Dinitrophenol				
0	0.20	0.49	0.30	1.2
1×10^{-4}	0.23	0.46	0.29	1.3
3.3×10^{-4}	0.17	0.32	0.18	1.1
1×10^{-3}	0.09	0.18	0.10	1.1
Ammonium sulfate				
0	0.21	0.59	0.36	1.2
1×10^{-4}	0.35	0.62	0.32	1.0
3.3×10^{-4}	0.62	0.74	0.12	0.32
1×10^{-3}	0.67	0.71	0.02	0.057

[a] From Jagendorf, 1959.

[b] The complete reaction mixtures contained a chloroplast preparation (30 μg chlorophyll per tube), Mg^{2+} (10 μmoles), ^{32}P-phosphate (10 μmoles), ferricyanide (2 μmoles), NaCl (70 μmoles), Tris buffer at pH 7.8 (40 μmoles), and ADP (0.1 μmole) in a final volume of 3.0 ml. After 2 minutes in the light at 25°, ^{32}P-ATP formation and ferricyanide reduction (determined spectrophotometrically at 400 nm) were measured. (Since ferricyanide is a one-electron acceptor, μmoles reduced are divided by 2 to calculate the P/2e observed.)

Tris buffer probably uncouples photophosphorylation to some extent under these experimental conditions (Winget *et al.*, 1965). However, a less precise method of measuring ^{32}P-ATP was used in these experiments compared with more recent work thus perhaps accounting for the relatively high P/2e ratios observed in the controls.

Although dinitrophenol inhibits electron flow above pH 7.0, it also functions as an uncoupler at pH 7.0 or below (Neumann and Jagendorf, 1964).

be sufficient noncyclic photophosphorylation to accommodate CO_2 fixation as well as other ATP-requiring reactions without requiring any of the cyclic type.

The dependence on O_2 and the results with inhibitors indicate that noncyclic photophosphorylation provides most of the ATP synthesis in the light in photosynthetic tissues of higher plants. When tobacco leaf disks were floated on glucose or sucrose solution in the light, no starch was produced (a reaction that requires at least two ATP's for each glucose, Chapters 4.K and 5.A) after 24 hours of continuous illumination in a N_2 atmosphere, but starch was readily synthesized in an atmosphere containing O_2 (Krall and Bass, 1962). The inhibitor CMU, which affects noncyclic but not cyclic photophosphorylation, completely blocked starch formation in the light, again suggesting that ATP from noncyclic photophosphorylation is used primarily for starch synthesis. Additional evidence comes from experiments with isolated Class I spinach chloroplasts which were supplied with concentrations of phlorizin, a phenolic dihydrochalcone glucoside, at concentrations that inhibited cyclic photophosphorylation with PMS as cofactor. Nevertheless, CO_2 fixation continued to proceed at rapid rates by present standards suggesting that the noncyclic pathway was providing the needed ATP (Vose and Spencer, 1967).

There are suggestions that the coupling sites where cyclic and noncyclic photophosphorylation occur in the electron-transport chain may differ (unlike Fig. 3.5 which indicates a single site) because certain inhibitors often have different effects on these two types of phosphorylation in isolated chloroplasts (Avron and Neumann, 1968). For example, the inactivation of photophosphorylation by treatment of chloroplasts with *n*-heptane resulted in a greater inhibition of noncyclic than cyclic phosphorylation, indicating that multiple sites for photophosphorylation do occur in isolated chloroplasts (Black, 1967). Low concentrations of carbonylcyanide *m*-chlorophenylhydrazone (CCCP), which acts as an uncoupler as defined below, inhibited the ferricyanide-dependent noncyclic phosphorylation 90% while the cyclic reaction was inhibited only 10% when carried out with PMS as cofactor (Avron and Shavitt, 1965). Other workers have also obtained different effects on cyclic and noncyclic photophosphorylation in a variety of electron acceptor systems (Arnon *et al.*, 1967).

The phosphorylation sites in the energy-transport chain have only been partly determined. No sites for photophosphorylation apparently exist between Photosystem I and the reduction of $NADP^+$, because this portion of the electron-transport chain carries out a light-dependent electron transfer without ATP synthesis. There is positive experimental evidence that a site for photophosphorylation does occur between Photosystems II and I (Fig. 3.5). One method involved observing the oxidation and reduction of cytochrome f in isolated spinach chloroplasts in a dual wavelength spectrophotometer by measuring the difference in spectrum between 554 and 540 nm (Table 3.1) under conditions where the rate of electron flow is limited by the absence of phosphorylating reagents (Avron and Chance, 1967; Ben Hayyim and Avron, 1970). When light of 640 nm (Photosystem II) is provided, the addition of an uncoupling agent should accelerate electron flow to cytochrome f and cause its reduction if the coupling site for ATP formation is rate-limiting for electron transport and phosphorylation occurs between Photosystem II and cytochrome f. On the other hand, uncoupling should cause an oxidation of cytochrome f if the coupling site were located between cytochrome f and Photosystem I. The addition of the uncoupler NH_4Cl to a chloroplast suspension caused a sharp reduction of cytochrome f, indicating that ATP is synthesized prior to cytochrome f reduction:

Photosystem II $\longrightarrow \longrightarrow \longrightarrow$ cytochrome f $\longrightarrow$ Photosystem I $\longrightarrow$ strong reductant
$\searrow$ ATP

Another phosphorylation may occur between the region where water serves as the electron donor in Photosystem II and that portion of the energy-transport chain where ascorbate serves as the electron donor (Böhme and Trebst, 1969).

3. INHIBITORS

Although ATP synthesis by photophosphorylation requires a photosynthetic electron flow in the chloroplast, the two processes are entirely distinct. The inhibition of electron transport will also slow ATP formation. but ATP synthesis occurs by a series of unknown reactions that can be disengaged from electron transport. Thus a substance that permits electron transport to continue while blocking ATP synthesis is said to be an uncoupler, and this mode of inhibition is frequently detected by showing an enhancement of electron transport without producing a corresponding increase in phosphorylation. A number of compounds have been found to uncouple photophosphorylation, and these have been useful in studies of the mechanism of this process. N. Good et al. (1966) listed many uncouplers, and the list of such substances is increasing constantly. The nature of the activities of only a few representative inhibitors is mentioned here.

Ammonium salts such as ammonium sulfate act as uncouplers of noncyclic photophosphorylation (Table 3.2). At a concentration of 10^{-3} M this salt stimulated the rate of ferricyanide reduction threefold, while ATP synthesis was almost completely abolished (Jagendorf, 1959). In contrast, 2,4-dinitrophenol, when added in increasing concentrations at pH 7.8 inhibited both ferricyanide reduction and ATP synthesis equally and did not change the observed P/2e ratio. Hence dinitrophenol did not act as an uncoupler but as an inhibitor of electron transport. Dinitrophenol does uncouple photophosphorylation from electron transport when the reaction is carried out at pH 7.0 or below instead of at pH 7.8 (Neumann and Jagendorf, 1964).

The essentiality of some membrane integrity for photophosphorylation was shown in studies with the detergent Triton X-100. At a concentration of 0.007%, Triton X-100 increased the rate of ferricyanide reduction by isolated chloroplasts in the light about twofold, while photophosphorylation was almost completely inhibited (Neumann and Jagendorf, 1965). The osmotic shock of frozen and thawed spinach chloroplasts also uncoupled photophosphorylation in isolated fragments (Uribe and Jagendorf, 1968).

The antibiotic Dio-9 acts as an energy-transfer inhibitor by still another mechanism. Dio-9 inhibits electron transfer in noncyclic photophosphorylation much as the antibiotic oligomycin inhibits oxidative phosphorylation in mitochondria (McCarty *et al.*, 1965; K. R. West and Wiskich, 1969). Ferricyanide reduction was inhibited 65% in one experiment in the presence of phosphate acceptor (ADP), while there was only a 24% inhibition when the phosphate acceptor was omitted from the reaction mixture. The inhibition brought about by Dio-9 in the presence of ADP was reversed by the addition of μM concentrations of CCCP, hence Dio-9 acted as an energy-transfer inhibitor and CCCP functioned as an uncoupler of photophosphorylation.

4. MECHANISM

In both oxidative phosphorylation and photophosphorylation some workers have assumed that a high-energy intermediate (X $\sim$ I) (read X squiggle I; the squiggle draws attention to the high free energy of hydrolysis of such compounds, the hydrolysis scale being a convenient means of thermodynamic bookkeeping) was produced as a result of the activity of the energy-transport chain, and that this intermediate then produced a phosphorylated intermediate which subsequently yielded ATP.

$$
\begin{array}{c}
\text{Energy-transfer} \\
\text{inhibitors} \\
\downarrow
\end{array}
$$

Energy-transport chain $\longrightarrow$ X $\sim$ I $\leftrightarrows$ X $\sim$ P $\quad + $ I $\leftrightarrows$ ATP + X

High-energy $\; + P_i \quad$ Phosphorylated $\qquad + $ ADP
intermediate $\qquad\qquad$ intermediate

$$\uparrow\downarrow$$

Cation production
+
H^+ uptake

SCHEME 3.1. Chemical coupling hypothesis.

Although Scheme 3.1 appeared to fit many of the known facts about phosphorylation, including cation uptake and H^+ production as a side reaction, no investigator isolated any high-energy chemical intermediate that would serve as a direct precursor of ATP synthesis. In addition, this hypothesis does not consider that reconstitution of the phosphorylating systems requires the integrity of membrane structure.

Certain cations may be moved across the membrane in exchange for protons in a process that is linked to energy transport in both mitochondria and chloroplasts, and this suggested the hypothesis that the energy transport chain may serve as a proton pump (Mitchell, 1961; Robertson, 1968). According to this hypothesis, the ATP is synthesized as a result of the light-activated passage of an equal number of hydrogen atoms and electrons in opposite directions across the grana membranes,

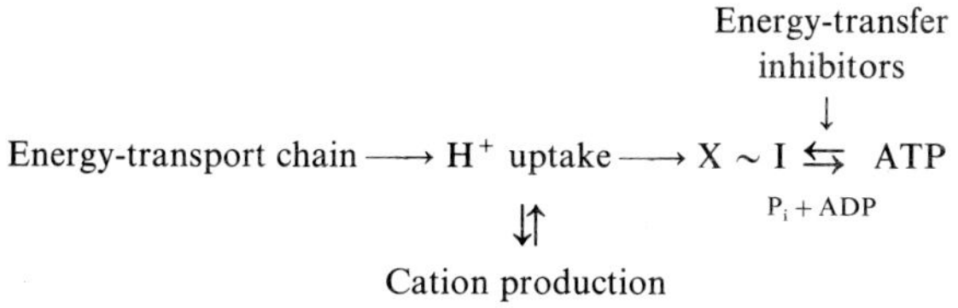

$$\text{Energy-transport chain} \longrightarrow \text{H}^+ \text{ uptake} \longrightarrow \text{X} \sim \text{I} \overset{\text{Energy-transfer}}{\underset{\text{}}{\rightleftarrows}} \text{ATP}$$

Energy-transfer
inhibitors
↓
Energy-transport chain ⟶ H$^+$ uptake ⟶ X ∼ I ⇆ ATP
⇕   P$_i$ + ADP
Cation production

SCHEME 3.2. Chemiosmotic hypothesis.

Scheme 3.2 shows that the proton gradient creates energy, and coupling to ATP synthesis occurs by means of a reversible membrane-bound ATPase. During ATP synthesis hydroxyl ions move to the inside and hydrogen ions to the outside of the membrane (Fig. 3.7). Comparing Schemes 3.1 and 3.2 shows that, in the chemical coupling hypothesis, cation production and hydrogen ion uptake result from the formation of a high-energy intermediate, while in the chemiosmotic hypothesis the high-energy intermediate is produced as a result of hydrogen ion uptake and cation production. The concentration gradient produces an electrical potential difference across the membrane for a transient period of time and this gradient is the "intermediate."

Mitchell proposed (Fig. 3.7) that if ADP and P$_i$ were present at the active site of the membrane ATPase (discussed below), and if a potential were formed across a membrane, water could be extracted from the complex (hence "chemiosmotic") with OH$^-$ moving to the positive or acid side and H$^+$ moving to the alkaline side. Thus ATP could be synthesized without formation of chemical intermediates. The chemiosmotic hypothesis also requires that the membrane must be relatively impermeable to protons since if protons diffused outward readily there would be no concentration gradient. The introduction of these two alternative hypotheses, and modifications of them, has greatly stimulated research on the mechanism of phosphorylation. Neither hypothesis can be dismissed entirely at present (Deamer, 1969). In order to disprove the chemiosmotic formulation, one must (*1*) find a true X ∼ I chemical intermediate that functions as an electron

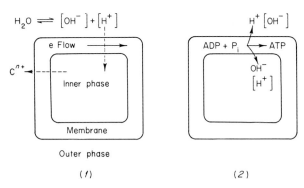

FIG. 3.7. Diagrammatic representation of the chemiosmotic hypothesis of photophosphorylation. (*1*) Photosynthetic electron flow results in an inward transport of hydrogen ions (H^+) across the grana membrane of the chloroplast. The inward movement produces a concentration gradient of protons when this flow is not immediately matched by the outward movement of cations (C^{n+}). This gradient serves as a high-energy intermediate. (*2*) A reversible ATPase within the grana membranes removes a proton and a hydroxyl ion (OH^-) from ADP and P_i, ATP is synthesized and the OH^- is released to the inside of the membrane and H^+ is returned to the outside.

carrier, or (*2*) devise a treatment that would totally remove the ability of chloroplasts to translocate protons yet not interfere with photophosphorylation (Jagendorf, 1967).

Among the interesting experiments resulting from Mitchell's hypothesis was the illumination of broken chloroplasts in an unbuffered solution at pH 6.5 or below. A rise in the pH of the medium in the light took place which was reversed in the dark (Jagendorf and Neumann, 1965). This pH change occurs without addition of P_i or ADP, and presumably results from the movement of protons to the inside of double membranes in the light (Fig. 3.7). The pH rise is inhibited by DCMU, showing that electron flow is needed, and by detergents, suggesting that intact membranes are also essential. Energy-transfer inhibitors do not affect the uptake of H^+ ions in the light. Uncouplers when added prior to illumination also inhibit the pH increase although they stimulate electron flow, also suggesting that this phenomenon is related to the production of a high-energy intermediate in the light.

If the high-energy intermediate in photophosphorylation consists of a proton gradient, then light should not be required for ATP synthesis in chloroplast preparations providing the inner space of the grana membranes were loaded with protons in some alternative manner. This was accomplished (Jagendorf and Uribe, 1966) by placing chloroplast fragments in succinate solution at pH 3.8 in the presence of DCMU for 60 seconds in the absence of light or O_2. The acid medium was then quickly raised to pH 8, and ADP, P_i and Mg^{2+} were added for 15 seconds. ATP was produced, and it was clearly caused by the acid–base transition. The synthesis of ATP was also inhibited by the addition of uncouplers in the acid stage, and the presence of DCMU and the absence of O_2 and light assured that ATP synthesis occurred without electron transport. Although raising the pH undoubtedly produced a proton gradient from the inside to the outside of the membranes, as seen from the schemes of the two hypotheses for ATP formation shown above, these interesting experiments do not by themselves prove that the proton gradient resulted in a direct synthesis of ATP without first producing a chemical intermediate.

A characteristic of all phosphorylating systems is the presence of membrane-bound ATPases, and these enzymes usually function in the synthesis of ATP *in vivo* rather than in its hydrolysis. Chloroplasts, unlike mitochondria, do not generally hydrolyze ATP at an appreciable rate, but, in the presence of a strong sulfhydryl reducing agent such as dithiothreitol, chloroplasts show a light-triggered and Mg^{2+}-dependent ATPase activity (McCarty and Racker, 1968). This enzyme can be extracted from chloroplasts with ethylenediaminetetraacetate (EDTA) and it then functions as a cold-labile Ca^{2+}-ATPase. The enzyme has been purified and called "chloroplast factor 1" (CF_1) (McCarty and Racker, 1967). The addition of CF_1 to subchloroplast particles obtained from sonically disrupted chloroplasts or to EDTA-treated chloroplasts greatly stimulated photophosphorylation in a reconstructed system. If the high-energy intermediate reacts with P_i to form a phosphorylated intermediate, then CF_1 would catalyze the reactions,

$$X \sim I + P_i \rightleftharpoons X \sim P + I$$
$$X \sim P + ADP \rightleftharpoons ATP + X$$

thus the chloroplast ATPase, as expected from the above equations, also catalyzes an exchange between $^{32}P_i$-ATP.

A second protein chloroplast factor (CF_2) was extracted by treatment of chloroplasts with dilute solutions of NH_4OH and enhances photophosphorylation in the presence of CF_1 with subchloroplast particles treated with the proteolytic enzyme trypsin (Livne and Racker, 1968). Both CF_1 and CF_2 when added together with plastocyanin-deficient subchloroplast particles, stimulated noncyclic electron flow to $NADP^+$ somewhat and increased the production of ATP greatly. Undoubtedly other soluble factors will also be found that stimulate photophosphorylation in chloroplasts as they have for oxidative phosphorylation in mitochondria (Racker, 1970). By analogy with mitochondria, membrane components will probably be essential in reconstructed photophosphorylation systems. Indeed it has already been shown that both a lipid component and a protein fraction are essential for the binding of CF_1 in the chloroplast membrane (Livne and Racker, 1969).

5. VOLUME CHANGES IN CHLOROPLAST MEMBRANES

Membranes of isolated spinach chloroplasts contract in the light, as shown by an increase in light scattering, under conditions where photophosphorylation occurs (Packer, 1963). Shrinking of the membranes was inhibited by DCMU and NH_4Cl. Similar observations were made by measurement of the packed volumes and of the chloroplast axial ratios by electron microscopy (Itoh *et al.*, 1963), and a volume decrease of 50 to 80% during illumination was observed. Shrinkage in the light occurred much more rapidly than the return of the dimensions to their dark state. Conformational changes in the thickness of the grana membranes are not limited to isolated chloroplasts but also occur *in vivo* in leaves of spinach and other species (Kushida *et al.*, 1964; U. Heber, 1969). Uncouplers supplied to intact leaves completely eliminated this shrinkage of chloroplasts in intact tissues. These changes in the lamellar membranes probably occur because weak acids in their undissociated form move freely across the membrane in both directions. In the light, an increase in H^+ uptake inside the membrane results in the association of a weak anion, and the unassociated acid then diffuses outside of the grana membranes resulting in an osmotic collapse of lamellar membranes, hence a change in volume (Robertson, 1968).

The light-dependent loss of water from chloroplasts was nearly always accompanied by both K^+ and Mg^{2+} effluxes, and the H^+ consumed by the chloroplast preparations was about equal to the amount of cations released during the light-induced membrane shrinkage (Dilley and Vernon, 1965). Consistent with this result is a study of the ionic concentration of pea chloroplasts, which were isolated within 2 minutes after the breaking of the leaf cells, from plants maintained in the light for 4 hours or in the dark for 4 to 12 hours (Nobel, 1969). The volume of isolated chloroplasts from leaves kept in the light was about 20% lower than those kept in the dark, confirming observations on chloroplasts *in vivo*, and the K^+, Mg^{2+}, and Ca^{2+} concentrations were about 30% less in chloroplasts obtained in the light (Nobel, 1969). When an uncoupler was supplied through the stems, the light-dependent decrease in the K^+ content of the chloroplasts was eliminated, hence the membrane shrinkage and accompanying K^+ loss appears to be related to the production of a high-energy intermediate or to ATP synthesis. The physiological significance of these light-induced contractions of chloroplast membranes remains to be determined.

4 Biochemical Pathways of CO_2 Fixation in Photosynthesis

A. EARLIER HYPOTHESES

The first attempts to envisage the pathway of CO_2 to carbohydrate (starch) during photosynthesis were based largely on intuition. Thus J. von Liebig in 1843 believed that there must be a stepwise reduction of CO_2 through such common organic acids as oxalate, tartrate, and malate. Since organic acids are likely the first intermediates in CO_2 reduction, although they are not the ones that were originally proposed, this suggestion was somewhat prophetic. In 1870 A. von Baeyer suggested that formaldehyde was the first product of CO_2 assimilation, presumably because it is a one-carbon compound at the same oxidation state as carbohydrate, and that the HCHO condensed to produce hexoses in a manner analogous to the reaction that occurs in the presence of alkali. Several variations of the formaldehyde hypothesis were later introduced by others, although the fact was largely ignored that the conditions by which hexoses are produced from formaldehyde in the laboratory differ greatly from those available to a living leaf. Nevertheless, Kostychev (1931), referring to Baeyer's hypothesis, wrote that ". . . this theory is now more than 50 years old and still no directly contradictory facts have been uncovered." The importance of formaldehyde as an intermediate in photosynthesis faded when experiments with radioactive carbon became possible.

B. USE OF RADIOACTIVE CARBON AS TRACER

Fortunately for progress in the field of photosynthesis, the use of isotopes as tracers of metabolic pathways, and the development of rapid methods of separating compounds by paper chromatography occurred about the same time. Ruben *et al.* (1939) attempted to identify intermediates in photosynthesis with the short-lived $^{11}CO_2$ (22 minute half-life). The labeled CO_2 was supplied to barley leaves in light or dark, but little progress was made in isolating intermediates because of the instability of the isotope and the cumbersome fractionation procedures that were required. The stable isotope of carbon—^{14}C—was discovered, and its production in nuclear reactors began about the same time that modern techniques used in paper chromatography were being perfected by A. J. P. Martin and R. L. M. Synge. Several groups of investigators initiated experiments with $^{14}CO_2$ as a tracer in photosynthesis in the late 1940's, and separated the radioactive products by paper and ion-exchange chromatography. The most rapid progress in tracing the path of carbon in photosynthesis was made by Calvin and his colleagues, and their earlier experiments are summarized in two little books (Bassham and Calvin, 1957; Calvin and Bassham, 1962). This highly original work provided entirely new insights into the biochemical reactions of carbon compounds associated with photosynthesis and therefore merits careful analysis.

Several approaches were used to deduce the sequences of reactions in the path of carbon in photosynthesis, and it is well to review some of the methods used and the underlying assumptions on which they are based. Calvin and Massini (1952) assumed that if a chain of intermediates exists in a sequence, and if during steady-state photosynthesis the uptake of $^{14}CO_2$ occurs at a constant rate, then "we should find the label appearing successively in time in the chain of intermediates . . ." and ". . . we may also determine the order in which the various carbon atoms within each compound acquire the label." Fig. 4.1 shows a theoretical distribution of ^{14}C in the carbon atoms of three hypothetical compounds that arise in the

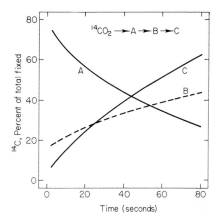

Fɪɢ. 4.1. Hypothetical distribution of percent of ¹⁴C fixed with time in three successive compounds arising in sequence from ¹⁴CO₂ during steady-state photosynthesis. The assumption is made that a constant rate of ¹⁴CO₂ uptake takes place and that only a single carboxylation reaction occurs.

sequence A → B → C when the enzymatic activities of the cells are rapidly inactivated at various times after supplying ¹⁴CO₂. At the shortest time, only the first intermediate (compound A) would be radioactive and thus account for nearly 100% of the total ¹⁴C fixed. The proportion of radioactivity in the first intermediate would decrease while the percentage of ¹⁴C in later products of the sequence would subsequently increase (with increasing time). After short intervals, compound B would contain a higher proportion of the total radioactivity than the third compound in the sequence (compound C). Since the photosynthetic experiments were carried out in the steady state, the concentrations of intermediates did not change during the course of the experiment, and the pool sizes were small compared with the quantity of the carbon being assimilated. Thus intermediates in the sequence should in time acquire the same specific activity as that of the ¹⁴CO₂ entering the tissues, the later intermediates reaching their maximum specific activity more slowly (Fig. 4.2).

The kinetic approach as outlined is sound when applied to a simple sequence, although it has some possible pitfalls when applied to intact tissues. Bassham and Calvin (1957) appreciated the importance of carrying out experiments under steady-state photosynthesis so as not to change the CO₂ concentration or other environmental factors that could alter the concentration of intermediates, but they performed many of the experiments at a CO₂ concentration of 1 to 4%. This concentration is greatly in excess of that normally found in air (0.03% or 300 ppm), and a distorted

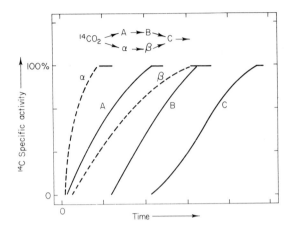

Fɪɢ. 4.2. Rate of change in the specific activities of compounds in a reaction sequence beginning with the fixation of $^{14}CO_2$ supplied in excess at zero time to tissues carrying out steady-state photosynthesis. It is assumed that two carboxylation reactions occur in parallel, and that compound α is present in concentrations too low to be readily detected.

result was possible if more than one carboxylation reaction occurred *in vivo* and they differed in their affinity for CO_2. There is now good evidence that multiple carboxylations do occur in photosynthetic tissues (Chapter 4.F).

They also recognized that intermediates may be present that are too unstable or too volatile to be isolated or that may be present in such low concentrations as to escape detection (Calvin and Bassham, 1962). Nevertheless, if unstable or volatile intermediates were present, their concentration was believed to be no greater than the quantity of carbon fixed in 5 seconds of photosynthesis (Bassham, 1964). Determining the importance of an intermediate from its pool size, or eliminating the possibility of an unknown pathway by this parameter is difficult, since the most active intermediates would probably turn over rapidly and be present in very low concentrations.

The ability to isolate intermediates quantitatively is implicit in the kinetic approach to determining sequences in a metabolic chain since the percent of the total ^{14}C in a given compound must be found. Chromatographic methods are generally satisfactory for this purpose, but this may not be true for all substances. For example, free glycolic acid, a compound of considerable interest in photosynthesis and photorespiration (Chapter 6), is volatile in the acidic solvents frequently used to develop paper chromatograms and "although 20 to 85% may evaporate from the paper during development of the chromatogram, the remainder disappears only very slowly from the papers" (Calvin and Bassham, 1962). This difficulty may affect the interpretation of the role of glycolic acid in photosynthesis in some experiments.

Another possible pitfall would occur *in vivo* if exchange reactions as measured with ^{14}C occur at rates that are more rapid or much slower than the net rate of carbon synthesis. For example, the net transamination reaction between glutamate and ^{14}C-oxaloacetate to yield ^{14}C-aspartate and α-ketoglutarate occurs only one-third as rapidly as the exchange reaction between ^{14}C-oxaloacetate and unlabeled aspartate to produce ^{14}C-aspartate without net synthesis (Hiller and Walker, 1961). Therefore ^{14}CO$_2$ may appear in aspartate by an exchange reaction in the absence of net synthesis and may cause erroneous conclusions. Since the overall discrimination against ^{14}CO$_2$ as compared with ^{12}CO$_2$ in photosynthesis by leaves is only about 2.0% (Yemm and Bidwell, 1969), this isotope effect is negligible in assessing photosynthetic assimilation. Nevertheless, a succession of isotope effects on individual reactions may discriminate against the distribution of ^{14}C by certain pathways more than others and therefore have an appreciable influence in ^{14}C-labeling experiments *in vivo*.

Stiller (1962) emphasized that one cannot necessarily determine the relative contributions of alternate carboxylation reactions when several occur simultaneously merely by observing the rate of appearance of radioactivity in various compounds from ^{14}CO$_2$. The apparent demonstration of a logical sequence by this method does not exclude the importance of other parallel pathways unless their existence is excluded in some independent manner. Such results may be especially misleading if the early intermediates in an alternate pathway are unstable or are present in concentrations too low to be detected easily. The consequence of a kinetic experiment under such circumstances is illustrated in Fig. 4.2, which shows how an alternate and perhaps equally important pathway (through compounds α and β), in terms of the quantities of carbon involved, might not be detected. Unless one is certain that only a single carboxylation reaction occurs, one cannot from such kinetic experiments determine how much carbon passes through one pathway in comparison with another merely by determining the "first products" formed in the shortest time, since several "first products" are produced by such methods (Chapter 4.E–4.H).

Many of these precautions were fully appreciated by Calvin and Benson (1949), although they have frequently been overlooked since then. Calvin and Benson wrote that the nonappearance of radioactivity in a given compound "does not necessarily preclude the possibility of its playing a part as an intermediate in a given sequence. For example, the reservoir of this compound in the sequence may be extremely small, or the compound

TABLE 4.1   *Percent Distribution of ^{14}C in Various Carbon Atoms after Photosynthesis in $^{14}CO_2$ by Barley Seedlings* [a,b]

Seconds in $^{14}CO_2$	Phosphoglyceric acid			Percent Distribution of ^{14}C Hexose			Glycolic acid	
	—COOH	—CHOH	—CH$_2$OP	C—3,4	C—2,5	C—1,6	—COOH	—CH$_2$OH
4	87	6.5	6.8					
15	56	21	23				50 ± 5	50 ± 5
30	44	30	25				48	52
60				52	25	24		

[a] From Calvin and Massini, 1952.
[b] The $^{14}CO_2$ was supplied at an initial concentration greater than 1% in air after a preliminary period of photosynthesis in unlabeled CO_2. The ^{14}C distribution within the various compounds was determined after separating them from an alcoholic extract by means of paper chromatography.

may never exist as a free compound in solution but rather as an enzyme-substrate complex, so that the amount of radioactivity present in that particular compound may be so small as to be missed. Conversely, the appearance of radioactivity in a particular compound does not necessarily prove its part as an intermediate in a direct sequence. It can be, and often is the result of a side reaction."

C. THE PHOTOSYNTHETIC CARBON REDUCTION CYCLE

When *Chlorella pyrenoidosa* cells were supplied $^{14}CO_2$ in the light for 30 seconds and the cells were at once killed in boiling 80% alcohol, Calvin and Benson (1948) found that over 70% of the total $^{14}CO_2$ fixed was present in triose phosphate and phosphoglyceric acid. If, however, the exposed cells were transferred to darkness, 10 to 35 times as much ^{14}C was found in phosphoglyceric acid. They deduced that the major reduction step in the light involved phosphoglyceric acid to yield triose phosphate. When the exposure time to $^{14}CO_2$ was reduced to about 5 seconds, 3-phosphoglyceric acid was the only compound labeled, and nearly all of its radioactivity was found in the carboxyl-carbon atom (Table 4.1). The other two carbon atoms were generally equally labeled suggesting that they were not formed *de novo* from CO_2 but arose from the carboxyl-carbon atom by a cyclic process. The amount of label approached the ^{14}C concentration of the carboxyl-carbon after about 60 seconds of photosynthesis. Hexoses had the highest concentration of radioactivity in carbons 3 and 4, suggesting that the sugar is produced by a condensation of two triose phosphate molecules catalyzed by the well-known aldolase reaction.

Later Kandler and Gibbs (1956) found that C-4 had higher radioactivity than C-3 in hexoses, and this probably results because a small pool of glyceraldehyde-3-phosphate reacts with a large pool of dihydroxyacetone phosphate in the aldolase step. The intermediate ribulose diphosphate was discovered on paper chromatograms and was shown to function in an early stage of CO_2 fixation; and the novel enzymatic carboxylation of ribulose diphosphate to produce two molecules of phosphoglyceric acid was established in cell-free preparations (Quayle *et al.*, 1954).

Thus by (*1*) experiments involving successive ^{14}C-labeling of intermediates at changing time intervals with results as expected from Fig. 4.1; (*2*) causing reciprocal changes in the pool sizes of intermediates in the cycle when the CO_2 concentration was decreased or the light was turned off (A. T. Wilson and Calvin, 1955); and (*3*) observing the changes in pool sizes of intermediates caused by inhibitors (Krause and Bassham, 1969), a cyclic process for CO_2 fixation in photosynthesis known as the photosynthetic carbon reduction cycle (or the Calvin cycle) was deduced. By extrapolating to zero time, 70 to 85% of the CO_2 uptake beginning with the carboxylation

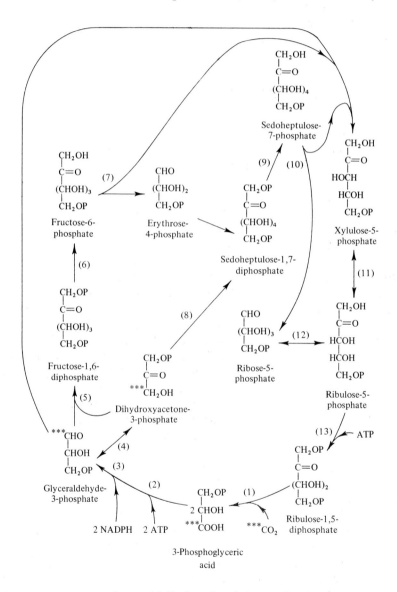

SCHEME 4.1 (See legend on facing page.)

of ribulose diphosphate was accounted for by this cycle in experiments with *Chlorella* carried out at 1 to 2% CO_2 (Bassham and Kirk, 1960).

The photosynthetic carbon reduction cycle in Scheme 4.1 consists of three phases: (*a*) a carboxylation phase in which ribulose diphosphate accepts CO_2 to yield two molecules of 3-phosphoglyceric acid; (*b*) a reductive phase in which the phosphoglyceric acid is reduced to triose phosphate in the presence of ATP and NADPH; and (*c*) a regenerative phase in which five triose phosphates are converted to three pentose phosphates and the CO_2 acceptor, ribulose diphosphate, is produced. The overall equation for the cycle is,

$$3CO_2 + 9ATP + 6NADPH + 6H^+ + 5H_2O \rightarrow \text{triose-P} + 9ADP + 8P + 6NADP^+ \quad (4.1)$$

Thus for each CO_2, two molecules of reduced nicotinamide adenine dinucleotide phosphate (NADPH), and three molecules of ATP are needed (Chapter 3.E). The three phases of the cycle can be expressed as follows:

$$3C_5 + 3CO_2 \rightarrow 6PGA$$
$$6PGA + 12H \rightarrow 6C_3$$
$$2C_3 \rightarrow C_6$$
$$2C_3 + C_6 \rightarrow C_5 + C_7$$
$$\underline{C_3 + C_7 \rightarrow 2C_5}$$
$$\text{Net:} \quad 3CO_2 + 12H \rightarrow C_3$$

Abbreviations: 3-phosphoglyceric acid (PGA); triose phosphates (C_3); hexose phosphate (C_6); pentose phosphates (C_5); sedoheptulose phosphate (C_7); reducing equivalents (H).

SCHEME 4.1. The enzymatic reactions of the photosynthetic carbon reduction cycle (after Bassham, 1964). The asterisks show the order of labeling after a short exposure (10 seconds) to $^{14}CO_2$. (1) Ribulose diphosphate in the presence of ribulose diphosphate carboxylase adds CO_2 at the C-2 position to produce two molecules of phosphoglyceric acid. (2) In the presence of phosphoglycerate kinase and ATP and (3) of triosephosphate dehydrogenase and reduced nicotinamide adenine dinucleotide phosphate (NADPH) glyceraldehyde phosphate is produced. (4) Triosephosphate isomerase produces dihydroxyacetone phosphate. (5) Aldolase condenses C-1 of glyceraldehyde phosphate with C-1 of dihydroxyacetone phosphate to form fructose diphosphate. (6) Phosphatase hydrolyzes the phosphate ester from C-1. (7) Transketolase transfers the glycolyl group (C-1,2) from fructose phosphate to glyceraldehyde phosphate to form xylulose phosphate and erythrose phosphate. (8) Aldolase condenses C-1 of erythrose phosphate with C-1 of dihydroxyacetone phosphate to produce sedoheptulose diphosphate. (9) A phosphatase hydrolyzes the phosphate ester on C-1 to form sedoheptulose phosphate. (10) Transketolase transfers the glycolyl group (C-1,2) from sedoheptulose phosphate to C-1 of glyceraldehyde phosphate, leaving ribose phosphate. (11) Epimerase catalyzes the epimerization of C-3 of xylulose phosphate to form ribulose phosphate. (12) Pentosephosphate isomerase catalyzes an equilibrium between ribose phosphate and ribulose phosphate. (13) Phosphoribulokinase esterifies C-1 in the presence of ATP to produce ribulose diphosphate.

The enzymes required for the thirteen steps of the carbon cycle (Scheme 4.1) are present in photosynthetic tissues, and the complete cycle has been reconstructed with isolated enzymes. However, low activities of ribulose diphosphate carboxylase [(1) in Scheme 4.1], sedoheptulose-1,7-diphosphatase (9), and fructose-1,6-diphosphatase (6) were obtained in spinach leaf homogenates compared with the rate of CO_2 fixation by leaves (Peterkofsky and Racker, 1961). Similar deficiencies in these enzymatic activities compared with photosynthetic rates of CO_2 fixation have been observed in isolated spinach chloroplasts (Table 4.2) as well as in other photosynthetic tissues. Bassham (1964) believes that the enzymes of the reduction cycle are present as a multifunctional enzyme and that by disrupting this organized system to assay the enzymatic activities some may become lost. Thus although there can be no doubt about the functioning of the photosynthetic carbon reduction cycle in a variety of photosynthetic tissues, the amount of carbon fixed by this pathway in a given organism under normal conditions is still uncertain because of the limitations of the methods used as discussed earlier.

TABLE 4.2 $^{14}CO_2$ *Fixation and Specific Activity of Spinach Chloroplasts Isolated in Different Media* [a,b]

	μmoles Substrate consumed mg chlorophyll^{-1} hr^{-1}		
Activity	Sucrose-	Salt-	Sorbitol-chloroplasts
$^{14}CO_2$ fixation	5	7	123
(1) Ru-1,5-diP carboxylase	7.8	8.5	28
(2) 3-PGA kinase	1850	1420	2800
(3) G-3-P dehydrogenase (NADP$^+$)	104	210	505
(4) Triose-P isomerase	3500	4250	9650
(5) FDP-aldolase	84	79	189
(6) FDPase	43	46	101
(7), (10) Transketolase	115	122	267
(9) SDPase	0.9	—	4.1
(11) Xu-5-P epimerase	180	550	1610
(12) R-5-P isomerase	290	276	565
(13) Ru-5-P kinase	168	230	480

[a] From Latzko and Gibbs, 1968.
[b] The numbers in parentheses refer to the enzymatic reactions illustrated in Scheme 4.1. Salt- and sucrose-chloroplasts (0.35 M in the medium) were isolated by the methods of Allen *et al.* (1955) and Gibbs and Calo (1959). More nearly intact sorbitol-chloroplasts were prepared according to Jensen and Bassham (1966) as discussed in Chapter 2.A.

D. PHOTOSYNTHESIS IN ISOLATED CHLOROPLASTS

Studies on the reconstruction of photosynthetic reactions in isolated chloroplasts are of considerable importance because the problem of photosynthetic carboxylation may be simpler to analyze in cell-free systems. Early studies on photosynthesis with isolated chloroplasts were frequently suspect, however, because of the low rates of CO_2 incorporation obtained. Recently, more efficient methods of isolating intact chloroplasts have been developed (Chapter 2.B). Some workers state that whole leaves at saturating light and CO_2 concentrations are capable of fixing only about 200 μmoles of CO_2 mg chlorophyll^{-1} hr^{-1} (Walker, 1964; Jensen and Bassham, 1966). Leaves of a standard variety of tobacco fix 69 mg CO_2 dm^{-2} hr^{-1} at close to the optimal temperature, saturating light intensities, and higher than normal CO_2 concentrations (J. D. Hesketh, 1963; Clendenning, 1957), and this value is closer to 500 μmoles of CO_2 mg chlorophyll^{-1} hr^{-1}. Rates of approximately 650 μmoles CO_2 mg chlorophyll^{-1} hr^{-1} have been obtained with maize. [One dm^2 = about 2.0 gm fresh weight of tobacco lamina with small veins = about 3 mg chlorophyll.]

Fixation rates of 24 to 37 μmoles CO_2 mg chlorophyll^{-1} hr^{-1} were obtained in the presence of catalytic concentrations of ribose-5-phosphate when essentially intact-appearing pea chloroplasts were isolated at pH 6.8 and photosynthetic $^{14}CO_2$ uptake was measured at pH 7.5 (Walker, 1964). Walker (1965) also observed that Class I chloroplasts, which contain intact envelope membranes (Chapter 2.B), fix CO_2 at more than 10 times the rate of Class II chloroplasts. However, cyclic photophosphorylation with pyocyanine as cofactor was considerably faster with membrane-free chloroplasts than with intact chloroplasts, presumably because either the cofactor or ADP does not readily pass through the outer membranes. Jensen and Bassham (1966, 1968a) isolated spinach chloroplasts rapidly at pH 6.1, suspended them in buffer in a medium at pH 6.7, and carried out photosynthesis at pH 7.6. The chloroplast suspension was illuminated for 3 minutes because there was a lag of 3 to 5 minutes before linear rates of CO_2 fixation occurred; $H^{14}CO_3^-$ was added at zero time and the reaction was stopped by the addition of acid or methanol after about 6 minutes. The rates of CO_2 fixation by isolated spinach chloroplasts decrease after 20 minutes, but fixation can be sustained for a longer period of time by the addition of metabolites such as fructose-1,6-diphosphate.

Rates of assimilation of 122 to 197 μmoles of CO_2 mg chlorophyll^{-1} hr^{-1} were obtained, a great improvement over the rates of fixation 10 years earlier. In occasional experiments rates as high as 240 μmoles of CO_2 mg chlorophyll^{-1} hr^{-1} have been observed (Plaut and Gibbs, 1970), and with a somewhat simpler medium than Jensen and Bassham's, spinach chloroplasts fixed 50 μmoles of CO_2 mg chlorophyll^{-1} hr^{-1} (Kalberer *et al.*, 1967).

Results with spinach chloroplasts were unlike the experience with pea chloroplasts; photosynthesis by isolated spinach chloroplasts was not enhanced by the addition of intermediates of the Calvin cycle. Pyrophosphate at concentrations between 1 and 5 $\times$ 10^{-3} M was required (Jensen and Bassham, 1968a). Maximal rates of photosynthesis were obtained with about 2000 ft-c of illumination and at bicarbonate concentrations of 3 mM, and half-maximal rates at 0.6 mM. From the Henderson–Hasselbach equation, pH = pK + log(HCO_3^-)/(CO_2), and with the value of 6.3 for the pK, a K_m of 0.6 mM for HCO_3^- is equivalent to 3.2 $\times$ 10^{-5} M free CO_2 at pH 7.6. A similar calculation from the data of Plaut and Gibbs (1970) yields a K_m for CO_2 of about 1.3 $\times$ 10^{-5} M. Thus the affinity of isolated chloroplasts for CO_2 does not seem to be as great as in intact leaves (Chapter 8.C) and it is greater than with isolated carboxylating enzymes (Chapter 4.E, 4.F).

The products of $^{14}CO_2$ fixation by isolated chloroplasts have been determined by many investigators and consist of intermediates of the Calvin cycle (phosphoglyceric acid, dihydroxyacetone phosphate, fructose and sedoheptulose diphosphate, and ribose-5-phosphate). Glycolic acid is formed in rather large amounts under some conditions (Chapter 6).

Finding ^{14}C in phosphoglyceric acid does not by itself provide evidence that the Calvin cycle is functioning (Havir and Gibbs, 1963), although there is good evidence that the photosynthetic carbon reduction cycle occurs in isolated chloroplasts. These workers demonstrated that intact spinach chloroplasts carry out the cyclic sequence of reactions because the α- and β-carbon atoms of phosphoglyceric acid became increasingly labeled with ^{14}C with increasing time of $^{14}CO_2$ fixation (as in Table 4.1). However, when $^{14}CO_2$ fixation was studied with fragmented chloroplasts in the presence of an extract of chloroplasts supplemented with ribulose diphosphate, almost all of the radioactivity was found exclusively in the carboxyl-carbon atom of phosphoglyceric acid even after a considerable period of $^{14}CO_2$ fixation (40 minutes). Thus only the carboxylation reaction, but not the Calvin cycle, was operating in such a reconstructed system although CO_2 fixation readily took place.

Latzko and Gibbs (1968) have examined the enzymatic activities associated with the photosynthetic carbon reduction cycle in chloroplasts isolated by three different procedures and compared these activities with the rates of photosynthetic $^{14}CO_2$ assimilation by the same chloroplast preparations (Table 4.2). The rate of CO_2 fixation by chloroplasts prepared in sorbitol (Jensen and Bassham, 1966) was about 20 times higher than with chloroplasts prepared in a medium containing NaCl or sucrose. The specific activities of the enzymes on a chlorophyll basis were about twice as high in the sorbitol chloroplasts although this increase was not found when the specific activities were expressed on a protein basis. The ribulose diphosphate carboxylase activity did not account for more than 23% of the CO_2 fixation in the better preparations, and sedoheptulose diphosphatase and fructose diphosphatase reaction rates were also insufficient to account for the observed rates of photosynthesis (Scheme 4.1). These are troublesome results if the Calvin cycle is the primary carboxylation system. The data demonstrate that low rates of CO_2 fixation obtained with chloroplasts isolated in 0.35 M NaCl in place of sorbitol were not the result of the leaching out of enzymes of the photosynthetic cycle. A major difference between the two preparations was the higher protein/chlorophyll ratio in the more active chloroplasts, suggesting that proteins associated with the membrane envelope were extracted during isolation and this limited the CO_2 fixation of isolated chloroplasts.

Additional direct evidence for the functioning of the photosynthetic carbon reduction cycle in isolated chloroplasts comes from measurements of O_2 evolution. The cycle (Scheme 4.1) shows that the strong reductant produced in photosynthesis (Chapter 3.D) eventually reduces phosphoglycerate to triose phosphate. Walker and Hill (1967) showed the dependence of O_2 evolution on the presence of CO_2 with isolated spinach and pea chloroplasts. In the absence of CO_2 phosphoglycerate substituted as a substrate to initiate O_2 evolution, whereas triose phosphate (the product of the reduction step) is only effective when CO_2 is also present (Walker *et al.,* 1967).

Attempts have been made to isolate factors from chloroplasts that stimulate the rate of CO_2 fixation. This approach may yet be fruitful in obtaining higher rates since isolated chloroplasts still do not fix CO_2 as efficiently as intact leaves. Evidence was obtained for the presence of both stimulatory as well as inhibitory substances in spinach leaf extracts (Moore *et al.,* 1969). An inhibitory fraction that functions in the presence of pyrophosphate has been purified from spinach chloroplasts that were disrupted sonically, and the inhibitor appears to be identical with the enzyme fructose-1,6-diphosphatase (Springer-Lederer *et al.,* 1969). The inhibitory action of the diphosphatase on CO_2 fixation was greatly enhanced when the fructose diphosphatase was combined with another factor, which is presumably a small molecule. These factors are believed to play a role in regulating the flow of intermediates in the photosynthetic carbon reduction cycle through the chloroplast outer membrane as well as the flow of carbon from the cycle to other biosynthetic reactions that take place in the cytoplasm.

The assimilation of CO_2 by isolated spinach chloroplasts is inhibited by increasing the O_2 concentration in the atmosphere, and the inhibition is greater at lower concentrations of CO_2 (Ellyard and Gibbs, 1969), as is observed in intact leaves. Various consequences of this phenomenon (sometimes known as the Warburg effect) are discussed further in Chapters 5.D, 6.A, and 8.E.

It is now well established that certain plant species such as maize, sorghum, and sugarcane are more efficient in their net photosynthetic CO_2 fixation on a leaf area basis than most crop species (Chapters 5, 8, and 9). Accordingly, some attempts have been made to isolate and study chloroplasts from leaves of maize and sugarcane. A light-stimulated carboxylation of phosphoenolpyruvate (Chapter 4.F) by sugarcane chloroplasts has been observed (Baldry *et al.*, 1969), but the highest rates of CO_2 fixation so far obtained were only about 2 μmoles of CO_2 mg chlorophyll^{-1} hr^{-1}. Low rates of CO_2 fixation from chloroplasts of maize, sorghum, and sugarcane have also been found by Hew and Gibbs (1969), although their maize preparations showed relatively better rates of $NADP^+$ reduction in the light. Isolated maize chloroplasts had fixation rates as high as 45 μmoles CO_2 mg chlorophyll^{-1} hr^{-1} when a sulfhydryl reducing agent was present in the medium (Gibbs *et al.*, 1970), and surprisingly the first product labeled by $^{14}CO_2$ was phosphoglyceric acid. It is not stated whether chloroplasts from the bundle-sheath or from the mesophyll (Chapter 2.A) were obtained during isolation, although mesophyll chloroplasts would be isolated more easily. Edwards *et al.* (1970) have separated mesophyll cells from bundle-sheath cells in crabgrass (*Digitaria sanguinalis*), an efficient photosynthetic species, and showed that the mesophyll cells had rapid rates of $^{14}CO_2$ fixation when supplemented with phosphoenolpyruvate (Chapter 4.F). The bundle-sheath cells, where ribulose diphosphate carboxylase activity was apparently concentrated, produced low rates of CO_2 fixation even when ribose-5-phosphate or ribulose diphosphate were added, but the rates were higher in the presence of phosphoenolpyruvate. Investigations on isolated chloroplasts and cells from the more efficient photosynthetic species are in their infancy, and determining whether the greater efficiency of CO_2 fixation exhibited by certain species is related in part to differences in the biochemical activity of their chloroplasts should provide a productive area of research.

E. RIBULOSE DIPHOSPHATE CARBOXYLASE

Wood and Utter (1965) recognized the diversity of carboxylating systems in heterotrophic organisms and reasoned that "it seems unlikely that there would be only a single mechanism, that of ribulose diphosphate carboxylase, for the autotrophic utilization of CO_2." A number of carboxylating enzymes do occur in higher plants, and several of these carry out CO_2 fixation in photosynthesis. Perhaps the most studied of these reactions, that catalyzed by ribulose diphosphate carboxylase or "carboxydismutase," functions in the photosynthetic carbon reduction cycle. Its activity was first demonstrated in cell-free preparations from *Chlorella* (Quayle *et al.*, 1954). When $^{14}CO_2$ was supplied to extracts containing ribulose diphosphate, all of the radioactivity appeared in the carboxyl carbon atom of 3-phospho-D-glyceric acid:

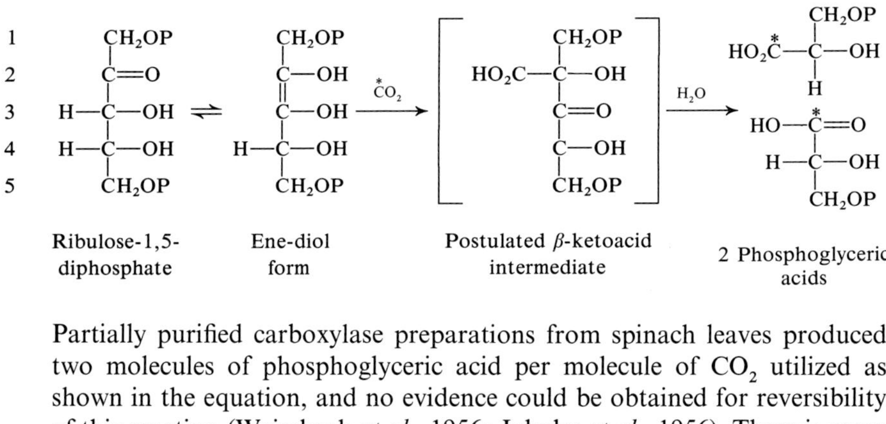

$$(4.2)$$

	Ribulose-1,5-diphosphate	Ene-diol form	Postulated β-ketoacid intermediate	2 Phosphoglyceric acids

Partially purified carboxylase preparations from spinach leaves produced two molecules of phosphoglyceric acid per molecule of CO_2 utilized as shown in the equation, and no evidence could be obtained for reversibility of this reaction (Weissbach *et al.*, 1956; Jakoby *et al.*, 1956). There is some evidence that in intact tissues in the light the splitting of ribulose diphosphate occurs by a reductive reaction leading to the production of only one molecule of phosphoglyceric acid and one molecule at the oxidation level of triose phosphate (Calvin and Pon, 1959). The cleavage of ribulose diphosphate could conceivably occur between the C-2 and C-3 or between the C-3 and C-4 bond. By using variously ^{14}C-labeled ribulose diphosphate and carrying out the enzymatic carboxylation in the presence of deuterium-enriched water, it was clear from the nature of the labeled phosphoglyceric acid produced that the CO_2 became affixed to the carbon atom that was originally in the C-2 position of ribulose diphosphate. Hence cleavage must have occurred between the C-2 and C-3 bond (Mullhöfer and Rose, 1965).

When extracts of chloroplasts are subjected to ultracentrifugation, a large portion of the soluble protein appears as a high molecular weight fraction which has been referred to as Fraction I protein. A useful review about Fraction I protein has been published (Kawashima and Wildman, 1970). This fraction has a high molecular weight, with a sedimentation coefficient of 18 to 21 S, and contains some of the enzymatic activities of the photosynthetic carbon reduction cycle including ribulose diphosphate carboxylase (Trown, 1965). Between 85 and 100% of the ribulose diphosphate carboxylase activity of leaves is found in the chloroplast fraction in many species of higher plants (A. Heber *et al.*, 1963) when nonaqueous methods of chloroplast isolation were used.

The isolated enzyme also requires the presence of a metal ion such as magnesium for maximal activity, in addition to the substrates shown in Equation (4.2). Magnesium ions also serve as an allosteric activator of the enzyme. Thus the pH optimum changed from 8.5 without magnesium to pH 6.5 in the presence of 1×10^{-2} M magnesium ions, and the K_m for CO_2 was also lowered slightly (Sugiyama *et al.*, 1968). This carboxylase may be more active in the light than in the dark in isolated chloroplasts, since spinach chloroplasts do not fix $^{14}CO_2$ appreciably in the dark even when the appropriate substrates are supplied and the concentration of ribulose diphosphate in the dark compares with that in the light (Jensen and Bassham, 1968b). High rates of CO_2 fixation in darkness can be induced in isolated chloroplasts providing magnesium ions are added and the intact chloroplasts are swollen by changing the osmotic conditions of the medium (Jensen, 1969). These experiments indicate that the activity of this enzyme in the light increases *in vivo* because of changes in the ionic environment. Björkman (1968) observed that sun-adapted species generally have a greater carboxylase activity than species that normally grow in the shade (Table 4.3), a further demonstration of the control of the activity of this enzyme by light.

A highly purified and apparently homogeneous preparation of ribulose diphosphate carboxylase was obtained from spinach leaf extracts (Paulsen and Lane, 1966). This purified enzyme was characterized by having a low turnover rate, 1300 moles HCO_3^- fixed per mole of enzyme per minute at pH 7.9 and 30°. The enzyme was present in large amounts, accounting for about 16% of the protein initially present in the leaf homogenate, and had a molecular weight of about 557,000. The large protein had two distinct kinds of subunits which consisted of noncovalently linked polypeptide chains. Each enzyme molecule contained about eight heavy subunits,

TABLE 4.3   *Ribulose Diphosphate Carboxylase Activity in Leaf Extracts from Species with an Active Photosynthetic Carbon Reduction Cycle*

Species	Activity[a] μmoles CO_2 fixed mg chlorophyll^{-1} hr^{-1}	Source
Spinach	118	Weissbach et al. (1956)
Spinach	150	Peterkofsky and Racker (1961)
Spinach	187	Paulsen and Lane (1966)
Various	252–282	Johnson and Hatch (1968)
Various	312–504	Björkman and Gauhl (1969)
Various shade-adapted	40	Björkman (1968)
Various sun-adapted	200–330	Björkman (1968)

[a] Where the activity was expressed on a fresh weight basis only, the assumption was made that 1.0 gm fresh weight contains 1.5 mg chlorophyll and when the activity was given on a leaf area basis it was assumed that 1.0 dm² of lamina was equal to 2.0 gm fresh weight.

molecular weight 55,800, and eight to ten lighter subunits, molecular weight 12,000 (Rutner and Lane, 1967; Rutner, 1970). These different types of subunits differed in their amino acid composition and their physical behavior. Recent experiments suggest that only the large subunits are enzymatically active (Akazawa and Sugiyama, 1969), thus the smaller subunits presumably have a regulatory action on the enzyme. Alterations in the subunit structure are found in some tomato mutants, and this accounts for higher or lower specific activities of the carboxylase in the mutant plants (Andersen et al., 1970).

Wishnick and colleagues (1969) have made an exciting discovery about ribulose diphosphate carboxylase that stemmed from the observation that cyanide binds to the purified enzyme stoichiometrically, but only in the presence of ribulose diphosphate (Wishnick and Lane, 1969). This suggested that the ribulose diphosphate alters the enzyme conformation and thereby allows a heavy metal in the enzyme to become available for reaction with the cyanide. This hypothesis led to an analysis of the enzyme for heavy metals, and copper was found to be a tightly bound component of ribulose diphosphate carboxylase (Wishnick et al., 1969). There was approximately 1.0 gm-atom of copper per mole of carboxylase as determined by several different methods. This unexpected finding raises interesting questions about the mechanism of action of the enzyme.

When extracts of leaves that possess an active photosynthetic carbon reduction cycle, as evidenced by phosphoglyceric acid production after

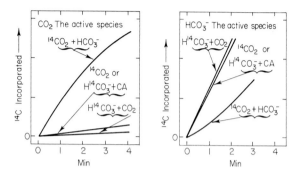

FIG. 4.3. Theoretical incorporation in a radiochemical assay for carboxylation when the active species is either free CO_2 or bicarbonate (from T. G. Cooper *et al.*, 1969). It is assumed that the rate of ^{14}C incorporated is directly proportional to the concentration of the active species at pH 8 and 10°, and that under these conditions the rate of hydration of CO_2 to produce bicarbonate to CO_2 is slow in the absence of added carbonic anhydrase (CA).

short periods of $^{14}CO_2$ fixation, are assayed for ribulose diphosphate carboxylase activity under optimal conditions, rates of enzymatic CO_2 fixation of from 120 to 500 μmoles mg chlorophyll^{-1} hr^{-1} are usually obtained (Table 4.3). These rates of reaction are generally insufficient to account for maximal rates of photosynthetic CO_2 fixation by intact leaves except for the highest rates (Chapter 4.D). Leaves of species that normally grow in open sunlight have considerably higher rates of net photosynthesis on a leaf area basis than those which are adapted to growing in the shade (Chapter 8.B), and the low ribulose diphosphate carboxylase activity found in shade-adapted species (Table 4.3) may be partly responsible for their low photosynthetic CO_2 assimilation (Björkman, 1968). Nevertheless, in one sun plant, *Plantago lanceolata*, net photosynthesis at 300 ppm of CO_2 (a nonsaturating concentration) was 840 μmoles CO_2 dm^{-2} hr^{-1} or about 280 μmoles CO_2 mg chlorophyll^{-1} hr^{-1}, while the carboxylase activity measured in extracts with near-saturating CO_2 concentrations was only 200 μmoles CO_2 dm^{-2} hr^{-1}. Hence there is insufficient ribulose diphosphate carboxylase activity (usually measured at saturating CO_2 concentration) to account for the rates of net photosynthetic CO_2 assimilation even when the CO_2 concentration is limiting (Chapter 4.D).

The purified carboxylase has a reasonably low K_m for ribulose diphosphate (1.2×10^{-4} M) and for magnesium ions (1.1×10^{-3} M) (Paulsen and Lane, 1966). Bicarbonate is generally used as the substrate, and its K_m is high, about 2.2×10^{-2} M (Weissbach *et al.*, 1956; Racker, 1957; Paulsen and Lane, 1966), indicating this carboxylase does not have a very strong affinity for bicarbonate. Until recently there was uncertainty whether bicarbonate or free CO_2 was the active species fixed in this reaction, but this question has now been resolved. The hydration of CO_2 to form bicarbonate ($CO_2 + H_2O \rightleftharpoons H_2CO_3 \rightleftharpoons H^+ + HCO_3^-$) is relatively slow below 15° in the absence of the enzyme carbonic anhydrase. Thus if free CO_2 is the active species and $^{14}CO_2$ is supplied under these conditions it will be taken up more rapidly than $H^{14}CO_3^-$, whereas if bicarbonate is the active species the reverse would be true. The rate of ^{14}C incorporation that would be obtained under these two circumstances, in the presence and absence of carbonic anhydrase, is illustrated in Fig. 4.3. T. G. Cooper *et al.* (1969) used this method and clearly showed that free CO_2, and not bicarbonate, was the active species utilized by ribulose diphosphate carboxylase. Thus the K_m for free CO_2 should be revised downward to about 4.5×10^{-4} M. Still, the affinity of isolated chloroplasts (Chapter 4.D) and intact leaves (Chapter 8.C) for CO_2 appears to be greater than the affinity of ribulose diphosphate carboxylase for CO_2. The affinity for CO_2 is also not as great as that shown by isolated phosphoenolpyruvate carboxylase (Chapter 4.F).

F. PHOSPHOENOLPYRUVATE CARBOXYLASE AND THE C₄-DICARBOXYLIC ACID PATHWAY

The enzyme that catalyzes the carboxylation of phosphoenolpyruvate to produce oxaloacetate was discovered in extracts of spinach leaves by Bandurski and Greiner (1953) and Bandurski (1955):

$$
\begin{array}{ccc}
\text{COOH} & & \text{COOH} \\
| & & | \\
\text{C}-\text{OPO}_3\text{H}_2 + \overset{*}{\text{C}}\text{O}_2 + \text{H}_2\text{O} \xrightarrow{\text{Mg}^{2+}} & \text{C}{=}\text{O} & + \text{H}_3\text{PO}_4 \\
\| & & | \\
\text{CH}_2 & & \text{CH}_2 \\
& & | \\
& & \overset{*}{\text{C}}\text{OOH}
\end{array}
\qquad (4.3)
$$

Phosphoenolpyruvate   Oxaloacetate

Until recently this enzyme was known to function only in plants exhibiting a crassulacean-type metabolism in spite of its high activity in leaf extracts (Table 4.4). Kortschak and co-workers (1965) supplied $^{14}CO_2$ for 10 seconds during an investigation of photosynthetic CO_2 fixation in sugarcane leaves and found that most of the radioactivity was present in C_4 compounds such as malate and aspartate rather than in phosphoglyceric acid as had previously been observed in other species. They compared $^{14}CO_2$ fixation in soybean leaves as a control and found that under the same experimental conditions most of the ^{14}C was present in phosphoglyceric acid. Leaves have small pools of oxaloacetate and high concentrations of

TABLE 4.4 *Phosphoenolpyruvate Carboxylase and Phosphopyruvate Synthetase Activities in Leaf Extracts from Species with the C_4-Dicarboxylic Acid Pathway.*

Species	Activity (μmoles mg chlorophyll^{-1} hr^{-1})		Source
	Phosphoenolpyruvate carboxylase	Phosphopyruvate synthetase	
Tropical grasses	900	72–210	Johnson and Hatch (1968)
Amaranthus palmeri	1260	222	Johnson and Hatch (1968)
Atriplex semibaccata	1080	84	Johnson and Hatch (1968)
Maize (130,000 lux 6 days)	1080	180	Hatch *et al.* (1969)
Maize (8600 lux 6 days)	288	30	Hatch *et al.* (1969)

malate dehydrogenase and transaminase that would convert the oxalo-acetate to malate and aspartate:

$$
\begin{array}{c}
\text{COOH} \\
| \\
\text{C}=\text{O} \\
| \\
\text{CH}_2 \\
| \\
\text{*COOH} \\
\text{Oxaloacetate}
\end{array}
+
\begin{array}{c}
\text{NADH} \\
\text{or} \\
\text{NADPH}
\end{array}
+ \text{H}^+
\underset{\substack{\text{Malate} \\ \text{dehydrogenase}}}{\rightleftharpoons}
\begin{array}{c}
\text{COOH} \\
| \\
\text{HO}-\text{C}-\text{H} \\
| \\
\text{CH}_2 \\
| \\
\text{*COOH} \\
\text{L-Malate}
\end{array}
+
\begin{array}{c}
\text{NAD}^+ \\
\text{or} \\
\text{NADP}^+
\end{array}
\qquad (4.4)
$$

$$
\begin{array}{c}
\text{COOH} \\
| \\
\text{C}=\text{O} \\
| \\
\text{CH}_2 \\
| \\
\text{*COOH}
\end{array}
+
\begin{array}{c}
\text{COOH} \\
| \\
\text{H}_2\text{N}-\text{C}-\text{H} \\
| \\
\text{CH}_2 \\
| \\
\text{CH}_2 \\
| \\
\text{COOH} \\
\text{L-Glutamate}
\end{array}
\underset{\text{Transminase}}{\rightleftharpoons}
\begin{array}{c}
\text{COOH} \\
| \\
\text{H}_2\text{N}-\text{C}-\text{H} \\
| \\
\text{CH}_2 \\
| \\
\text{*COOH} \\
\text{L-Aspartate}
\end{array}
+
\begin{array}{c}
\text{COOH} \\
| \\
\text{C}=\text{O} \\
| \\
\text{CH}_2 \\
| \\
\text{CH}_2 \\
| \\
\text{COOH} \\
\alpha\text{-Ketoglutarate}
\end{array}
\qquad (4.5)
$$

Thus the pattern of labeling from $^{14}\text{CO}_2$ showed that in certain species CO_2 is fixed by the phosphoenolpyruvate carboxylase reaction in preference to that catalyzed by ribulose diphosphate carboxylase. These experiments were confirmed and extended by Hatch and Slack (1966) who found that in sugarcane the malate produced from $^{14}\text{CO}_2$ came from ^{14}C-oxaloacetate and that malate was labeled almost exclusively in the C-4 carbon atom as expected from Reaction (4.3). Even after 90 seconds of photosynthesis in $^{14}\text{CO}_2$, this carbon atom contained more than 50% of the ^{14}C in the malate. On the other hand, the carbon atoms of phosphoglyceric acid became uniformly labeled more rapidly than those of malate, suggesting that the C-4 of malate is readily transferred to phosphoglyceric acid and the photosynthetic carbon reduction cycle is also active in these species. This alternative mechanism of photosynthetic CO_2 fixation has come to be known as the C_4-dicarboxylic acid pathway, and it appears to be most prominent in several families of plants that are characterized by having high rates of net photosynthesis per unit leaf area (Chapters 5 and 8.A), low rates of photorespiration (Chapter 5), and bundle-sheath chloroplasts (Chapter 2.A).

The first reaction in this pathway involves the synthesis of phosphoenolpyruvate, and it occurs in plants possessing this pathway by a novel enzymatic reaction that is active in efficient photosynthetic species such as sugarcane, maize, sorghum, and *Amaranthus*, but not in plants lacking the C_4 pathway (Slack, 1968; Hatch and Slack, 1968, 1969a). The enzyme, phosphopyruvate synthetase (also called pyruvate, phosphate dikinase) catalyzes Reaction (4.6):

$$
\begin{array}{l}
\text{COOH} \\
| \\
\text{C}\!=\!\text{O} \\
| \\
\text{CH}_3
\end{array}
+ \text{ATP} + \text{Pi} \underset{\substack{\text{Phosphopyruvate}\\ \text{synthetase}}}{\overset{\text{Mg}^{2+}}{\rightleftharpoons}}
\begin{array}{l}
\text{COOH} \\
| \\
\text{C}\!-\!\text{OPO}_3\text{H}_2 \\
\| \\
\text{CH}_2
\end{array}
+ \text{AMP} + \text{PPi} \xrightarrow{\text{Pyrophosphatase}} 2\,\text{Pi} \quad (4.6)
$$

Pyruvate   Phosphoenolpyruvate

$$\text{Phosphoenolpyruvate} + \text{CO}_2 \xrightarrow[\substack{\text{Phosphoenolpyruvate}\\ \text{carboxylase}}]{} \text{Oxaloacetate} + \text{Pi} \qquad (4.7)$$

$$\text{Oxaloacetate} + \text{NADPH} + \text{H}^+ \xrightarrow[\substack{\text{Malate}\\ \text{dehydrogenase}}]{} \text{malate} + \text{NADP}^+ \qquad (4.8)$$

$$[\text{Malate} + \text{NADP}^+ \xrightarrow[\text{``Malic'' enzyme}]{} \text{pyruvate} + \text{NADPH} + \text{H}^+ + \text{CO}_2] \qquad (4.9)$$

Reaction (4.6) converts pyruvate to the CO_2 acceptor. After leaves have been kept in light for 20 minutes the activity of the synthetase in leaf extracts increases at least tenfold in comparison with extracts prepared from leaves stored in the dark. To some extent, the light activation of this enzyme can be reproduced by treating the inactive enzyme with magnesium ions and dithiothreitol. The phosphoenolpyruvate is carboxylated to produce oxaloacetate in Reaction (4.7) of the C_4-pathway, and this compound is reduced to malate by malate dehydrogenase in Reaction (4.8). The C-4 of malate eventually becomes the C-1 of phosphoglycerate because the C-4 carbon atom is first converted to CO_2 by the "malic" enzyme, Reaction (4.9), and the CO_2 is then refixed by the ribulose diphosphate carboxylase reaction as part of the photosynthetic carbon reduction cycle. Thus both pathways of CO_2 fixation exist in the same leaf, and there is considerable evidence that the enzymes of the C_4-dicarboxylic acid pathway are located in the mesophyll chloroplasts while the reactions of the photosynthetic carbon reduction cycle take place primarily in the bundle-sheath chloroplasts in species that have both types although some of the enzymes are difficult to extract from such chloroplasts (Johnson and Hatch, 1968; Björkman and Gauhl, 1969; Slack *et al.*, 1969; Berry *et al.*, 1970; Hatch and Slack, 1969b, 1970). In *Atriplex*, one species gives phosphoglyceric acid in short-time experiments with $^{14}CO_2$ and another produces C_4 compounds (Osmond *et al.*, 1969). Intermediate examples may occur since plants are not necessarily only highly efficient in CO_2 fixation or very much less efficient and examples between these extremes are already known (Chapter 5.J).

Even in species that do not show the C$_4$-pathway in kinetic experiments, such as spinach, phosphoenolpyruvate carboxylase activity is present and is associated with the chloroplasts (L. L. Rosenberg *et al.*, 1958). In the tropical grasses and dicotyledons that exhibit the C$_4$-pathway, the phosphoenolpyruvate carboxylase activity in leaf extracts is greater than the rate of photosynthetic CO$_2$ fixation (Table 4.4) (cf. Chapter 4.E), but the phosphopyruvate synthetase activity is still somewhat less than that required to account for the functioning of this pathway in intact tissues. The activities of phosphoenolpyruvate carboxylase, as well as the phosphopyruvate synthetase, are considerably higher in leaves that have been kept for a few days at high light intensities (Table 4.4).

Free CO$_2$ rather than bicarbonate was shown to be the active substrate for phosphoenolpyruvate carboxylase activity in maize extracts (Waygood *et al.*, 1969) when methods were used similar to those described in Fig. 4.3. The K_m for HCO$_3^-$ in this system was 9×10^{-5} M, a figure similar to that previously observed with extracts of *Kalanchoe* (Walker and Brown, 1957) and in germinating peanut cotyledons (Maruyama *et al.*, 1966). The K_m for CO$_2$ should be revised downward to about 1.5×10^{-5} M (or about one-tenth of the K_m of ribulose diphosphate carboxylase, Chapter 4.E) since the reaction rate was initially six-fold faster with free CO$_2$ than with bicarbonate. The affinity of the isolated enzyme for CO$_2$ is still less than that which must occur in chloroplasts in intact leaves (Chapter 8.C).

Three molecules of ATP and two molecules of NADPH are required for each mole of CO$_2$ fixed in the photosynthetic carbon reduction cycle (Reaction 4.1) when ribulose diphosphate is the substrate for carboxylation. The production of dicarboxylic acid requires two additional molecules of ATP in the C$_4$-dicarboxylic acid pathway, one to produce phosphoenolpyruvate from ATP (Reaction 4.6) and a second to synthesize ADP from the AMP formed in the synthetase reaction. The C$_4$-dicarboxylic acid pathway would thus apparently be less efficient than the Calvin cycle. This may seem surprising since C$_4$-cycle plants show higher rates of net photosynthesis at high light intensities, and the possible explanation for this paradox may be that the more efficient species do not evolve much CO$_2$ in the light by the process of photorespiration, whereas the Calvin cycle plants in general exhibit high rates of CO$_2$ evolution in the light, which counteracts their apparently more efficient mechanism (Chapter 5).

It should be emphasized that just as short-time kinetic experiments with [14]CO$_2$ do not tell us how much carbon is fixed in a leaf by the photosynthetic carbon reduction cycle, one cannot tell how much carbon is fixed by the phosphoenolpyruvate carboxylase reaction simply because oxaloacetate or malate appear as the "first product." Such results do, however, provide convincing evidence that multiple carboxylation reactions occur in photosynthetic tissues, and considerable ingenuity will be required to determine how much carbon is fixed by one pathway compared with another.

G. FERREDOXIN-LINKED CARBOXYLATIONS

Two additional carboxylation reactions involving reduced ferredoxin have been discovered in photosynthetic bacteria, and they appear to play a special role in CO_2 assimilation (Buchanan and Arnon, 1965; Buchanan and Evans, 1965):

$$\text{Acetyl CoA} + CO_2 + Fd_{red} \xrightarrow[\substack{\text{Pyruvate} \\ \text{synthetase}}]{} \text{pyruvate} + Fd_{ox} + CoA \qquad (4.10)$$

$$\text{Succinyl CoA} + CO_2 + Fd_{red} \xrightarrow[\substack{\alpha\text{-Ketoglutarate} \\ \text{synthetase}}]{} \alpha\text{-ketoglutarate} + Fd_{ox} + CoA \qquad (4.11)$$

The pyruvate synthetase [Reaction (4.10)] and α-ketoglutarate synthetase [Reaction (4.11)] accomplish reversals of essentially irreversible steps of the tricarboxylic acid cycle (Chapter 5.A). These reactions have been studied in partially purified preparations from *Chromatium* (purple sulfur bacteria) and *Chlorobium* (green sulfur bacteria) and the requirements shown by the equations have been met. Together with two other carboxylation reactions, those catalyzed by isocitrate dehydrogenase [Reaction (4.13)] and phosphoenolpyruvate carboxylase, a cyclic mechanism may occur in these bacteria in which four molecules of CO_2 are fixed in one turn of a reductive carboxylic acid cycle to produce a four-carbon acid, or possibly only two molecules of CO_2 may be fixed to produce acetate. When $^{14}CO_2$ was supplied to *Chlorobium* cells for 30 seconds, 75% of the total ^{14}C was found in glutamate, which is undoubtedly derived from α-ketoglutarate synthesized by Reaction (4.11) and followed by a transamination reaction (M. C. W. Evans *et al.*, 1966). Other experiments with inhibitors also suggest that these cells fix CO_2 by a reversal of the tricarboxylic acid cycle (Sirevåg and Ormerod, 1970). Thus this organism shows still another "first product" of CO_2 fixation, although it is known that these cells also possess enzymes of the photosynthetic carbon reduction cycle and hence are capable of producing phosphoglyceric acid during photosynthesis. The relative role of each of these carboxylation mechanisms in these bacteria is still not known.

H. CARBOXYLATION REACTION YIELDING GLYCOLATE

The biochemical reactions involved in the synthesis of glycolic acid, CH_2-OH—$COOH$, are still unknown (Chapter 6.A), although this metabolite is an early product of photosynthesis (Table 4.1). When $^{14}CO_2$ of known specific activity at low concentrations was supplied to tobacco leaf tissue in the light, glycolic acid was synthesized with a specific activity similar to that of the added $^{14}CO_2$ and in excess of the specific activity of the carboxyl-carbon atom of phosphoglyceric acid (Zelitch, 1965a). These experiments provide some of the evidence that glycolic acid is synthesized by a carboxylation reaction different from ribulose diphosphate carboxylase. Further evidence that an unknown carboxylation reaction may be involved in the synthesis of glycolate comes from kinetic $^{14}CO_2$ experiments carried out with the photoheterotrophically grown photosynthetic bacterium *Rhodospirillum rubrum* (L. Anderson and Fuller, 1967). About 70% of the ^{14}C from $^{14}CO_2$ fed to cells for 3 seconds was found in glycolate, greatly in excess of the ^{14}C present in phosphate esters. The percentage of radioactivity in glycolate decreased with a negative slope as if it were a "first product" of photosynthesis, and phosphate esters showed a positive slope (Fig. 4.1). Thus there is a strong possibility that glycolate is synthesized by a mechanism involving a novel carboxylation reaction that remains to be discovered.

I. PYRIDINE NUCLEOTIDE-LINKED CARBOXYLATIONS

Three reactions that require NADPH for CO_2 fixation are known to occur in higher plants:

$$
\begin{array}{c}
\text{COOH} \\
| \\
\text{C}\!=\!\text{O} \\
| \\
\text{CH}_3
\end{array}
+ CO_2 + NADPH + H^+
\underset{\substack{\text{``Malic''}\\\text{enzyme}}}{\overset{Mg^{2+}}{\rightleftharpoons}}
\begin{array}{c}
\text{COOH} \\
| \\
\text{HO}\!-\!\text{C}\!-\!\text{H} \\
| \\
\text{CH}_2 \\
| \\
\text{COOH}
\end{array}
+ NADP^+ \qquad (4.12)
$$

Pyruvate   L-Malate

$$
\begin{array}{c}
\text{COOH} \\
| \\
\text{C}=\text{O} \\
| \\
\text{CH}_2 \\
| \\
\text{CH}_2 \\
| \\
\text{COOH}
\end{array}
\; + \; CO_2 + NADPH + H^+ \;
\underset{\substack{\text{Isocitrate}\\ \text{dehydrogenase}}}{\overset{Mn^{2+}}{\rightleftharpoons}}
\;
\begin{array}{c}
\text{COOH} \\
| \\
\text{H}-\text{C}-\text{OH} \\
| \\
\text{HOOC}-\text{C}-\text{H} \\
| \\
\text{CH}_2 \\
| \\
\text{COOH}
\end{array}
\; + \; NADP^+
\qquad (4.13)
$$

a-Ketoglutarate   *threo*-D_s-Isocitrate

$$
\begin{array}{c}
\text{CH}_2\text{OH} \\
| \\
\text{C}=\text{O} \\
| \\
\text{H}-\text{C}-\text{OH} \\
| \\
\text{H}-\text{C}-\text{OH} \\
| \\
\text{CH}_2\text{OPO}_3\text{H}_2
\end{array}
\; + \; CO_2 + NADPH + H^+ \;
\underset{\substack{\text{6-Phosphogluconate}\\ \text{dehydrogenase}}}{\overset{Mg^{2+}}{\rightleftharpoons}}
\;
\begin{array}{c}
\text{COOH} \\
| \\
\text{H}-\text{C}-\text{OH} \\
| \\
\text{HO}-\text{C}-\text{H} \\
| \\
\text{H}-\text{C}-\text{OH} \\
| \\
\text{H}-\text{C}-\text{OH} \\
| \\
\text{CH}_2\text{OPO}_3\text{H}_2
\end{array}
\; + \; NADP^+
\qquad (4.14)
$$

Ribulose-5-phosphate   6-Phosphogluconate

The "malic" enzyme was studied extensively by S. Ochoa and his colleagues in the 1950's. The equilibrium of this reaction, which requires manganese or magnesium ions for activity, favors oxidative decarboxylation [Reaction (4.9)] rather than CO_2 fixation. However, malate synthesis can be demonstrated in this reaction if large concentrations of CO_2 are supplied, or the malate produced is removed and a system for the continual regeneration of NADPH is also provided. As with the other photosynthetic carboxylation reactions thus far described, free CO_2 and not HCO_3^- is the active substrate (Dalziel and Londesborough, 1968). The K_m for CO_2 for the enzyme in extracts of *Kalanchoe* is about 7×10^{-3} M (Walker, 1962), and with the enzyme obtained from wheat germ, a K_m of 3.8×10^{-3} M was obtained (Dalziel and Londesborough, 1968). Although malate is frequently observed as a product of CO_2 fixation, it is probably synthesized by the combined action of the phosphoenolpyruvate carboxylase and malate dehydrogenase reactions rather than by the "malic" enzyme.

Isocitrate dehydrogenase [Reaction (4.13)] catalyzes one of the steps of the tricarboxylic acid cycle (Chapter 5.A), and the equilibrium of this reaction also favors decarboxylation rather than carboxylation. However, isocitrate can be synthesized with isolated enzymes from CO_2 and α-ketoglutarate by using high concentrations of the reactants. Methods similar to those shown in Fig. 4.3 showed free CO_2 to be the active species in this carboxylation reaction (Dalziel and Londesborough, 1968).

Reaction (4.14) catalyzed by 6-phosphogluconate dehydrogenase is associated with the oxidative pentose phosphate cycle (Chapter 5.A) and favors decarboxylation. Few quantitative data about the function of this carboxylation reaction are available although the carboxylation reaction has been demonstrated. Radioactive 6-phosphogluconate increased in pool size, but presumably not because of CO_2 fixation, when *Chlorella* cells were transferred from the light to dark; enzymes of the oxidative pentose phosphate cycle including this dehydrogenase were apparently present in chloroplasts (Bassham, 1968).

Although all three of these NADPH-linked enzymatic activities are present in photosynthetic tissues, their possible role in assimilating CO_2 has still not been fully evaluated.

J. THE WARBURG HYPOTHESIS

O. Warburg and his collaborators have evolved a very different view, compared with those that have been presented earlier, of the initial steps of CO_2 fixation and O_2 evolution during photosynthesis. His ideas are based upon experiments on the quantum efficiency of photosynthesis carried out largely with *Chlorella* and represent a minority viewpoint. They should be considered seriously, however, even though his hypothesis cannot be readily subjected to a critical experimental test. Some of the unresolved questions about the effects of CO_2 on the Hill reaction (Chapter 3.C) also suggest that Warburg's hypothesis requires further investigation. He believed that CO_2 reacts in air with a special form of chlorophyll to produce an active form of CO_2 known as the photolyte (Warburg *et al.*, 1963, 1969; Vennesland, 1963):

Light: $\quad\quad (H_2CO_3)^* + 1 \text{ quantum} \rightarrow (CH_2O) + O_2 + 1 \text{ chlorophyll} \quad (4.15)$

$\quad\quad\quad\quad\quad\quad$ Photolyte

Dark: $\quad\quad \frac{2}{3}(CH_2O) + \frac{2}{3}O_2 \rightarrow \frac{2}{3}H_2CO_3 + 76,000 \text{ cal} \quad\quad (4.16)$

Dark: $1\,H_2CO_3 + 1 \text{ chlorophyll} + 76,000 \text{ cal} \rightarrow (H_2CO_3)^* \quad\quad (4.17)$

Net: $\quad\quad \frac{1}{3}H_2CO_3 + 1 \text{ quantum} \rightarrow \frac{1}{3}(CH_2O) + \frac{1}{3}O_2$

or $\quad\quad 1\,H_2CO_3 + 3 \text{ quanta} \rightarrow 1\,(CH_2O) + 1\,O_2 \quad\quad (4.18)$

According to this scheme, in Reaction (4.15) the chlorophyll bound to CO_2 absorbs 1 quantum of light and is split into oxygen, reduced carbonic acid (carbohydrate), and free chlorophyll. The photolyte is then resynthesized in a dark reaction [Reaction (4.17)] from carbonic acid and free chlorophyll with the aid of energy derived from the reoxidation of two-thirds of the reduced carbon formed during the light reaction [Reaction

(4.16)]. These equations add up to a quantum requirement of three and indicate that the oxygen evolved in photosynthesis is derived from CO_2 rather than from the photooxidation of H_2O as suggested by the Hill reaction. The photolyte is difficult to isolate because it is rapidly converted to O_2 in the light and is decomposed to CO_2 in the dark.

Evidence for the existence of the photolyte comes from experiments in which photosynthesis is carried out with algae maintained in concentrations of 10 to 20% CO_2. A typical experiment is terminated by adding enough acidic fluoride to inhibit respiration completely. The resulting pressure in the manometer from the release of CO_2 is corrected by suitable controls and is believed to correspond to the aerobically bound photolyte. Warburg deduced by such experiments that when the cells are incubated in air and CO_2 in the dark, they form a CO_2 compound which is readily converted to O_2 in the light but which quickly decomposes to CO_2 in the dark as expected from the postulated properties of the photolyte. Warburg states that only 5 to 25% of the total chlorophyll is present as the photolyte during steady-state photosynthesis, and that high concentrations of CO_2 are used in the experiments so that the CO_2 is not consumed too rapidly. He says that concentrations of 2 to 5% CO_2 may also be used to demonstrate the existence of such an active CO_2 compound. Others who have attempted to duplicate these experiments on the effect of fluoride on the photolyte have confirmed the existence of aerobically bound CO_2, but unlike Warburg they could not show that its magnitude was affected by illumination and they observed that a variety of acids could cause CO_2 evolution in addition to acidic fluoride solutions (Miyachi et al., 1968).

K. THE PRODUCTS OF PHOTOSYNTHESIS

All of the carbon compounds found in green plants are ultimately derived from the CO_2 assimilated in photosynthesis. Varying quantities of the fixed CO_2 find their way into carbohydrates, lipids, and proteins in proportions that differ with the species and the environmental conditions under which the plants are grown. The study of factors that influence the flow of carbon into one pathway rather than another constitutes an important area of research, and the biochemical manipulation of the products of photosynthesis should become more feasible with increasing knowledge. Since most plants accumulate carbohydrate as the initial storage product of photosynthesis, the reactions responsible for carbohydrate synthesis are first discussed, and some methods of altering the flow of carbon into other compounds are also considered.

Starch and sucrose are the two main storage forms of carbohydrate in leaves, and in some species starch makes up 40% of the dry weight of the leaves. The biosynthesis of starch was shown in L. F. Leloir's laboratory to take place by the transfer of glucosyl residues from nucleoside diphosphate sugars (Chapter 1.C):

$$\text{ATP} + \alpha\text{-glucose-1-P} \xrightarrow[\substack{\text{ADP-Glucose} \\ \text{pyrophosphorylase}}]{} \text{ADP-glucose} + \text{PPi} \qquad (4.19)$$

$$\substack{\text{ADP-Glucose} \\ \text{or} \\ \text{UDP-glucose}} + \alpha\text{-1,4-glucan primer} \xrightarrow[\substack{\text{Starch} \\ \text{synthetase}}]{} \alpha\text{-1,4-glucosyl glucan} + \substack{\text{ADP} \\ \text{or} \\ \text{UDP}} \qquad (4.20)$$

The pyrophosphorylase activity is believed to control the rate of starch synthesis in leaves, since the activity of this enzyme in leaf extracts may be increased from 9- to 80-fold in the presence of low concentrations of phosphoglyceric acid (Preiss and Kosuge, 1970), and the activation occurs with enzyme obtained from plants that show the Calvin type or the C_4-dicarboxylic acid pathway of photosynthesis. This enhancement of enzyme activity is accomplished in part by phosphoglyceric acid decreasing the apparent affinity of the enzyme for its substrates, ATP and glucose-1-phosphate. The starch synthetase [Reaction (4.20)] in leaves is located in the chloroplasts, and it is also found in other plant tissues where starch is stored (Preiss *et al.*, 1967).

At one time starch synthesis was believed to occur by means of enzymes called phosphorylases, but the phosphorylase reaction is probably responsible for the breakdown of starch because of the high cellular concentration of orthophosphate:

$$\text{Starch} + \text{Pi} \underset{\text{Phosphorylase}}{\rightleftharpoons} \text{glucose-1-P} + \alpha\text{-1,4-glucan} \qquad (4.21)$$

Amylases, which carry out the hydrolysis of starch rather than its phosphorolysis as indicated above, also function in the breakdown of starch to yield soluble carbohydrate in plant tissues.

Sucrose is the primary storage form of carbohydrate in many plants and it is usually the main substance translocated from the leaf to other organs. Sugarcane provides an extreme example of sucrose storage, since

90% of the soluble material present in the stalk may be present as sucrose.
Leaves contain enzymes that can carry out the biosynthesis of sucrose by
two separate mechanisms:

$$\text{UDP-Glucose} + \text{fructose} \underset{\substack{\text{Sucrose}\\\text{synthetase}}}{\rightleftharpoons} \text{sucrose} + \text{UDP} \qquad (4.22)$$

$$\text{UDP-Glucose} + \text{fructose-6-P} \xrightarrow[\substack{\text{Sucrose-phosphate}\\\text{synthetase}}]{} \text{sucrose-P} + \text{UDP} \qquad (4.23)$$

$$\text{Sucrose-P} + \text{H}_2\text{O} \xrightarrow[\text{Phosphatase}]{} \text{sucrose} + \text{Pi} \qquad (4.24)$$

Experiments based on both the rate of sucrose labeling during $^{14}\text{CO}_2$
fixation and by providing glucose-^{14}C to leaf tissues suggest that the mechanism in Reactions (4.23) and (4.24) involving sucrose-phosphate synthetase
and phosphatase predominates. Equal labeling of the glucose and fructose
moieties of sucrose is found, and this would not be expected if the sucrose
synthetase [Reaction (4.22)] were the primary mechanism for sucrose
biosynthesis. The sucrose-phosphate synthetase activity in leaf extracts of
several species is about the same or somewhat greater than the activity of
sucrose synthetase, but both activities are low compared with the rate of
sucrose production in intact leaves (Hawker, 1967). Because the sucrose
synthetase reaction, unlike the sucrose-phosphate synthetase system, is
thermodynamically freely reversible, it may function primarily in the breakdown of sucrose to provide nucleoside diphosphate sugars that are substrates
of a number of reactions involving the biosynthesis of carbohydrate polymers (Hassid, 1967).

Certain external factors are known to influence the nature of the carbohydrate produced during photosynthesis. For example, under conditions
of a nitrogen deficiency, the starch content of leaves is frequently increased
(H. K. Porter, 1966), and at low temperatures in some tissues the sucrose
concentration is high. During photosynthesis by isolated spinach chloroplasts, from 5 to 20% of the assimilated CO_2 is found as starch (Gibbs *et al.*,
1967), but sucrose is found as a product only in rare instances. Hydrolysis
of the sucrose which was formed established that there was isotopic equilibrium between the two hexose units of sucrose, suggesting that sucrose
was not synthesized from UDP-glucose and the unlabeled fructose that
was available from the large pool of sucrose present in the medium. Both
mechanisms for the synthesis of sucrose [Reactions (4.22), and (4.23) plus
(4.24)] exist in chloroplasts, but the sucrose-phosphate synthetase reaction
probably predominates. The reasons for the variability in sucrose synthesis
by isolated chloroplasts are still not understood.

The concentration of oxygen in the atmosphere also greatly influences the products of CO_2 assimilation during photosynthesis (Chapter 6.A). Sunflower leaves were supplied with $^{14}CO_2$ for 3 minutes at 600 ppm of CO_2; in 2% O_2, 9% of the radioactivity was present in amino acids and 42% in carbohydrates; in 21% O_2, the amino acids increased to 31% and the ^{14}C in carbohydrates decreased to 21% of the total carbon fixed (Fock *et al.*, 1969a). Thus increasing the oxygen concentration resulted in an increased rate of synthesis of amino acids and a decreased synthesis of carbohydrate. The incorporation of $^{14}CO_2$ in bean leaf disks into glycolate and glycine was suppressed about 50% in 0.5% O_2 compared with normal air, and this occurred at low and at high light intensities (Voskresenskaya *et al.*, 1970a).

Likewise the CO_2 concentration in the atmosphere affects the products of photosynthetic CO_2 assimilation. *Chlorella* cells were grown in 5% CO_2, then $^{14}CO_2$ was given to the cells in light for 6 minutes at a CO_2 concentration of 0.3% or at 0.003%. The ^{14}C incorporation (as percent of the total) was: total phosphate esters, 18% in 0.003% CO_2, 54% in 0.3% CO_2; malate plus aspartate, 69% in 0.003% CO_2 and 25% in 0.3% CO_2 (Graham and Whittingham, 1968). These results may be explained in terms of a competition between the phosphoenolpyruvate carboxylase and the ribulose diphosphate carboxylase reaction, the latter enzyme having a lower affinity for CO_2 (Chapter 4.E). Studies of $^{14}CO_2$ fixation in tomato plants grown in air or in a high CO_2 concentration (0.1% CO_2) showed that there was a consistent decrease in the ^{14}C in the glycine-serine fraction from leaves assimilating CO_2 at the higher CO_2 concentration (Chapter 6.A) (P. M. Bishop and Whittingham, 1968). This occurred whether the plants were grown at high or low light intensities, and the result again emphasizes that the CO_2 concentration can alter the equilibrium between various pathways and thus change the products of CO_2 assimilation.

L. CRASSULACEAN-TYPE METABOLISM

Although most species accumulate carbohydrates in their leaves during photosynthesis, many plants also accumulate large concentrations of organic acids. Malic and citric acids are the most common, but in certain species large amounts of oxalic acid, isocitric acid, tartaric acid, or malonic acid are found. Lesser quantities of other organic acids associated with the tricarboxylic acid cycle are always present. Tobacco leaves in various positions on the stalk may differ in the composition of specific organic acids by as much as fourfold, and the total nonvolatile organic acids varied

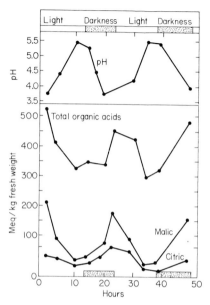

Fɪɢ. 4.4. Diurnal variation of organic acids and pH in excised leaves of *Bryophyllum calycinum* (from Vickery, 1952).

from 240 meq/kg fresh weight (over 0.1 *M*) in younger leaves to 145 meq in the older leaves (Vickery, 1961). Since organic acids are intermediates in the biochemical interconversions between carbohydrates, lipids, and proteins, their metabolism has received considerable attention and certain aspects of organic acid metabolism in leaves have been reviewed (Zelitch, 1964b; Ranson, 1965).

Of special interest is the organic acid metabolism exhibited by many species in the family Crassulaceae and a number of other succulent plants which accumulate large quantities of malic acid at night, and lose most of the malic acid in the subsequent light period (Fig. 4.4). Plants possessing this metabolism have a reciprocal diurnal variation in starch. This striking diurnal fluctuation in the organic acid content of the leaves was first recorded by B. Heyne in 1815 who observed that leaves of *Bryophyllum calicynum* taste strongly acid at sunrise and merely bitter in the afternoon. Changes in organic acid composition were described in 1875, and some of its quantitative aspects were established by T. A. Bennet-Clark in 1933. In the dark, 200 meq malic acid and 50 meq citric acid per kg of fresh leaf may accumulate while these organic acids largely disappear in the subsequent light period. The pH of the leaf frequently changes by 2.0 units as expected from the alterations in the organic acid concentration (Fig. 4.4).

In crassulacean-type metabolism, most of the net CO$_2$ is fixed in the dark rather than in the light, while in the light there is generally little CO$_2$ uptake or net output. Thus these species store their reducing power needed for CO$_2$ fixation primarily in the form of starch, which may comprise 30% of the dry weight (Vickery, 1954), but actually carry out their net CO$_2$ uptake in the dark. This unusual metabolism has attracted considerable attention and the subject has been discussed in excellent articles by Ranson and Thomas (1960) and Walker (1962).

Plants carrying out such a metabolism fix CO$_2$ in the dark primarily by means of the phosphoenolpyruvate carboxylase reaction. However, the following experiment indicates that ribulose diphosphate carboxylase also plays a role. When leaves of *Kalanchoe* were supplied with ^{14}CO$_2$ in the dark, the malic acid isolated after exposures of from 4 seconds to 20 hours still showed the same distribution of ^{14}C in the carbon atoms, and subsequently placing the leaves in the light for 3 to 9 hours did not change the ^{14}C distribution. About two-thirds of the ^{14}C was in the C-4 atom, one-third in the C-1 carbon, and none in the middle two carbons (Bradbeer *et al.*, 1958). Such results could be explained if the ribulose diphosphate carboxylase reaction also functioned to produce one mole of labeled and one mole of unlabeled phosphoglyceric acid, and these by the usual sequence produced one mole of phosphoenolpyruvate labeled in the carboxyl-carbon atom and one unlabeled molecule. The fixation of ^{14}CO$_2$ by the phosphoenolpyruvate carboxylase reaction would then produce the isotopic distribution found:

$$
\begin{array}{c}
\text{*COOH} \\
| \\
\text{C—OP} \\
\| \\
\text{CH}_2 \\
+ \\
\text{COOH} \\
| \\
\text{C—OP} \\
\| \\
\text{CH}_2
\end{array}
\quad + \; 2\,\overset{*}{\text{CO}_2} \;\xrightarrow[\text{carboxylase}]{\text{Phosphoenolpyruvate}}\; 2
\begin{array}{c}
\text{*COOH} \\
| \\
\text{HO—C—H} \\
| \\
\text{CH}_2 \\
\text{**COOH}
\end{array}
\qquad (4.25)
$$

Phosphoenolpyruvate

L-Malate

Whether this is the mechanism has not been fully established, although ribulose diphosphate carboxylase is present in these leaves. These isotope-distribution results clearly emphasize that the malic acid produced must be quickly and completely isolated or compartmented from other metabolic events since even after 20 hours the middle carbon atoms are still unlabeled and the distribution of ^{14}C in the carboxyl-carbon atoms is also unequal.

Another interesting aspect of crassulacean acid metabolism involves organic acids other than malic acid. Leaves of these species frequently accumulate large amounts of isocitric acid, an important intermediate in the tricarboxylic acid cycle. For some time an unusual organic acid was known to be present in such plants, and because its properties resembled those of malic acid it had sometimes been referred to as "crassulacean malic acid." Pucher (1942) identified this substance as isocitric acid, a compound present in concentrations of from 10 to 15% on a dry weight basis in various crassulacean species (Vickery and Wilson, 1958; Soderstrom, 1962). In *Bryophyllum* leaves provided with isocitrate-^{14}C, the labeled organic acid was rapidly metabolized, although the isocitric acid concentration remains fairly constant during a 24 hour period. Thus there is a small pool of metabolically active isocitrate together with a large inactive pool which remains in storage in the vacuole (A. O. Klein, 1964).

It is difficult to understand how a plant can efficiently store its reducing power in the form of starch in the light and respire this in the dark to provide ATP and the CO_2 acceptor. Some theoretical assumptions on how this may be achieved have been suggested (Walker, 1966). Crassulacean metabolism results from an adaptation which evolved under arid conditions, and it may permit survival under such adverse conditions where productivity is not of prime consideration. In these succulent plants, the stomata are usually closed in the light and open in the dark as a further means of conserving water (Nishida, 1963; Ekern, 1965). Those who have grown crassulacean plants will have observed how slowly they grow. A notable exception is the pineapple plant which presumably carries out reasonable rates of net CO_2 assimilation yet possesses a crassulacean-type metabolism (Sideris *et al.*, 1948). Thus further work would seem to be of the utmost importance to explain the detailed mechanism that permits a plant to fix CO_2 at appreciable rates in the dark while conserving water at the same time, an attribute that would seem to have widespread utility in many parts of the world.

M. COMPARTMENTATION AND MOVEMENT OF METABOLITES WITHIN LEAF CELLS

If compounds that participate in biochemical sequences are present in different organelles or within the cytoplasm as well as the vacuole, the turnover rates of such substances will differ in the various compartments and this will strongly affect reaction rates. Evidence exists that many metabolites are so compartmented that they do not rapidly equilibrate (see below), and this also has consequences in attempting to understand the role of different biochemical pathways in intact cells. The vacuole of leaf cells (Chapter 1.E) constitutes a particularly large pool in which the turnover of substances is very slow. For example, when pyruvate-2-^{14}C at a concentration of 0.2 M was fed to excised tobacco leaves in the dark, the specific activity (cpm per mole) of the citrate formed was the same as that of the pyruvate supplied (Vickery and Zelitch, 1960). Therefore, one-half of the carbon found in the citrate must have come from some source other than the added pyruvate-^{14}C. Since the specific activity did not change for 48 hours at least two pools of citrate must be present in the leaf, one that is rapidly metabolized to intermediates of the Krebs tricarboxylic acid cycle, and the other that is metabolically inert and is presumably present in the vacuole.

Similar conclusions were obtained when acetate-1-^{14}C was supplied to wheat leaves in the dark and the specific activity of the ^{14}CO$_2$ released was compared with that of the carboxyl-carbon atoms of several of the intermediate compounds that make up the tricarboxylic acid cycle and are precursors of the CO$_2$ (MacLennan *et al.*, 1963). When a constant specific activity in the respired ^{14}CO$_2$ was achieved, the specific activity in the carboxyl-carbon atoms of citric and malic acids was very much lower, again indicating that large pools of the individual acids are physiologically remote from the respiratory centers.

The outer chloroplast envelope constitutes a considerable barrier to the penetration of certain compounds in the leaf cell and creates metabolic compartments. The concentration of NADP$^+$, for example, is greatly different in the chloroplast and the cytoplasm, and when leaves are illuminated the NADP$^+$ present in the chloroplast is reduced after 15 to 30 seconds but not the NADP$^+$ in the cytoplasm (U. W. Heber and Santarius, 1965). Isolated intact spinach chloroplasts are also relatively impermeable to pyridine nucleotides, whereas ADP and ATP readily move out through the plastid envelope. Phosphoglyceric acid is rapidly taken up by isolated chloroplasts, and triose phosphate synthesized within the chloroplast in the light by reduction of phosphoglyceric acid easily moves out through the boundary membranes (Stocking and Larson, 1969).

The rate of migration of products of photosynthesis from chloroplasts has been studied in intact tobacco leaves supplied with ^{14}CO$_2$ for periods of from 5 seconds to 10 minutes. The tissue was rapidly frozen in liquid nitrogen and the chloroplasts and supernatant fractions were separated in nonaqueous media. Such experiments showed that the assimilated CO$_2$

migrates rapidly from the chloroplasts, and after 60 seconds about 50% of the newly formed carbon compounds are already outside of the chloroplasts (Stocking *et al.*, 1963). After 10 minutes of photosynthesis in $^{14}CO_2$, the distribution of radioactivity among various classes of labeled compounds was generally similar in the chloroplast fraction and in the cytoplasm, except that phosphate esters and starch accounted for a higher percentage of the total ^{14}C in the chloroplast (Ongun and Stocking, 1965b). Similar experiments in which photosynthesis in $^{14}CO_2$ was followed by periods of photosynthesis in normal air showed that sugar phosphates moved readily from the chloroplast to the cytoplasm where they appear to be converted to other products including glycine and serine (Whittingham and Keys, 1969). Thus various metabolites move at different rates between organelles and the cytoplasm and changes in these rates can be expected to influence the leaf composition greatly.

Section II

*Respiration Associated with
Photosynthetic Tissues*

5   Dark Respiration and Photorespiration

The photosynthetic assimilation of CO_2 is partly counterbalanced by the loss of CO_2 by respiration:

$$C_6H_{12}O_6 + 6\ O_2 \longrightarrow 6\ CO_2 + 6\ H_2O + 672{,}000\ cal \qquad (5.1)$$

The overall reaction of respiration of carbohydrate is the same but opposite in direction to the overall equation for CO_2 uptake in photosynthesis [Reaction (3.1)]. Net CO_2 assimilation (sometimes called "apparent" photosynthesis) is equal to the gross photosynthesis (sometimes referred to as "true" photosynthesis) minus the loss resulting from respiration. Photosynthesis and respiration do not represent reversals of the same biochemical reactions, however, although both processes have enzymes and intermediates in common. Moreover, photosynthesis and respiration occur in different cell organelles and therefore these processes are largely separated physically. Respiration is needed to provide energy for the essen-

127

tial biochemical reactions concerned with growth and the synthesis of ATP. Some portion of the respiration may be nonessential for these purposes and, if it could be controlled or eliminated, a net gain of plant dry weight should ensue provided that photosynthesis was unaffected. Factors other than respiration that regulate the rates of gross and net photosynthesis are discussed separately in Chapters 7, 8, and 9.

Research since about 1960 has strongly supported the view that respiratory losses are perhaps the most important single factor affecting net photosynthesis. This is especially true of the vast quantities of CO_2 lost by many species in the light by reactions that differ from the usual dark respiration observed in nonphotosynthetic as well as in photosynthetic plant tissues. Respiration of CO_2 in the light, usually called photorespiration, results from the oxidation of photosynthetic products and often occurs at rates three to five times greater than dark respiration. However, even dark respiration causes a considerable decrease of dry weight gain, amounting to 50% or more of net photosynthesis in a crop (Chapter 9.E), hence its potential role in affecting plant productivity must also be considered.

According to Equation (5.1), the photosynthetic quotient (ratio of net O_2 released/net CO_2 uptake) should be 1.0. If O_2 uptake and CO_2 evolution resulting from respiration are large compared with photosynthesis and have a respiratory quotient (CO_2/O_2) differing from 1.0, the photosynthetic quotient may also vary. Measurements of the photosynthetic quotient have frequently been made (Rabinowitch, 1945, p. 33), and the experiments of R. Willstätter and A. Stoll in 1918 are typical. Very high rates of photosynthesis were desired to minimize the error caused by respiration, thus high light intensities and 5 to 6.5% CO_2 in air were used in experiments with several species. A photosynthetic quotient of 1.0 was obtained, but it will become evident from the data to be presented below that such high CO_2 concentrations also completely abolish photorespiration (Chapters 5.E and 6.A), hence their results may represent an artifact.

A sensitive infrared CO_2 analyzer has been developed that permits accurate measurements of small changes in CO_2 concentration in an open system in which the gas phase is passed over a leaf at a known rate and the CO_2 concentration in the entering and leaving streams is analyzed [in a closed system, the gas is recycled past the leaf and the change in CO_2 concentration is measured at a single point.] This instrument has been used in conjunction with a paramagnetic O_2 analyzer to measure the photosynthetic quotient in a range of CO_2 concentrations at low (2%) and high (42%) O_2 concentrations at a light intensity of 15,000 lux, temperatures between 20° and 25°, and CO_2 concentrations up to 600 ppm (Fock *et al.*, 1969b). At low ambient concentrations of O_2, leaves of *Amaranthus paniculatus* (low photorespiration) and bean (*Phaseolus vulgaris*) (high photorespiration) had photosynthetic quotients of 1.0 at all CO_2 concentrations examined. At the high O_2 concentrations, *Amaranthus* leaves had a photosynthetic quotient of 0.50 and bean leaves 0.33. These quotients were essentially independent of the CO_2 concentration, and indicate that high concentrations of O_2 in the atmosphere can greatly lower the ratio of the net O_2 evolved to the net CO_2 taken up. This resulted because either the rate of O_2 uptake was enhanced in the light or the rate of O_2 evolution was inhibited or both. A greater O_2 uptake in the light compared with CO_2 evolution also appears from measurements of $^{18}O_2$ and $^{13}CO_2$ in a mass spectrometer (Chapter 5.C). A biochemical explanation for a relative enhancement of O_2 uptake induced by increasing the O_2 concentration is suggested in Chapter 6.B.

The question whether respiration in the light is the same, greater than, or less than the easily measured CO_2 evolution in the dark in photosynthetic tissues has been raised by research workers for at least 100 years. The uncertainty results because the gross CO_2 evolution or O_2 uptake cannot be measured directly during photosynthesis at high light intensities when the reverse processes, net CO_2 uptake and O_2 evolution, take place from 2 to 30 times faster than the dark respiration. Therefore information about respiration in the light can be obtained only by indirect methods, and all of these methods underestimate photorespiration by various amounts. For this reason the discussion of photorespiration has been arranged according to the assay method used so that the conclusions derived from different experiments can more easily be evaluated.

Since a considerable body of recent evidence now points to the existence and importance of photorespiration in affecting net photosynthesis, it may prove instructive to examine why the widespread recognition of this process was so long delayed. Rabinowitch (1945, 1951, 1956) discussed the possibility that an acceleration of "normal" (meaning "dark") respiration might occur in the light ("photorespiration") and cited apparently contradictory results. He vividly concluded that "the possibility of such an effect is a nightmare oppressing all who are concerned with the exact measurement of [gross] photosynthesis" (Rabinowitch, 1945, p. 569). Photorespiration is a biochemical process very different from dark respiration, and the earlier attempts to detect it did not recognize this fact so that experiments were generally carried out under conditions that tended completely to obscure or eliminate photorespiration: high concentrations of ambient CO_2, low concentrations of O_2 in the atmosphere, weak light, and low temperatures. Also many workers failed to consider the occurrence of a rapid reutilization (recycling) of CO_2 and O_2 within photosynthetic tissues in preference to the assimilation of CO_2 and O_2 provided from the gas phase, an oversight which confused the interpretation of some experiments.

An additional obstacle to appreciating the importance of photorespiration was the finding that certain photooxidation processes occur in photosynthetic tissues kept in the light at 10 to 100% O_2, and that these photooxidations occurred for a considerable period of time in the dark and even when the tissue was killed by heating. Thus Gaffron and Fager (1951) stated that "the extra oxygen uptake can be explained as a pathological photooxidation," although Franck (1951) pointed out that such non-enzymatic photooxidations were too slow by a factor of twenty to account for the inhibition of photosynthesis caused by O_2 in the atmosphere (Chapters 5.D and 8.E). Clearly photorespiration is very different in magnitude and in many other characteristics from the relatively slow photooxidation process that had previously been observed.

"Photorespiration" is defined as the respiration (especially the CO_2 evolution) that differs biochemically from normal dark respiration and is specifically associated with substrates produced during photosynthesis. Normal dark respiration also occurs in the light and "photorespiration" is sometimes used to mean the total CO_2 evolved by the processes of dark respiration plus photorespiration. Since photorespiration usually greatly exceeds dark respiration, there is consequently little difference between the more precise and the all-inclusive definition. Photorespiration was first used in its current sense by Decker and Tió (1959) in explaining the post-illumination CO_2 outburst shown by leaves of many species (Decker, 1955) (Chapter 5.E). They recognized the importance of photorespiration and suggested that the limitation of photorespiration by genetic means may increase net photosynthesis and hence net dry weight increment. The sad history of the failure of many scientists to recognize the significance of the post-illumination CO_2 outburst has recently been described (Decker, 1970).

It is now clear that Rabinowitch's "nightmare" has come to reality and that photorespiration greatly decreases net photosynthesis in many species. But when the process is better understood so that it may be controlled, diminishing photorespiration may be one of the easiest methods of obtaining large increases in net photosynthesis.

A. OXYGEN UPTAKE AND CO_2 EVOLUTION BY DARK RESPIRATION

Dark respiration provides ATP which is synthesized by the reactions of oxidative phosphorylation in addition to causing O_2 uptake and CO_2 evolution. Reduced pyridine nucleotides and other electron carriers are also byproducts, and these products of respiration are utilized in specific reactions to synthesize the carbon compounds that make up proteins, lipids, nucleic acids, organic acids, aromatic compounds, and other cell constituents as well as to move ions and carry out cell division. The energy of respiration is made available in stages by several biochemical sequences that are usually cyclic in nature so that the product of the series of reactions functions in catalytic concentrations to start the entire sequence all over again. Several excellent books (W. O. James, 1953; Beevers, 1961) and a number of articles (Ducet and Rosenberg, 1962; Zelitch, 1964b; Lieberman and Baker, 1965) are available that discuss various aspects of plant respiration. Most writers have generally been concerned with dark respiration although a comprehensive review on photorespiration has recently appeared (Jackson and Volk, 1970).

Dark respiration is usually measured by manometric or polarographic determination of O_2 uptake, assessing the rate of loss of dry weight, or by measuring the rate of CO_2 evolution manometrically or by means of an infrared gas analyzer while air is passing rapidly over the leaf. The rates of dark respiration of leaves vary from about 0.5 to 4.0 mg CO_2 dm^{-2} hr^{-1} at about 25°, and the rate approximately doubles for every 10° rise from 0 to 35° (Altman and Dittmer, 1968, p. 507). For example, soybean leaves released 0.3 mg CO_2 dm^{-2} hr^{-1} at 15° and 0.7 at 25° in darkness, and the rates were the same with about 1% O_2 or 21% O_2 in the atmosphere (Hofstra and Hesketh, 1969a). Other typical values for dark respiration are given in Tables 5.2 and 5.4.

The respiratory quotient (RQ), the ratio of the CO_2 evolved to the O_2 taken up, is usually close to 1.0 for leaves. This would be expected if hexose were the primary respiratory substrate and it were completely oxidized [Reaction (5.1)]. However, hexose could be respired and result in a different RQ if some of the CO_2 evolved were consumed again, if hexose were not completely oxidized, or if diffusion of O_2 into the tissues were hindered. If the primary substrate of respiration is a fatty acid such as stearic acid, the RQ would be less than 1.0 (0.69), and if the substrate were an organic acid such as citric acid the RQ would be greater than 1.0 (1.33).

The rate of dark respiration decreases in leaves with age, and is generally higher after a regime of high illumination (Chapter 5.A.5). In a number of instances dark respiration is considerably lower in shade-tolerant plants than in sun-plants even when the former are grown in full sunlight (Grime, 1965). Perhaps this characteristic enables these species to survive the adverse light conditions prevailing in a dense forest. Respiration of leaves in darkness shows a maximal rate of CO_2 evolution at ambient concentrations of 2 to 3% O_2, considerably less than the 21% of O_2 present in normal air (Forrester *et al.*, 1966a). Thus the oxygen consumption reaction occurs by means of an oxidase with a high affinity for O_2, such as cytochrome oxidase.

Dark respiration in leaves is usually the same in CO_2-free air as in normal air, although some exceptions can be found, as indicated below.

1. ANAEROBIC GLYCOLYSIS

The breakdown of hexose does not go to completion in the absence of O_2, and the end product of anaerobic glycolysis is either lactic acid or ethanol. In leaves, the product is more likely ethanol, since the enzyme lactate dehydrogenase required in the last step is usually present in very small amounts. The sequence of reactions which normally takes place by means of soluble enzymes (shown in parentheses) located in the cytoplasm is given in Scheme 5.1.

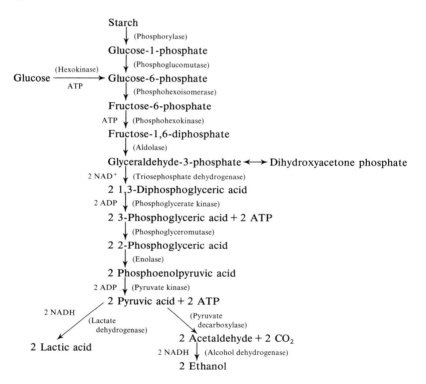

SCHEME 5.1. Transformation of starch or glucose by anaerobic glycolysis.

A number of the intermediates, such as 3-phosphoglyceric acid, and enzymes are also involved in the photosynthetic carbon reduction cycle (Chapter 4.C). Although the complete oxidation of 1 mole of glucose results in the release of 672,000 cal, only about 57,000 cal are produced when 2 moles of lactic acid are synthesized during glycolysis. Beginning with starch, 1 mole of ATP is consumed (at the phosphohexokinase-catalyzed reaction) while 4 moles of ATP are synthesized, for a net gain of three "energy rich" phosphate bonds. Since each ATP contains about 8000 cal/mole in its pyrophosphate bond, a gain of 24,000 represents 24,000/57,000 or about 42% of the potentially available energy conserved as chemical energy. When starting with glucose, an additional ATP is consumed in the hexokinase step, so the net gain is then only two ATP's.

2. TRICARBOXYLIC ACID CYCLE

The pyruvic acid produced during glycolysis may be completely oxidized to CO_2 and H_2O with the accompanying synthesis of much more ATP in aerobic tissues. This occurs in the mitochondria (Chapter 1.F) by means of a cyclic sequence of reactions involving tricarboxylic acids which was first demonstrated in animal muscle by H. A. Krebs in 1937. Its functioning in plants was surmised by A. C. Chibnall in 1939. The first product of pyruvate oxidation, acetyl coenzyme A, reacts with oxaloacetic acid to synthesize citric acid. Oxaloacetic acid is eventually regenerated as a product of the cycle. This sequence of reactions (also known as the Krebs cycle or citric acid cycle) accounts for 5 moles of O_2 uptake and 6 moles of CO_2 per mole of hexose oxidized (one more mole of O_2 uptake results from the triosephosphate dehydrogenase reaction of glycolysis). The details of the tricarboxylic acid cycle are readily available, thus only the oxidative steps of the cycle are summarized in Scheme 5.2.

		Per mole of glucose			
Oxidative reaction	Reduced product	Atoms O_2 uptake	Moles CO_2	Moles ATP	Ratio P/O
Pyruvate → acetyl CoA + CO_2 (pyruvate dehydrogenase)	NADH	2	2	6	3
Isocitrate → α-ketoglutarate + CO_2 (isocitrate dehydrogenase)	NADPH	2	2	6	3
α-Ketoglutarate → succinate + CO_2 (α-ketoglutarate dehydrogenase)	NADH	2	2	8	4
Succinate → fumarate (succinate dehydrogenase)	Flavin	2	—	4	2
Malate → oxaloacetate (malate dehydrogenase)	NADH	2	—	6	3
	Net:	10	6	30	

SCHEME 5.2. Summary of oxidative reactions of the tricarboxylic acid cycle concerned with ATP formation.

Reduced pyridine nucleotides are produced in four of the reactions shown in this scheme. The electrons are transported from NADH or NADPH in the mitochondria to flavoprotein and cytochrome carriers. Eventually they are transferred to O_2 by the enzyme cytochrome oxidase. The oxidation of each pair of electrons from NADH or NADPH results in the synthesis of 3 moles of ATP by the process of oxidative phosphorylation which occurs in the mitochondria, hence these reactions are said to have a P/2e or P/O ratio of 3.0. An additional phosphorylation, not connected with the oxidation of reduced pyridine nucleotides, occurs during the oxidation of α-ketoglutarate so that a P/O ratio of 4.0 may be obtained in this step of the cycle. However, the oxidation of succinate begins at the flavoprotein-oxidation level and results in a P/O ratio of only 2.0. Thus the cycle accounts for 5 moles of O_2 consumed, 6 moles of CO_2 produced, and 30 moles of ATP synthesized during glucose oxidation. In terms of conversion of chemical energy, the tricarboxylic acid cycle is therefore about 12 times more productive than anaerobic glycolysis.

Reactions of the tricarboxylic acid cycle were demonstrated in tobacco leaves in a number of experiments conducted by H. B. Vickery in which various organic acids or inhibitors were supplied to excised tobacco leaves in darkness and the products of respiration were quantitatively analyzed. Reactions of the Krebs cycle were also shown when glutamate metabolism was investigated (Vickery, 1963). Isolated particulate preparations from pea leaves were capable of actively oxidizing substrates of the tricarboxylic acid cycle at good rates and with P/O ratios as expected from Scheme 5.2 (Smillie, 1956). Mitochondrial preparations from spinach leaves (Zelitch and Barber, 1960) and tobacco leaves (Pierpoint, 1959, 1960) with similar activities have also been obtained.

The oxidation of organic acids of the cycle by mitochondria isolated from pea leaves is tightly coupled (oxygen uptake and ATP formation are not rapid unless ADP is also present to facilitate ATP synthesis) (Geronimo and Beevers, 1964), as they are in mitochondria isolated from most other tissues. The close association of phosphorylation with oxidations of the tricarboxylic acid cycle implies that much of the energy obtained by such oxidations is retained as useful chemical energy rather than being wasted in the form of heat. The overall stoichiometry during glycolysis, the tricarboxylic acid cycle, and oxidative phosphorylation can be written as:

$$C_6H_{12}O_6 + 6\ O_2 + 32\ ADP + 32\ H_3PO_4 \longrightarrow 6\ CO_2 + 32\ ATP + 38\ H_2O \qquad (5.2)$$

The mechanism of oxidative phosphorylation is under intense biochemical investigation as is photosynthetic phosphorylation (Chapter 3.E), and the two processes have many features in common. Both types of phosphoryla-tion clearly involve membrane organization (Hall and Palmer, 1969), and recent work carried out with animal mitochondria indicates that changes in membrane conformation produce an energized state that results in ATP formation (Green and MacLennan, 1969). Soluble coupling factors (com-pounds of mitochondrial origin that stimulate ATP formation when added to suitably depleted mitochondrial membrane preparations) that are analogous to those required for photosynthetic phosphorylation have also been found for oxidative phosphorylation (Racker, 1965).

3. OXIDATIVE PENTOSE PHOSPHATE PATHWAY

In addition to the mechanism for the breakdown of glucose by glycolysis, many plant tissues also utilize the pentose phosphate pathway, a mechanism that has been thoroughly reviewed by Axelrod and Beevers (1956). Most plant tissues have an abundant supply of glucose-6-phosphate dehydro-genase activity, an enzyme that catalyzes the reaction:

$$\text{Glucose-6-P} + \text{NADP}^+ \longrightarrow \text{6-phosphogluconate} + \text{NADPH} + \text{H}^+ \qquad (5.3)$$

A second enzyme, 6-phosphogluconate dehydrogenase, carries out an oxidative decarboxylation at the C-1 atom of phosphogluconate to produce pentose phosphate:

$$\text{6-Phosphogluconate} + \text{NADP}^+ \longrightarrow \text{ribulose-5-P} + \text{CO}_2 + \text{NADPH} + \text{H}^+ \qquad (5.4)$$

This is the same as Reaction (4.14) written in the reverse direction. Addi-tional enzymes are present in many tissues, for example in pea (Gibbs, 1954) and tobacco leaves (Clayton, 1959), that convert the pentose into hexose and triose. The latter may then be transformed to pyruvic acid by the usual reactions of anaerobic glycolysis. The net result of the pentose phosphate pathway is thus to provide an alternative mechanism of con-verting hexose to triose. The CO_2 released is derived from the C-1 of hexose, while CO_2 evolved during anaerobic glycolysis comes first from the C-3 and C-4 atoms of hexose, and during operation of the tricarboxylic acid cycle CO_2 would arise next from C-2 and C-5 and lastly from C-1 and C-6 of hexose.

Hence if tissues are supplied with radioactive glucose-^{14}C labeled either in the C-1 or in the C-6 positions the relative rates at which the ^{14}C appears in the respired CO_2 from these two kinds of labeled molecules provide a rough estimate of the relative contributions of the glycolytic and pentose phosphate pathways. If the ratio of the $^{14}CO_2$ produced from C-6/C-1 equals 1.0, only glycolysis is presumed to occur, but if this ratio is less than 1.0, it suggests that some of the respired CO_2 is derived from the alternative pathway. In leaves of several species that were examined by this technique, the C-6/C-1 ratio was less than 1.0 (Gibbs and Beevers, 1955), suggesting that the pentose phosphate pathway was functioning, and younger tissues generally had higher values for the ratio indicating the pentose phosphate mechanism increases with age. This technique depends on the slow inter-conversion of C-1 and C-6 of hexose, but in wheat seedlings there was a rapid equilibration of these carbon atoms (Edelman *et al.*, 1955). If such interconversions take place, an underestimation of the part played by the pentose phosphate pathway would occur. Evidence that the oxidative pentose phosphate cycle occurs in chloroplasts comes from experiments in which phosphogluconate accumulated in isolated spinach chloroplasts in the presence of an inhibitor (Krause and Bassham, 1969).

4. OXIDASES

Cytochrome oxidase catalyzes the major reaction by which electrons are transferred to O_2 during dark respiration in plant tissues (Beevers, 1961). This carrier is located in the mitochondria and is thus closely connected with the reactions of oxidative phosphorylation. Cytochrome oxidase contains an iron porphyrin associated with its protein, and the iron is alternately reduced and oxidized during electron transport to O_2. Its activity is severely inhibited by cyanide (which reacts with the ferrous form of the enzyme) and by carbon monoxide (which forms a stable complex with the ferric cytochrome oxidase that is characteristically dissociated by visible light). The inability to inhibit dark respiration by cyanide or carbon monoxide has frequently been cited as evidence that cytochrome oxidase was not functioning in a particular tissue, but reasonable explanations can in most cases account for the experimental difficulties. Leaf tissues do, however, contain a number of other oxidases, although their function *in vivo* is not always clear.

The phenol oxidases are active in leaves, for example, and these enzymes contain copper as the prosthetic group and are inhibited by the same in-hibitors as cytochrome oxidase, except that inhibition by carbon monoxide is not reversed by light. Phenol oxidase activity is responsible for the famil-iar "browning" observed in plant cells after injury, a reaction caused by the oxidation of phenolic compounds. During oxidation of the phenolic substrate, it is converted to the corresponding quinone, the cupric form of the enzyme is reduced to the cuprous form, and the reduced copper is then reoxidized by O_2. Phenolase activity is found in chloroplasts and other cell fractions (Mayer and Friend, 1960), and this oxidase may function in the synthesis of lignin and in providing disease-resistance.

Ascorbic acid oxidase is another copper-containing enzyme that is widely distributed in plant tissues, frequently being associated with cell walls. Some functions of this enzyme have been reviewed by Mapson (1958). The substrate is reduced to dehydroascorbic acid by the enzyme with the uptake of O_2. Reconstructed systems have been prepared with enzymes isolated from plant tissues whereby the ascorbic acid oxidase reaction can be linked with the oxidation of NADPH produced by respiratory dehydrogenase reactions. However, such a role for this oxidase has not yet been demonstrated in intact tissues.

Glycolate oxidase is a flavoprotein which catalyzes the oxidation of glycolic acid, an early product of photosynthesis, and this enzyme appears to be universally present in higher plants. The reaction it catalyzes is most important in photorespiration, and this aspect is discussed in detail in Chapter 6.B.

5. EFFECT OF LIGHT ON DARK RESPIRATION

The rate of dark respiration is frequently affected by the previous intensity of illumination, and respiration usually increases following periods of increasing light intensity. Although the post-illumination outburst typically related to photorespiration has been thoroughly examined (Chapter 5.E), some unexplained post-illumination effluxes of CO_2 are observed in certain species.

Thus G. Heichel (1970) found that the rate of CO_2 released into a stream of CO_2-free air by attached maize leaves after a period of illumination increased slowly to a maximum 20 minutes after darkening and returned to a steady minimum rate only after 2 to 3 hours. The maximal rate of dark respiration was quantitatively related to the intensity during the light period, but appeared to be independent of net CO_2 fixation and required only a low concentration of O_2 in the ambient atmosphere to produce the substrates utilized for the dark respiration. Increasing the light intensity in leaves of *Rumex acetosa* also subsequently produced a higher rate of CO_2 efflux in darkness. The rate of the dark respiration differed little whether 2 ppm or 369 ppm of CO_2 were present in the ambient atmosphere during exposure to high irradiance (Holmgren and Jarvis, 1967), suggesting that net photosynthesis was not directly involved. Although one would not ordinarily expect CO_2 evolution in the dark to be affected by the CO_2 concentration in the atmosphere, in lucerne leaves dark respiration was twice as great in CO_2-free air as in normal air (Begg and Jarvis, 1968), indicating that higher concentrations of CO_2 may inhibit dark respiration in some tissues.

Several interesting effects of blue light on respiratory systems have been described. Low intensities of blue light stimulated dark respiration in a chlorophyll-free mutant of *Chlorella*, and the resulting twofold increase in respiration, observed only in starved cells, was not inhibited by DCMU (Kowallik and Gaffron, 1967). Blue light may also stimulate respiration in nitrogen-deficient leaves (Chapter 5.I). The blue light probably does not stimulate respiration directly, but serves to release a substrate by changing the permeability of cell membranes, thus making substrates more readily available. There is some evidence that the blue light acceptor is probably riboflavin mononucleotide (Schmid and Schwarze, 1969).

High intensities of blue light inhibited respiration in the colorless alga *Prototheca*, and the photoinhibition of respiration has been correlated with the destruction of cytochrome oxidase (Epel and Butler, 1969). The mechanism of this photodestruction is still unknown, but if this phenomenon is widespread in photosynthetic organisms it could result in a diminished dark respiration at high light intensities.

The view that the normal dark respiration is inhibited in photosynthetic tissues in the light comes from several indirect sources. B. Kok in 1947 observed that the curve of CO_2 uptake or O_2 evolution versus light intensity showed an abrupt change in slope at low light intensities, and this phenomenon became known as the Kok effect. Because the upper segment of the curve extrapolated to an intercept on the ordinate with a value smaller than dark respiration, the Kok effect was widely interpreted as indicating that dark respiration is slower during photosynthesis than in darkness. The Kok effect can also be explained as resulting from an *increase* in respiration with increasing light intensity (photorespiration), and this alternative interpretation is just as plausible.

The possibility that dark respiration is inhibited in the light has also been suggested because the CO_2 compensation point (Chapter 5.I) in soybean leaves was linearly related to the concentration of O_2 in the ambient air, and apparently extrapolated to zero when the O_2 concentration was zero (Forrester *et al.*, 1966a). However, Forrester *et al.* underestimated the rate of respiration in the light because of the preferential reutilization of CO_2 within the tissues, and such an extrapolation is based on the assumption that the mechanism of respiration in the light is identical to dark respiration, a point of view that seems untenable in view of considerable other evidence. Moreover, not all workers observe a "zero" CO_2 exchange when the CO_2 compensation point is extrapolated to zero O_2 in the ambient atmosphere (Bunt, 1969).

Products of the tricarboxylic acid cycle such as glutamic and isocitric acids were not labeled during short periods of photosynthesis in experiments on $^{14}CO_2$ fixation by *Chlorella* and barley leaves (Benson and Calvin, 1950), and inhibition of the cycle by light was therefore suggested. Such an interpretation may be incorrect as was shown in experiments with mung bean leaves supplied with $^{14}CO_2$ or with various ^{14}C-labeled intermediates of the tricarboxylic acid cycle in the light and dark (Graham and Walker, 1962). All metabolites of the Krebs cycle were readily labeled, although the rates of labeling were somewhat different in light and darkness. The change in the flow of carbon into specific compounds probably occurred because pyridine nucleotides are more completely reduced in the light. Accordingly, more ^{14}C is found in aspartate than malate in the dark while the reverse is true in the light, and although the order of labeling may be slightly different, the cycle functioned in the light as well as in the dark.

Glyoxylate, a product of the glycolate oxidase reaction in the light [Reaction (6.1)], is a very effective inhibitor of the tricarboxylic acid cycle because it interferes with several of the steps (Adinolfi *et al.*, 1969), and therefore it has been suggested that glyoxylate interferes with the Krebs cycle during photosynthesis. However, glyoxylate is probably not synthesized in the mitochondria (Chapter 6.A) and it is a reactive substance that may not be available to mitochondria in inhibitory concentrations.

Isotopic equilibrium among succinic, malic, and citric acids was found when $^{14}CO_2$ was supplied to tobacco, barley, and rhubarb leaves in the light (Stutz and Burris, 1951), suggesting that dark respiration occurs during illumination. When aspartic acid U-^{14}C was fed to detached wheat leaves, the percentage distribution of ^{14}C metabolized in malic, citric, succinic, and glutamic acids was similar after 1 hour in light or in darkness (Naylor *et al.*, 1958), again supporting the view that the tricarboxylic acid cycle functions during photosynthesis.

Perhaps the most convincing evidence for the existence of the tricarboxylic acid cycle in photosynthetic tissues in the light comes from the experiments of Marsh *et al.* (1964, 1965). They supplied ^{14}C-labeled acetate and pyruvate to *Scenedesmus* cells and isolated succinate, α-ketoglutarate, malate, and citrate after different periods in the light or in darkness. The percentage distribution of ^{14}C in the various metabolites and the specific activity of the individual carbon atoms were consistent with the functioning of the cycle in the light, and moreover light did not affect the rate of equilibration of the isotope in the intermediates of the Krebs cycle.

B. FACTORS AFFECTING PHOTORESPIRATION

Use of an electrical analog is helpful in examining the relation between photosynthesis and respiration in the light (Fig. 5.1) because CO_2 assimilation during photosynthesis may be considered as a diffusion process and is thus subject to certain specific limitations (Chapter 7). Many workers including Gaastra (1959), D. N. Moss (1966), Lake (1967), El-Sharkawy *et al.* (1967), Bravdo (1968), and Samish and Koller (1968) have found similar representations useful in explaining differences in net photosynthesis and transpiration caused by various factors. A more inclusive extension of such a model has been used to simulate closely the net photosynthesis observed under different conditions by a single leaf and a community of leaves in a stand (Waggoner, 1969a, b). Some of the algebraic consequences of such an electrical representation with respect to photorespiration are discussed later (Chapter 5.H), and the possible magnitude of the different resistances concerned with photosynthesis is treated in Chapter 7.

Figure 5.1 indicates that CO_2 is taken up by a leaf because of a current

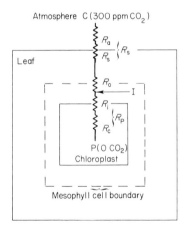

FIG. 5.1. Diagrammatic representation of the relation between net photosynthesis and respiration in a leaf (from Waggoner, 1969a). Photosynthesis is shown as a flux of CO_2 caused by the higher concentration of CO_2 (300 ppm) in the atmosphere and the zero concentration at some point within the chloroplast. The total CO_2 evolved in the light, I (dark respiration plus photorespiration), joins the main photosynthetic current outside the chloroplast and thus diminishes net CO_2 uptake. The current of CO_2 assimilation is opposed by a series of resistors including R_a in the air, R_s in the stomata, R_o at the cell boundary, R_i inside the cell, and R_c caused by limitations of photochemistry and biochemistry on the carboxylation within the chloroplast. In the text R_a and R_s are sometimes combined as R_S; R_i and R_c are taken together as R_P.

resulting from the CO_2 concentration gradient that exists between the atmosphere (which contains 300 ppm of CO_2) and some point within the chloroplast where the CO_2 concentration is close to zero. By analogy with an electrical current, the CO_2 enters the leaf where its rate is measured as net photosynthesis, and this rate is controlled by a series of resistors and by the magnitude of an opposing current of respiratory CO_2 (I) within the leaf derived from photorespiration and dark respiration that joins the main stream at a point separated by a diffusive resistance from the site of fixation of CO_2 in the chloroplast. This implies that respiratory CO_2 in the light originates at a site outside the chloroplast or it could conceivably arise within the chloroplast but at a different location from the site of CO_2 fixation (in the stroma). R_a, R_s, R_o, R_i, and R_c are respectively the resistances to diffusion of CO_2 in the air, through the stomata, at the cell wall boundary, and inside the aqueous medium surrounding the chloroplast, and that caused by the limitations of photochemistry and biochemistry on CO_2 fixation in the chloroplast. For simplicity, R_a and R_s will sometimes be combined in this chapter as R_S and similarly R_i and R_c will be combined as R_P.

Figure 5.1 shows that a part of the photorespiratory CO_2 will always be refixed under photosynthetic conditions, and measurement of photorespiration by any method that requires an analysis of the atmosphere outside a leaf must underestimate it. Methods have been devised to try to minimize this internal recycling of CO_2 (Chapter 5.H).

Photorespiration frequently occurs at rates considerably greater than dark respiration, and the process is completely dependent on photosynthesis to provide its substrates. This alone clearly distinguishes it from other types of light responses upon respiratory processes. Other compelling data have been accumulated that show that photorespiration is different from dark respiration because (*a*) photorespiration is greatly stimulated by increasing concentrations of O_2 in the atmosphere, whereas dark respiration is fully operative at 2 to 3% O_2; (*b*) photorespiration increases more rapidly with increasing temperature than does dark respiration; (*c*) photorespiration is not much affected in the range between zero and 300 ppm of CO_2 in the atmosphere and is very much inhibited when the CO_2 concentration exceeds 1000 ppm CO_2, while dark respiration is generally not greatly affected by the CO_2 concentration; and (*d*) biochemical inhibitors can be shown to affect photorespiration more than net CO_2 assimilation and without inhibiting dark respiration. Differences in photorespiration between species and within varieties of a single species have now been found, and the effect of photorespiration on photosynthetic efficiency has been examined.

C. OXYGEN UPTAKE IN THE LIGHT MEASURED WITH $^{18}O_2$

Measurements of O_2 uptake during photosynthesis have been carried out by observing, by means of a mass spectrometer, the disappearance of the $^{18}O_2$ added to the atmosphere. Experiments by this technique have frequently been cited as evidence that respiration does not increase in the light. For example, the rate of $^{18}O_2$ uptake by tobacco leaf tissue in the light appeared to be the same as in darkness (A. H. Brown, 1953, Fig. 6). Although Brown recognized that the O_2 produced within the tissues by photosynthesis may be expected to have a significant advantage over the tracer oxygen diffusing in from the environment, he concluded that there was no evidence for a photostimulation of respiration. In order to conserve expensive isotopes such experiments with $^{18}O_2$ have frequently been conducted under unusual environmental conditions such as low concentrations of O_2; high concentrations of CO_2 have also been employed. In Brown's experiment cited above, the $^{18}O_2$ uptake was measured at a concentration of 1.9% O_2, a level at which there is little photorespiration.

Decker (1958) correctly pointed out that this widely accepted interpretation that respiration is the same in light as in darkness did not fully consider the magnitude of the dilution of $^{18}O_2$ inside the leaf by the large quantity of $^{16}O_2$ released from normal H_2O, and that the data could be masked by a respiration that accelerated as an approximately linear function of photosynthesis. "If, and only if one assumes that respiration does not change, the data show that photosynthesis and respiration can be measured independently. Or, if (and only if) one assumes that respiration and photosynthesis can be measured independently, the data show that the respiration rate does not change. But neither proposition is established independently of the other." Therefore, Decker concluded that in order to obtain the apparently constant rates of O_2 uptake in the light and in darkness as observed by Brown, the rate of respiration in the light must have actually increased several-fold. In later experiments, Good and Brown (1961) attempted to evaluate the contribution of endogenous oxygen in such measurements, but this was done by carrying out experiments at low light intensities with algae grown in 5% CO_2, conditions which would also tend to obscure photorespiration.

Hoch *et al.* (1963) have studied oxygen uptake in the light in *Anacystis* (blue-green algae) and *Scenedesmus* (green algae). They used an efficient system of sampling the gases dissolved in the liquid phase by allowing them to diffuse into the mass spectrometer through a membrane. The cells were grown in 3% CO_2, and nitrogen was bubbled through the cell suspensions for several minutes to remove dissolved O_2 (and CO_2) before supplying $^{18}O_2$ in the light. With increasing light intensity, $^{18}O_2$ uptake first decreased compared with the dark rate and then increased three- to fivefold at the higher light intensities. The $^{18}O_2$ uptake was inhibited by low concentrations of DCMU, indicating that the photorespiration was catalyzed by an enzyme system and photooxidations were not involved. They concluded that part of the O_2 consumption in the light occurs by the same mechanism that occurs in darkness, and a new type of O_2 uptake is superimposed on dark respiration at higher light intensities.

The difficulties in clearly interpreting experiments with labeled O_2 have been discussed by Jackson and Volk (1970). They have emphasized that corrections made for internal recycling must be minimal, and that isotopic O_2 measurements during illumination therefore underestimate O_2 consumption. However, unlike measurements with isotopic carbon, the opportunity for evolution of the isotopic O_2 species to the atmosphere is not so probable (because of dilution by the large pool of cell water) as that of $^{14}CO_2$. Because the recycling of $^{18}O_2$ inside the leaf will not be very great, the assay of photorespiration using labeled O_2 would seem to offer an important advantage. On the other hand, there may be oxygen consuming reactions under some conditions that have no stoichiometric relation to CO_2 evolution in the light (such as the oxidation of reduced ferredoxin, Chapter 6.A.6) so that "photorespiration" will be exaggerated by this method.

Ozbun *et al.* (1964) measured $^{18}O_2$ uptake and CO_2 evolution (in the presence of $^{13}CO_2$) by bean leaves in the light, and found that O_2 uptake was increased whereas CO_2 evolution appeared to be inhibited by light. The O_2 evolution was greater than the CO_2 uptake, suggesting that CO_2 was recycling within the tissue to a greater extent, and less was therefore released to the atmosphere where it could be detected. Moreover these experiments were carried out in an ambient atmosphere of 1.6% O_2 and 1.8% CO_2. This combination of low levels of O_2 and a high concentration of CO_2 would undoubtedly greatly inhibit photorespiration, and the high CO_2 concentration would certainly cause stomatal closure (Chapter 7.C) and thus cause a still greater internal recycling of CO_2 and permit less to be released from the leaf (Fig. 5.1).

Measurements of $^{18}O_2$ uptake have also been carried out at the CO_2 compensation point with illuminated maize leaves (Jackson and Volk, 1969), a species that shows low rates of photorespiration by all of the methods that measure CO_2 evolution that are described in this chapter. They observed an increased O_2 uptake in the light that was stimulated by increasing the O_2 concentration in the atmosphere from 2 to 6%, but the measured rates of respiration (though they are minimal rates) are about those that are usually observed for dark respiration in maize leaves. Hence the data do not support the view that there is an active photorespiratory process in maize although there may be a low rate of photorespiration as is also indicated by measurements of $^{14}CO_2$ released by this tissue (Chapter 5.H). Experiments with $^{18}O_2$ have thus far not provided as clear evidence as other methods about the nature of the respiratory process in the light, perhaps because of a number of limitations imposed by the technique and occasional poor (or unfortunate) choice of conditions.

D. EFFECT OF PHOTORESPIRATION ON INHIBITION BY OXYGEN OF NET CO₂ ASSIMILATION

Large increases in net photosynthesis are obtained in many species when the O_2 content of the atmosphere (normally 21%) is lowered to about 1 to 3% O_2. Several explanations for this phenomenon (Warburg effect) have been suggested (Chapter 8.E), but it seems likely that most of this effect is caused by the increase in CO_2 evolution in the light resulting from photorespiration. The percentage increase in net CO_2 assimilation when the O_2 concentration is decreased provides an indication of the maximal importance of photorespiratory CO_2 in the CO_2 budget of a leaf in the light, as indicated below.

Net photosynthesis in species with a low rate of photorespiration, such as maize, is not inhibited until the O_2 concentrations in the ambient atmosphere exceeds 50% at 300 ppm of CO_2 (Forrester *et al.*, 1966b). A comparison of net photosynthesis in several species has been made at high light intensities in 21% O_2 and low concentrations of O_2 at 300 ppm of CO_2 (Hofstra and Hesketh, 1969a). Leaves of the efficient photosynthetic species, maize and *Atriplex nummularia,* showed no stimulation of net photosynthesis at 25° in low O_2 concentrations, but *Atriplex hastata,* soybean, and sugar beet were stimulated 33, 53, and 40% respectively. Other workers have also frequently observed increases in net photosynthesis of from 33 to 50% in the less efficient photosynthetic species upon decreasing the O_2 concentration from 21% to very low levels. Thus the Warburg effect is observed in species with high rates of photorespiration, and cannot be detected in leaves having low rates of photorespiration even at O_2 concentrations up to 50%.

On the assumption that the inhibition of net photosynthesis at 21% O_2 is caused by increases in the release of photorespiratory CO_2, Bulley and co-workers (1969) have obtained an action spectrum for photorespiration. Radish and maize leaves were compared in these experiments, and net photosynthesis was measured at 300 ppm of CO_2 and 21% and 2% O_2 at light intensities where net photosynthesis was proportional to the light intensity. They found that O_2 inhibited net CO_2 uptake in radish leaves about 50% at all wavelengths from 402 to 694 nm, whereas in maize there was no inhibition over the visible spectrum. Thus photorespiration and photosynthesis are probably linked directly at all wavelengths in the visible portion of the spectrum, and increases in gross photosynthesis are accompanied by proportional increases in photorespiration.

E. POST-ILLUMINATION CO_2 OUTBURST

Commercial infrared CO_2 analyzers first became available in the early 1950's and their development greatly stimulated research on photosynthesis and respiration. Decker (1955) soon observed that when CO_2 uptake by tobacco leaves was measured in a chamber and after a period of illumination the chamber was darkened, there was a high rate of CO_2 evolution that decelerated for several minutes (6 to 9 minutes in other species) following

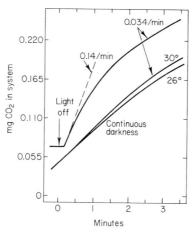

FIG. 5.2. The post-illumination CO_2 burst shown by a tobacco leaf (from Decker, 1955). The leaf was illuminated at 3800 ft-c in a chamber and the leaf temperature was maintained at 24 to 26° while the air surrounding the leaf in the closed system was rapidly recycled through an infrared CO_2 analyzer. When the CO_2 concentration reached the CO_2 compensation point, the light was turned off at zero time (upper curve). The initial rate of the CO_2 outburst is compared with the rate of CO_2 released in continuous darkness.

TABLE 5.1   *Effect of Light Intensity, Temperature, and CO_2 Concentration on CO_2 Outburst and Net Photosynthesis in Tobacco[a,b]*

Expt. No.	Light intensity (ft-c)	Net photosynthesis mg CO_2 dm^{-2} hr^{-1}	CO_2 outburst mg CO_2 dm^{-2} hr^{-1}	Relation of outburst to net photosynthesis %
1	500	5.8	1.4	24
	1300	11.5	3.6	31
	2500	16.9	6.5	38
	Temp. (°C)			
2	17.5	15.1	2.9	19
	25.5	16.9	7.6	45
	33.5	14.8	9.7	66
	CO_2 concentration (ppm)			
3	45	0	4.2	
	300	16.6	5.4	

[a] From Decker, 1959a.

[b] The experiments were carried out with attached tobacco leaves that were kept for 20 minutes at 2500 ft-c at 25° to 26° before measurements began. In Experiment 1, the measurement of net photosynthesis and post-illumination outburst were made on leaves at 25.5° and 300 ppm CO_2; each figure is the mean of 70 observations made on 14 different plants. In Experiment 2, the data were obtained at 2500 ft-c and 300 ppm CO_2; each figure is the mean of 35 observations. In Experiment 3, the results were obtained at 2500 ft-c and 25.5°; each figure is the mean of 25 observations.

darkening, after which the dark respiration resumed its normal and constant rate (Fig. 5.2). A similar so-called outburst in an open system had probably been observed earlier by Van der Veen (1949) but its importance was largely ignored. Decker found similar post-illumination CO_2 outbursts with leaves of bean (*Phaseolus vulgaris*), castor bean (*Ricinus communis*), tulip poplar (*Liriodendron tulipifera*), and white ash (*Fraxinus americana*). After determining that this was not caused by temperature changes resulting from darkening, he demonstrated the high rate of CO_2 evolution in twenty consecutive cycles of light and darkness and thus proved it was a reversible and reproducible phenomenon. He estimated the initial rate of CO_2 release by extrapolation to zero time of darkness, and in a typical experiment, such as that shown in Fig. 5.2, the initial rate of CO_2 released (0.14/minute) was 4.1 times greater than the steady dark respiration (0.034/minute).

Decker (1959a) suggested that the CO_2 outburst observed after a period of illumination resulted from a slight delay between the rate of deceleration of photosynthesis and the rate of deceleration of photorespiration. His well-documented data on the effect of light intensity, temperature, and CO_2 concentration on the initial rate of the CO_2 outburst on darkening are summarized in Table 5.1. The CO_2 outburst increased with light intensity during the period of illumination in a manner similar to net photosynthesis, and its magnitude was about 40% as great as the net photosynthesis at light saturation. The CO_2 outburst increased greatly with leaf temperature between 17.5 and 33.5°, increasing about threefold for every 10° rise. At the highest leaf temperature examined, 33.5°, the outburst accounted for 66% of the net photosynthesis (Chapter 6.C). Finally, the results showed that the magnitude of the CO_2 outburst was essentially unchanged whether the leaf was previously maintained at the CO_2 compensation point (45 ppm CO_2) or in air (300 ppm CO_2).

The CO_2 outburst was readily explained if photorespiration was being caught on the fly as it were. Photosynthesis diminishes more quickly than does respiration when the light is turned off, and this causes a momentarily rapid but rapidly decelerating evolution of CO_2 (Decker and Wien, 1958; Decker and Tió, 1959). This brief overshoot would be expected if the substrate for photorespiration was synthesized only in the light and occurred at low concentrations (Chapter 6.A).

Subsequently, Decker's initial observations have been confirmed and extended by a number of workers with a variety of plant tissues. The large initial CO_2 outburst in tobacco leaves was dependent on the previous light intensity and was not observed at 2% O_2 as it was in normal air after a period in the light at 6500 lux at the same O_2 concentration. The final steady rate of dark respiration was achieved within 5 to 6 minutes after the light was turned off, and the dark respiration was unaffected by O_2 concentrations around the leaf between 2 and 47% (Krotkov, 1963; Tregunna *et al.*, 1966). Also, no outburst was observed in soybean leaves kept in 1% O_2, but it occurred in 21% O_2 and was three times as large in an atmosphere of 100% O_2 (Forrester *et al.*, 1966a). Thus photorespiration as detected by this assay increases greatly with oxygen concentration.

The effect of CO_2 concentration during the photosynthetic period on the outburst was examined in attached rosettes of *Rumex acetosa*, and three CO_2 outbursts were observed with this tissue before the CO_2 efflux in the dark stabilized (Holmgren and Jarvis, 1967). They found the first outburst associated with photorespiration was greater when the leaves were in the light at an ambient concentration of 40 ppm CO_2 than at concentrations below 5 ppm, hence if photorespiration changes with ambient levels it would appear that photorespiration is depressed at concentrations of

CO_2 close to zero compared with CO_2 concentrations at the compensation point. The appearance of several consecutive peaks of CO_2 suggests that the CO_2 was derived from different substrates or from different cellular sites.

At very high atmospheric concentrations of CO_2 (1200 ppm) the CO_2 outburst was completely eliminated in tissues that showed the outburst in air, and such an inhibition of photorespiration by high levels of CO_2 was found in bean and sunflower leaves and with thalli of the liverwort *Concephalum conicum* (Egle and Fock, 1967). Thus the effect of CO_2 concentration on photorespiration in the normal range of CO_2 is small, and high concentrations of CO_2 inhibit photorespiration as reflected by the CO_2 outburst after illumination probably because synthesis of the substrate for photorespiration is inhibited (Chapter 6.A).

A greater CO_2 outburst with increasing temperature has also been observed in sunflower leaves previously maintained at 1800 ft-c and 21% O_2 (Hew *et al.*, 1969b). However, the rate of increase with temperature was not as rapid as those described by Decker (1959b), especially at the warmer temperatures, and this may have resulted in part because the stomata in the leaves in these experiments were not very widely open so that less of the photorespiratory CO_2 escaped to the atmosphere (Chapter 5.H).

Additional investigations have also been carried out on the effect of intensity and quality of light on the CO_2 outburst. In a study carried out with tobacco leaves (Tregunna *et al.*, 1961), the outburst in darkness increased with light intensity during the preliminary light period in a manner similar to net photosynthesis. But the response was not quantitatively as clear as that shown in Table 5.1, perhaps because the magnitude of the outburst was measured rather than its initial rate. In sunflower leaves, the CO_2 outburst occurred after illumination at wavelengths from 590 to 700 nm, and its magnitude was related to the light intensity in a manner similar to net photosynthesis (Semenenko, 1964). The outburst also increased with increasing light intensities between 350 and 1200 ft-c in soybean leaves, and disappeared after 100 seconds (Tregunna *et al.*, 1964). These workers first observed that the outburst did not occur in all species since maize leaves did not show the expected CO_2 efflux after illumination. Thus photorespiration must be slower in such species. The lack of a CO_2 outburst in maize leaves was confirmed by Forrester *et al.* (1966b), who showed there was no outburst in maize in the leaves of twelve different varieties that were examined two weeks after planting. They found no outburst even when 100% O_2 was present in the atmosphere. These results provide clear evidence that species vary in the magnitude of their photorespiration.

Some other and less understood post-illumination outbursts have also been observed (Chapter 5.A.5). In an open system, tobacco leaves showed several post-illumination CO_2 bursts (Heichel, 1971a). The first was over within 1 minute, and then a second burst was observed about equal in magnitude and with its maximum after about 5 minutes. Dark respiration then returned to its steady level within about 10 minutes. In a similar open system with maize, a post-illumination burst was also observed in CO_2-free air, but it had a much slower onset (maximum in 10 minutes) than the normal burst of photorespiration, and the dark respiration resumed its steady level in another 10 minute period.

F. EXTRAPOLATION OF NET PHOTOSYNTHESIS TO "ZERO" CO_2 CONCENTRATION

During an investigation of the effect of CO_2 concentration on net photosynthesis, Decker (1957) assumed that an extrapolation of the curve of net photosynthesis to zero CO_2 should indicate the dark respiration if there were no dark CO_2 fixation and if CO_2 production in the light was unchanged in the dark. Figure 5.3 shows that the extrapolation yielded a value for respiration 3.5-fold greater than the measured dark respiration $(3.4/0.97 = 3.5)$ for tobacco leaves, and this represents a minimal estimate of the magnitude of photorespiration. The results in Table 5.2 illustrate that photorespiration by this method was from 2.1 to 4.9-fold greater than the dark respiration in a number of species. Photorespiration in *Mimulus* also increased more rapidly than dark respiration as the temperature was raised (Decker, 1959b).

Bravdo (1968) used a model similar to Fig. 5.1 and showed algebraically why extrapolating CO_2 uptake to zero CO_2 underestimates photorespiration. The extrapolation is valid if the resistances in Fig. 5.1 do not change, the CO_2 compensation point is constant, and the line of net photosynthesis versus CO_2 concentration is straight. The intercept, however, is equal to the CO_2 compensation point (IR_P) divided by $(R_S + R_P)$ and not merely photorespiration plus dark respiration (I). In other words, the intercept is a function of stomatal resistance, resistance to CO_2 fixation in the chloroplast, as well as photorespiration. The same conclusion was reached independently in an algebraic analysis described by Samish and Koller (1968).

Difficulties in applying this method would occur when, as some investigators have observed, departures from linearity are found in experiments similar to Fig. 5.3. This may be caused by differences in the stomatal diffusive resistance (R_S) in some species as the CO_2 concentration is changed (Chapter 7.D). Thus departures from linearity at CO_2 levels close to the CO_2

TABLE 5.2   *Comparison of Dark Respiration and Minimum Photorespiration Estimated by the Extrapolation Method*[a,b]

Species	Dark respiration mg CO_2 dm^{-2} hr^{-1}	Minimum photorespiration mg CO_2 dm^{-2} hr^{-1}	Photorespiration/ dark respiration
Tobacco (*Nicotiana langsdorfii* × *N. sanderae*)	0.97	3.4	3.5
White ash (*Fraxinus americana*)	0.61	2.16	3.5
Tuliptree (*Liriodendron tulipifera*)	0.72	2.92	4.1
Geranium (*Pelargonium* sp.)	1.48	4.57	3.1
Grapefruit (*Citrus paradisi*)	0.90	3.74	4.2
Foxglove (*Digitalis* sp.)	1.91	4.00 —	2.1
Poinsettia (*Euphorbia pulcherrima*)	1.37	3.64	2.7
A fern (*Selliguea caudiforme*)	0.36	1.76	4.9
Ginkgo (*Ginkgo biloba*)	0.72	3.56	4.9

[a] From Decker, 1957.
[b] Attached leaves were sealed in a chamber for 30 minutes in 2500 ft-c of light at 25° to 26°. Net photosynthesis was measured at concentrations from 400 ppm of CO_2 to the CO_2 compensation point as in Fig. 5.3 for tobacco, and minimum photorespiration was determined by extrapolation to zero CO_2. Three runs at the different CO_2 concentrations were made, and then the dark respiration was measured. This was repeated with two or more plants.

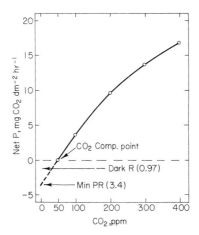

FIG. 5.3. Effect of CO_2 concentration on net photosynthesis of tobacco leaves at 2500 ft-c and 25° (from Decker, 1957). Each upper point was the mean of 45 observations made on attached leaves sealed in a chamber, and the CO_2 compensation point was the mean of 15 determinations. Dark respiration was measured after the other measurements were completed and the leaves had been in darkness for at least 10 minutes.

compensation point were observed in *Rumex acetosa* (Holmgren and Jarvis, 1967) and with Douglas fir seedlings, *Pseudotsuga menziesii*, at lower light intensities (Brix, 1968). With Scots pine seedlings, *Pinus silvestris*, at 10,000 lux the extrapolation method showed photorespiration to be 2.8 times greater than dark respiration (Zelawski, 1967). This assay method showed photorespiration in sunflower to be only 1.2 times dark respiration, perhaps because a lower temperature was used, but there was again no detectable photorespiration with maize (Hew *et al.*, 1969a).

G. DILUTION OF $^{14}CO_2$ SPECIFIC ACTIVITY IN THE AMBIENT ATMOSPHERE

If photosynthesis is permitted to occur in a closed system in the presence of an external supply of $^{14}CO_2$, the CO_2 arising from photorespiration will contain CO_2 originating from endogenous sources of carbon, and would thus have a lower specific activity (Fig. 5.1). A portion of this $^{12}CO_2$ arising from photorespiration will diffuse out of the leaf and thus lower the specific activity of the $^{14}CO_2$ in the atmosphere. If photorespiration is low in activity, the specific activity would not change. The demonstration of photorespira-

TABLE 5.3   *Uptake and Release of CO_2 by Wheat Leaves[a,b]*

Expt. No.	Experimental conditions	mg CO_2 hr^{-1} gm fresh wt.$^{-1}$	
		Uptake $^{14}CO_2$	Release $^{12}CO_2$
1	Dark	0.06 ± 0.37	0.06 ± 0.41
	Light	3.73 ± 0.29	0.95 ± 0.33
2	Dark	0.57 ± 0.23	0.64 ± 0.29
	Light	4.40 ± 0.18	1.67 ± 0.20

[a] From Krotkov *et al.*, 1958.

[b] Excised leaves from plants 14 to 16 days old were placed in light (800 ft-c) for 5 hours in a chamber at 20°. The chamber was closed, and $^{14}CO_2$ of known specific activity and quantity (initial concentration as high as 5% CO_2) was released. After 1 hour, the CO_2 remaining was collected and the quantity was again determined to permit a calculation of the $^{14}CO_2$ uptake. The extent of the $^{12}CO_2$ released from the leaves was then determined from the decrease in specific activity of the $^{14}CO_2$ in the atmosphere surrounding the leaves.

tion by such a method was first described in wheat leaves by Krotkov *et al.* (1958) and some of their data are given in Table 5.3. Even at 20°, more CO_2 was evolved in the light than in darkness, and although the results in Experiments 1 and 2 do not agree quantitatively, photorespiration was at least 2.6 times greater than dark respiration.

Photorespiration has been demonstrated by this method (Fig. 5.4) in tobacco (Krotkov, 1963) and in sunflower leaves. In the closed system in the light, the CO_2 concentration rapidly diminished to the CO_2 compensation point (about 50 ppm), which was reached in 5 minutes. The specific activity of the $^{14}CO_2$ supplied in the atmosphere also rapidly decreased indicating that $^{12}CO_2$ was being released by the leaf. It is important to note in Fig. 5.4 that the specific activity continued to decrease even after the CO_2 compensation point was reached. This demonstrates that a turnover or refixation of CO_2 occurs within the tissues even in the absence of net CO_2 uptake. In darkness there was a net efflux of CO_2, and the specific activity changed little after the end of the light period. In contrast, maize leaves decreased the CO_2 concentration to a compensation point close to zero ppm, while the specific activity of the $^{14}CO_2$ in the ambient atmosphere remained constant for at least 15 minutes, indicating that no photorespiration of $^{12}CO_2$ could be detected (Fig. 5.5). In the dark, CO_2 was evolved and the specific activity decreased slowly in time. Thus this method clearly indicates that efficient photosynthetic species like maize possess little photorespiration while it is readily detectable in less efficient species.

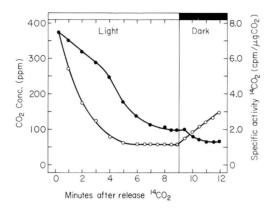

FIG. 5.4. Rate of change in the specific activity of $^{14}CO_2$ (●——●) and of the CO_2 concentration (○——○) in a closed system containing a sunflower leaf in air at an illuminance of 1800 ft-c and in darkness at 21° (from Hew *et al.*, 1969a). The $^{14}CO_2$ was introduced at zero time after a period in the light.

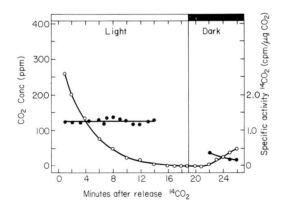

FIG. 5.5. Rate of change in the specific activity of $^{14}CO_2$ (●——●) and of the CO_2 concentration (○——○) in a closed system containing a maize leaf under the same conditions as Fig. 5.4 (from Hew *et al.*, 1969a).

A variation of this method has also been used to estimate photorespiration in chloroplasts obtained from the alga *Acetabularia mediterrania* (Bidwell *et al.*, 1969). They assumed that measurement of $^{14}CO_2$ uptake in short periods of time represents the gross photosynthesis, and $^{12}CO_2$ uptake measures the net CO_2 uptake. Thus the $^{14}CO_2$ uptake minus the $^{12}CO_2$ uptake should be equal to the CO_2 evolved by photorespiration. However, recently fixed $^{14}CO_2$ is respired so rapidly that the $^{12}CO_2$ released within the tissue can be detected as a diminished specific activity in the atmosphere surrounding sunflower leaves within 15 to 45 seconds after the introduction of $^{14}CO_2$ to the system (Ludwig and Krotkov, 1967). This indicates that detectable recycling must occur in less than 15 to 45 seconds within the tissue. Thus this method, like all others, will underestimate the extent of photorespiration, especially if times beyond 15 seconds are employed for the $^{14}CO_2$ fixation.

DISCRIMINATION AGAINST $^{13}CO_2$ DURING PHOTOSYNTHESIS

When plants are grown in CO_2 with a controlled $^{13}C/^{12}C$ ratio, the tissues have a ^{13}C concentration less than that present in the CO_2 that is assimilated (R. Park and Epstein, 1961). The ^{13}C concentration is usually expressed as the $\delta^{13}C$ per mil which indicates the difference per thousand of the $^{13}C/^{12}C$ ratio of a sample relative to a standard source of carbon, and is defined as:

$$\delta^{13}C \text{ per mil} = \left[\frac{^{13}C/^{12}C \text{ sample}}{^{13}C/^{12}C \text{ standard}} - 1 \right] \times 1000 \qquad (5.5)$$

Thus a sample with a $\delta^{13}C$ per mil of -10 has a $^{13}C/^{12}C$ ratio less than the standard by 10 per mil or 1%, and the larger the negative number the more the ^{13}C has been discriminated against. These ratios are obtained in a mass spectrometer with a precision of ±0.1 per mil.

Not all species discriminate equally against ^{13}C during photosynthesis (Bender, 1968). Grasses known to have high rates of photorespiration such as orchard grass, oats, wheat, and barley had a $\delta^{13}C$ per mil of between -26.6 and -27.8. Species with a low rate of photorespiration including maize, sugarcane, sorghum, and crabgrass (*Digitaria sanguinalis*) had $\delta^{13}C$ per mil values of between -12.3 and -13.7 although grown in a similar environment. Differences in $^{13}CO_2$ discrimination have also been obtained in dicotyledons with high photorespiration, *Atriplex hastata* (-25.5), and a species with low photorespiration, *Atriplex rosea* (-12.8), as well as in other related species (Tregunna *et al.*, 1970). Thus tissues with more efficient photosynthesis and low rates of photorespiration discriminate less against the heavier $^{13}CO_2$ than do the less efficient species.

The explanation for the above correlation is a matter of uncertainty. A significant discrimination against ^{13}C may occur in any chemical reaction in which the reactants and products are in a steady state. Therefore in a given tissue each active or inactive pool is likely to have a different $\delta^{13}C$ value, and indeed the $^{13}C/^{12}C$ ratio of the lipid in potato tuber tissue is considerably lower than that for starch or protein (Jacobson *et al.*, 1970). The values of $\delta^{13}C$ do not necessarily become more negative as one goes farther away from the CO_2 fixation reaction, since a strong discrimination against ^{13}C in a later reaction could lead to accumulation of ^{13}C in an earlier pool. Before useful conclusions about the above correlation can be drawn it will be necessary to have direct comparisons *in vitro* of the kinetic isotope effects of the primary CO_2 fixation reactions and of the CO_2 releasing reactions, especially those related to photorespiration.

H. PHOTORESPIRATION IN CO_2-FREE AIR AND WITH ^{14}C-LABELED TISSUES

Since photorespiration is not greatly affected when levels of CO_2 are close to "zero" or at the normal concentration in air (Chapter 5.E), it is apparent from Fig. 5.1 that if a leaf were placed in the light in a CO_2-free atmosphere, then the flux of CO_2 arising from photorespiration (I) would reach the main CO_2 current, and a portion of it would be refixed by the chloroplast and part would diffuse out to the atmosphere where it could be measured. The smaller the stomatal diffusive resistance (open stomata) and the more nearly the concentration of CO_2 in the atmosphere approached zero, the greater would be the fraction of CO_2 arising from photorespiration that would diffuse outside of the leaf. Bravdo (1968) showed that at zero CO_2 concentration outside the leaf the efflux of CO_2 is a function of the photorespiration, the diffusive resistance to CO_2 fixation in the chloroplast, and the stomatal resistance (Chapter 5.F). This technique therefore also measures only a fraction of the photorespiration. In addition there is one distinct disadvantage connected with the making of such measurements. The rate of CO_2 efflux is determined in an open system, and at high flow rates and concentrations of CO_2 close to zero, the sensitivity of most infrared CO_2 gas analyzers is not much better than 10 to 15 ppm. Hence the accuracy of the determinations of CO_2 efflux is limited.

The measurement of photorespiration in CO_2-free air was first described by El-Sharkawy and Hesketh (1965) who found photorespiration in cotton leaves was greater than dark respiration while CO_2 efflux from maize leaves in the light could not be detected (Table 5.4). D. N. Moss (1966) also used this technique to demonstrate an excess of photorespiration over dark respiration in several species, and confirmed the failure to obtain detectable CO_2 release from maize leaves. Results from several laboratories have shown that species can generally be separated by this method into groups

TABLE 5.4   *Comparison of Dark Respiration and Minimum Photorespiration Estimated in CO$_2$-Free Air*[a]

Leaf species	Temperature °C	Illuminance ft-c	Minimum photorespiration mg CO$_2$ dm^{-2} hr^{-1}	Dark respiration mg CO$_2$ dm^{-2} hr^{-1}	Photorespiration/ dark respiration	Reference
Maize	35	2500	—	—	0	D. N. Moss (1966)
Maize	25	2800	0	3.0	0	Hofstra and Hesketh (1969a)
Maize	35	—	0	2.0	0	El-Sharkawy and Hesketh (1965)
Atriplex nummularia	25	2800	0	2.5	0	Hofstra and Hesketh (1969a)
Atriplex hastata	25	2800	4.2	3.0	1.4	Hofstra and Hesketh (1969a)
Atriplex hastata	32	3000	8.7[b]	4.4	2.0	Osmond et al. (1969)
Atriplex spongiosa	32	3000	0	3.3	0	Osmond et al. (1969)
Soybean	25	2800	6.0	3.0	2.0	Hofstra and Hesketh (1969a)
Soybean	25	1800	2.6	1.6	1.6	Hew et al. (1969b)
Sunflower	20	2000	3.3	2.1	1.6	Hew et al. (1969a)
Sunflower, normal	24	1800	2.3	1.7	1.4[c]	Hew and Krotkov (1968)
Sunflower, chlorophyll mutant	24	1800	1.2	1.2	1.0[c]	Hew and Krotkov (1968)
Geranium	11–28	2500	—	—	0.95–2.5[d]	D. N. Moss (1966)
Tobacco	25	2500	—	—	1.4–1.5	D. N. Moss (1966)
Cotton	35	—	7.5	3.0	2.5	El-Sharkawy and Hesketh (1965)

[a] In some of the experiments the dark respiration was measured in a stream of normal air.

[b] Diminished to 0.97 in <2% O$_2$.

[c] Photorespiration in the normal sunflower leaf increased and dark respiration did not change in 100% O$_2$; thus the ratio of photorespiration/dark respiration increased to 2.9 in 100% O$_2$. The ratio did not change in the chlorophyll-deficient sunflower leaf when the O$_2$ concentration was raised to 100%.

[d] Ten different determinations were made at varying temperatures, and the mean ratio was 1.5.

that show high rates of photorespiration or low rates, and some examples are summarized in Table 5.4. In CO_2-free air, leaves of species showing a high light respiration doubled their rate of photorespiration in 100% O_2 compared with 21% O_2. The stimulation of photorespiration brought about by increasing O_2 occurred only in tissues capable of high rates of photosynthesis and not in a chlorophyll-deficient mutant or in variegated leaves (Hew and Krotkov, 1968) showing that photorespiration was dependent on photosynthesis.

Not all investigators have observed rates of photorespiration in excess of dark respiration in species that apparently do show high rates when tested by other methods. Usually this can be explained by one or more of the following: failure to ensure that stomata were wide open; not maintaining the CO_2 concentration in the air as near zero as possible; and failure to use a rapid flow of CO_2-free air. For example, increasing the flow rate of CO_2-free air over leaf tissue in the light continued to increase the outward flux of CO_2 even up to flow rates as great as seven chamber volumes per minute (Table 5.4) (El-Sharkawy et al., 1967; Zelitch, 1968; Hew et al., 1969a). Photorespiration was apparently only 30 to 40% of dark respiration in rice and hydrangea leaves, but a flow rate of only 0.036 flask volumes per minute was used (Nishida, 1962). Efflux of CO_2 was not detected in bean leaves when low light intensities (600 lux) were used (Doman, 1959). Rates of photorespiration frequently lower than dark respiration were obtained in spruce, wheat, and soybean leaves, but flow rates of 0.25 volumes per minute were employed (Poskuta et al., 1967).

The importance of the effect of high light intensities and high O_2 concentrations in the atmosphere on photorespiration has also been confirmed by the CO_2-free air method. Thus the ratio of the CO_2 released in the light compared with darkness increased from 0.83 to 1.5 at 20° when the illuminance on sunflower leaves was increased from 500 to 1000 ft-c (Hew et al., 1969a). The respiration in light relative to that in darkness also increased when the O_2 concentration was raised above 21% O_2 in sunflower and geranium (Hew and Krotkov, 1968) and in bean and sunflower leaves (Egle and Fock, 1967).

The interpretation of experiments on the effect of temperature on photorespiration as assayed in CO_2-free air appears to be confounded by failure to consider that stomata in many species begin to close at temperatures above 30°, especially when the demand for water by the leaf increases (Chapter 7.C). This increase in stomatal diffusive resistance would result in an apparent decrease in photorespiration because of greater refixation of CO_2 and would not affect dark respiration since the recycling and fixation of CO_2 is negligible in the dark. A twofold increase in the dark respiration was observed in several species when the leaves were warmed from 25 to

35° while the measured photorespiration decreased relative to dark respiration at the higher temperatures (Hofstra and Hesketh, 1969a). Similar results of an apparent decrease in photorespiration compared with dark respiration at warmer temperatures were obtained by Hew and co-workers (1969b), and again the possible role of stomatal diffusive resistance in these experiments was not fully examined.

The method of measuring CO_2 efflux in CO_2-free air appears to be one of the least sensitive methods of estimating photorespiration because of the shortcomings already described, but a variation of this technique has provided one of the most sensitive means of assaying for light respiration. Goldsworthy (1966) labeled tobacco leaf segments for about 6 hours in a constant stream of $^{14}CO_2$ (flow rate 0.28 volumes/minute) in air while the tissue was floated on water at 20°. The $^{14}CO_2$ released in the light and in periods of darkness was then collected at the same flow rate in a stream of CO_2-free air. The specific activity of the $^{14}CO_2$ released in light was 40% greater than in dark indicating that the substrates differed and photorespiration makes a greater use of recently assimilated CO_2. Under these experimental conditions the ratio of the quantity of $^{14}CO_2$ released in light compared to dark was 1.3, and the photorespiration increased 2.4-fold when the O_2 in the gas stream was increased from 21 to 100%.

A method similar to Goldsworthy's was developed independently by me (Zelitch, 1968), except that warmer temperatures and much faster flow rates of CO_2-free air were used. Leaf disks (six 1.6 cm diameter disks per flask, fresh weight about 240 mg) were floated on water at 30 to 35° with 1000 to 2000 ft-c of illuminance for at least 30 minutes, then the system was closed and a known quantity of $^{14}CO_2$ was released in amounts such that it was assimilated by the tissue in about 15 minutes. After waiting 45 to 60 minutes to insure complete $^{14}CO_2$ assimilation and maximal stomatal opening, CO_2-free air was passed through the flasks at a rate of from 3 to 7 flask volumes per minute and the $^{14}CO_2$ released was collected in alkali and its radioactivity determined. The results were generally expressed as the ratio of the $^{14}CO_2$ evolved in the light to that in the dark. This method of presenting the results has some advantages because it provides a more nearly constant basis for comparison since dark respiration does not vary so greatly and thus tissue sampling errors are minimized.

Figure 5.6 shows that the rate of $^{14}CO_2$ efflux in CO_2-free air is fairly constant for a considerable period of time and that photorespiration in tobacco tissue was at least three times greater than dark respiration. From 4 to 10% of the $^{14}CO_2$ fixed by the tissue was usually released in 30 minutes. The $^{14}CO_2$ efflux by maize disks in light was only 2% of tobacco, although the disks of both species initially fixed the same amount of $^{14}CO_2$ and the

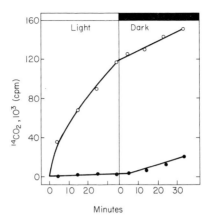

FIG. 5.6. Comparison of $^{14}CO_2$ released in the light and dark by tobacco (○———○) and maize (●———●) leaf disks at 30° (from Zelitch, 1968). After 45 minutes in air at 2000 ft-c the flasks (75 ml) were closed and 5 μmoles of $^{14}CO_2$ (1.08×10^6 cpm) were released. After another 45 minutes, at zero time CO_2-free air was passed through the vessels at a rate of 3 volumes/ minute, and the released $^{14}CO_2$ was collected and the radioactivity determined.

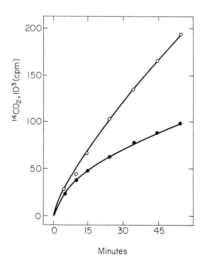

FIG. 5.7. The effect of light (○———○) and darkness (●———●) on the rate of $^{14}CO_2$ released from tobacco leaf disks (from Zelitch, 1968). The experiment was carried out as in Fig. 5.6, and 2.12×10^6 cpm of $^{14}CO_2$ was fixed. At zero time, one of the duplicate flasks was covered with aluminum foil.

rates of $^{14}CO_2$ released in the dark were similar. Unlike tobacco, the photorespiration by maize was much slower than its dark respiration, being only 10% as fast in this experiment.

If tobacco tissue was darkened at zero time in the ^{14}C assay, instead of after the usual period in the light in CO_2-free air, after 10 minutes (post-illumination outburst) the $^{14}CO_2$ released in the light was again about three times as great as the dark respiration, and both types of respiration continued at a constant rate for at least 55 minutes (Fig. 5.7). Standard varieties of tobacco showed a three- to fivefold greater photorespiration than dark respiration in many experiments, and although the specific activity of the $^{14}CO_2$ in the light may be somewhat greater than in the dark (Goldsworthy, 1966), the ^{14}C assay still probably underestimates photorespiration because of the high rate of refixation of CO_2 within the tissue (Chapter 6.C).

Considerably greater rates of photorespiration than dark respiration were also observed with this ^{14}C-assay with cells of *Chlorella pyrenoidosa* and *Chlamydomonas reinhardi* grown in normal air and with leaves of tomato and wheat (Zelitch and Day, 1968a). Variations in rates of photorespiration relative to dark respiration have been observed in leaves of tobacco varieties by us (Table 5.8) as well as by Kisaki and Tolbert (1970) suggesting that photorespiration is genetically controlled. Appreciable photorespiration was also detected in bean and geranium leaves by this method while hardly any was observed in maize (Yemm and Bidwell, 1969).

It was possible to demonstrate clearly some of the differences between photorespiration and dark respiration with this technique because of the sensitivity and reproducibility of this assay of photorespiration (Table 5.5). The $^{14}CO_2$ released by photorespiration is stimulated relative to the CO_2 efflux in the dark with open stomata, fast flow rates of CO_2-free air, higher temperatures, and higher concentrations of O_2. Photorespiration is also severely inhibited by compounds that do not affect dark respiration or stomatal opening and only partially inhibit net CO_2 uptake under the conditions of the experiments, thus the processes of photorespiration and dark respiration are biochemically clearly distinguished from each other. When the assay is carried out in normal air rather than CO_2-free air, the $^{14}CO_2$ released in photorespiration is diluted within the leaf by the large pool of normal CO_2 so that the CO_2 evolved is derived from less radioactive substrates that have recently been synthesized. Therefore even obtaining a light-to-dark ratio of 1.0 in normal air (Table 5.5) is possible only because photorespiration actually greatly exceeds dark respiration. An increase in photorespiration relative to dark respiration occurs at higher temperatures in this assay, but the increase in photorespiration may be twice as great as

TABLE 5.5 *Factors Affecting $^{14}CO_2$ Released from Tobacco Leaf Tissue in CO_2-Free Air in the Light and Dark*[a,b]

Factor changed	Mean stomatal width μ	Temp (°C)	Flow rate (vol/min)	$\dfrac{^{14}CO_2 \text{ released in light}}{^{14}CO_2 \text{ released in dark}}$
Control	6.1	30	3	5.1
Stomata closed	1.1	30	3	2.1
Control		35	7	5.2
Slower flow rate		35	3	3.3
Control	3.2	35	7	6.5
Lower temperature	3.4	25	7	4.9
Control		25	7	2.7
3% O_2		25	7	0.73
Control		27	3	2.9
Normal air		27	3	0.99
Control	5.2	35	3	3.1
0.1 mM CMU	5.2	35	3	1.6

[a] From Zelitch, 1968.
[b] The standard ^{14}C-assay of photorespiration was used as in Fig. 5.6 and the ratios of the $^{14}CO_2$ released in the light to that released in the dark were determined between 5 and 35 minutes in the light and between 15 and 45 minutes in darkness. To close the stomata, the leaf disks were floated on 3.3 mM phenylmercuric acetate solution (Chapter 7.C.4). The solution of CMU was added at zero time after $^{14}CO_2$ had been fixed by the tissue in the preliminary light period.

can be measured (in spite of the slightly higher specific activity of $^{14}CO_2$ released in the light) because refixation of CO_2 also greatly increases at warmer temperatures (Chapter 6.C). The underestimation of photorespiration because of the turnover of CO_2 has been calculated to be 148 to 233% in various species (Samish and Koller, 1968).

The 50-fold difference in photorespiration between a standard variety of tobacco and maize must occur either because maize is more efficient in its gross photosynthesis (a lower R_P in Fig. 5.1) so that evolved $^{14}CO_2$ is rapidly refixed and is thus not released to the intercellular space of the leaf, or because maize lacks an active system of photorespiration. The evidence at present supports the view that photorespiration is in fact slower in maize (Chapter 6.E).

I. THE CO_2 COMPENSATION POINT

The light compensation point is reached during photosynthesis when the light is decreased until assimilation and respiration rates compensate each other with a resulting zero net gaseous exchange (Fig. 9.1). This term was introduced by H. Plaetzer in 1917. At constant light intensities above the light compensation point, assimilation decreases the CO_2 concentration in the atmosphere until a CO_2 level is reached at which there is no net gas exchange (Fig. 5.3), the CO_2 compensation point. At both compensation points there is a steady state between CO_2 assimilation and respiration, but there is still a turnover of CO_2 (Chapter 5.G, Fig. 5.4) in the absence of net changes. Figures 5.8 and 5.9 show how this steady-state CO_2 concentration is attained in maize and tobacco starting at 300 ppm of CO_2 in the ambient atmosphere or at CO_2 concentrations close to zero. The CO_2 compensation point in bright light may differ greatly among various species and is affected by temperature and O_2 concentration. The CO_2 compensation point is a function of photorespiration as well as other factors, and because these other factors have not always been fully appreciated there is considerable misunderstanding in the literature concerning the significance of the CO_2 compensation point and about its limitations as a measure of photorespiration.

The manner in which the CO_2 concentration in the atmosphere should change from a given CO_2 concentration to the compensation point in a closed system has been described from a model similar to Fig. 5.1. Gross photosynthesis, respiration, and the diffusive resistance through stomata are assumed constant for leaves in a closed system in the light (Bravdo, 1968). When a plant is at the CO_2 compensation concentration and there is no net flux of CO_2, IR_P, the product of I (the dark respiration plus photorespiration) and the internal resistances to CO_2 assimilation, R_P (the R_i plus R_c in Fig. 5.1), is equal to the CO_2 compensation point. From this model an equation was derived that describes the expected rate of change of the CO_2 concentration in a closed chamber starting with a fixed CO_2 concentration:

$$\log_e(IR_P - C) = \log_e(IR_P) - \frac{t}{(R_S + R_P)V} \tag{5.6}$$

where C is the CO_2 concentration in the chamber at time t, R_S and R_P are diffusive resistances from the air plus the stomata and the internal resistances shown in Fig. 5.1, and V is the volume or capacity of the enclosure around

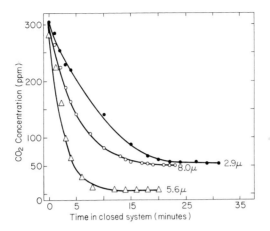

FIG. 5.8. Rate of CO_2 depletion in a closed system containing tobacco leaf disks with open (○———○) or closed stomata (●———●) or maize leaf disks (△———△) in light (from Zelitch, 1969b). Disks (5 disks, 3.2 cm in diameter) were cut from a tobacco leaf that had been sprayed 24 hours earlier in one portion with 5×10^{-5} M phenylmercuric acetate to close the stomata (Chapter 7.C.4). After the disks were exposed to air for at least 1 hour while floated upside down on water at about 25°, the CO_2 concentration was measured in a closed system with an infrared CO_2 analyzer. The CO_2 compensation points shown are 54 to 55 ppm for tobacco and 7 ppm for maize. Chamber volume, 0.95 liter. Gas flow, 8.5 liters/minute. Illuminance, 2000 ft-c. Mean stomatal widths were determined at the end of the experiment from silicone rubber impressions. (Reprinted from *Physiological Aspects of Crop Yield*, page 209, 1969, and reproduced with the permission of the publisher.)

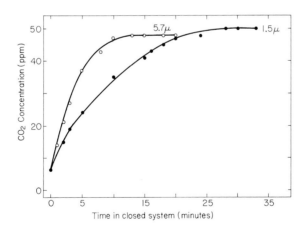

FIG. 5.9. Rate of CO_2 concentration increase in a closed illuminated chamber containing tobacco with open (○———○) or closed stomata (●———●) (from Zelitch, 1969b). Conditions as in Fig. 5.8. The CO_2 concentration in the closed system was depleted to 7 ppm of CO_2 by means of a CO_2 absorbent, and the absorbent was then removed from the system at zero time. (Reprinted from *Physiological Aspects of Crop Yield*, page 211, 1969, and reproduced with the permission of the publisher.)

163

TABLE 5.6   *Effect of Temperature on the CO_2 Compensation Point (ppm) of Various Species in Air*[a]

Species	CO_2 compensation point (ppm)				Reference
	20°C	25°C	30°C	35°C	
Maize	—	9	—	—	D. N. Moss (1962a)
Maize	—	2.7	—	5.5	Meidner (1962)
Sugarcane	—	7			D. N. Moss (1962a)
Wheat, barley, etc.	—	90		180	Thomas *et al.* (1944)
Wheat	32		55		Joliffe and Tregunna (1968)
Wheat (100 varieties)		52 ± 2			D. N. Moss *et al.* (1969)
Orchard grass		60			D. N. Moss (1962a)
Chlorella, Scenedesmus, Fucus		<3			D. L. Brown and Tregunna (1967)
Nitella		32			D. L. Brown and Tregunna (1967)
Amaranthus edulis		<5			Tregunna and Downton (1967)
Amaranthus albus		<5			Tregunna and Downton (1967)
Atriplex rosea		<5			Tregunna and Downton (1967)
Atriplex hastata		35			Tregunna and Downton (1967)
Atriplex hastata				109	Osmond *et al.* (1969)
Atriplex spongiosa				<5	Osmond *et al.* (1969)

Species					Reference
Tobacco			60		D. N. Moss (1962a)
Tobacco			48	80	Zelitch (1966)
Pelargonium	65	120	80		Egle and Schenk (1953)
Pelargonium			65	180	D. N. Moss (1962a)
Pelargonium			55		Heath and Orchard (1957)
Soybean (44 genotypes)		73 ± 0.9		120	Cannell et al. (1969)
Soybean			35		Forrester et al. (1966a)
Sunflower	42				Hew et al. (1969a)
Tomato			75		D. N. Moss (1962a)
Spinach			40		Tregunna and Downton (1967)
Norway maple			145		D. N. Moss (1962a)
Douglas fir	54	65			Brix (1968)
Coffee	80–100[b]				M. B. Jones and Mansfield (1970)
Bryophyllum sp.	145–200[b]				M. B. Jones and Mansfield (1970)
Kalanchoe sp.			<5		Tregunna et al. (1970)

[a] Lists of CO_2 compensation points of leaves in a number of species are given in Black et al. (1969), D. N. Moss et al. (1969), Tregunna and Downton (1967), and Widholm and Ogren (1969).
[b] Rhythmic changes in continuous light.

the leaf. Thus the natural logarithm of the difference between the CO_2 compensation point, IR_P, and the attained concentration, C, should be linearly proportional to time, t. The data in Fig. 5.8 and 5.9 when plotted in this fashion show that the leaf tissue did change C and $\log_e(IR_P - C)$ in the expected fashion, indicating that the rates of CO_2 evolution, stomatal diffusive resistance, and the physical and chemical resistances to CO_2 fixation were constant as the CO_2 concentration changed in the atmosphere to the CO_2 compensation point. The results in Fig. 5.8 and 5.9 also show that the stomata were functioning as variable resistances to diffusion, as expected from Fig. 5.1, since closing the stomata did not change the steady-state CO_2 concentration but only increased the time needed to achieve it.

The representation of the CO_2 compensation point in terms of its components of resistance is also helpful in understanding how the CO_2 compensation point might be expected to vary with increasing temperature in species with a high CO_2 compensation point. Estimates of the magnitude of the various resistances (Fig. 5.1) are given in Chapter 7.E, and the following substitutions may be helpful (Waggoner, 1969a):

$$CO_2 \text{ compensation point} = IR_P \tag{5.7}$$
$$= (\text{photorespiration} + \text{dark respiration})(R_i + R_c)$$

At 20°, for a 10° increase in temperature, assume that R_i is constant (2 cm sec^{-1}), photorespiration increases two or threefold (mg CO_2 dm^{-2} hr^{-1}), and dark respiration doubles (mg CO_2 dm^{-2} hr^{-1}). If gross photosynthesis doubles for a 10° increase, R_c will decrease from 2.4 to about 1.2 cm sec^{-1} at the warmer temperature. A constant, K, is needed as a units conversion; it depends on temperature and is about 1.5. Therefore substituting photorespiration (4.0), dark respiration (2.0), physical resistance (2.0), and carboxylation resistance (2.4);

At 20°, $IR_P = (4.0 + 2.0)(2.0 + 2.4)K = (6.0)(4.4)K = 26 K$ ppm CO_2
At 30°, $IR_P = (8.0 + 4.0)(2.0 + 1.2)K = (12)(3.2)K = 38 K$ ppm CO_2
or $= (12.0 + 4.0)(2.0 + 1.2)K = (16)(3.2)K = 51 K$ ppm CO_2

These substitutions show that even if photorespiration only doubled for a 10° rise in temperature, the CO_2 compensation point would still increase with temperature. If photorespiration tripled between 20° and 30° (Chapter 5.E), IR_P at 30° would equal 51 K CO_2, or be twice as large, a result more consistent with the data in Table 5.6. Changing the photosynthetic efficiency, R_c, which is in series with a constant physical resistance, R_i, is less effective than changing respiration on the CO_2 compensation point.

This method of evaluating photorespiration may not suffice to detect small changes that may occur in photorespiration in varieties within a given species. For example, if photorespiration alone were diminished by 25%, then the CO_2 compensation point would be about 16% smaller with the figures used above. Hence subtle differences in photorespiration might not be detected by seeking differences in the CO_2 compensation point, especially since the other parameters that make up this value may also change in an unpredictable manner, but large differences between species are observed by this method.

The CO_2 compensation points of a number of species obtained at light intensities well above the light compensation point are given in Table 5.6. The efficient photosynthetic species such as maize, sugarcane, some weeds like *Amaranthus* and *Atriplex*, as well as certain algae have CO_2 compensation points less than 10 ppm. Many common crop plants have CO_2 compensation points between 40 and 60 ppm at 25°, and the compensation point approximately doubles for a 10° rise in temperature. The CO_2 compensation point was fairly constant where investigators have examined large numbers of varieties of a single species such as wheat or soybean, suggesting that if differences in photorespiration were present they were not detectable by this method of assay. A rhythmic change in the CO_2 compensation point during continuous illumination has been described for two species and these observations are not yet explained in terms of changes in either respiration or gross photosynthesis that occurred in these experiments. It is noteworthy, however, that maize does not always have a CO_2 compensation point below 10 ppm (Table 5.8).

Photorespiration is greatly dependent on O_2 concentrations, and diminishing the ambient O_2 concentration to about 2% greatly decreases the CO_2 compensation point in species which have high values in normal air. This great change occurs also because at low O_2 concentrations R_c [a part of the R_P in Equation (5.7)] may decrease slightly because of a greater efficiency of carboxylation (Chapter 8.E). Thus the CO_2 compensation point in soybean leaves increased linearly with increasing O_2 concentration, and was less than 5 ppm in 2% O_2, 15 ppm at 10% O_2, 30 ppm in 20% O_2, and 60 ppm in 40% O_2 (Forrester *et al.*, 1966a). Similar results were obtained in tobacco (Krotkov, 1963). The CO_2 compensation point also increased with increasing O_2 concentration in wheat leaves, although in maize the CO_2 compensation point was low and unaffected by O_2 (Tregunna, 1966).

At low light intensities the CO_2 compensation point may be increased greatly, even in maize leaves. For example, in weak light a CO_2 compensation point of 40 ppm was obtained in 21% O_2, but this value did not change on decreasing the O_2 concentration to 2%, indicating that only dark respiration was occurring (Table 5.7). At low light intensities the CO_2 compensation point of wheat leaves was also increased greatly. In wheat,

TABLE 5.7   *Effect of O_2 Concentration on the CO_2 Compensation Point of Wheat and Maize Leaves at Limiting Light Intensities*[a]

| Species | Illuminance ft-c | CO_2 compensation point (ppm)[b] | |
		21% O_2	2% O_2
Wheat	65	170	27
Maize	18	40	39

[a] From Downton and Tregunna, 1968.
[b] The CO_2 compensation points were determined in a chamber at about 22°.

however, the value was affected by the ambient O_2 concentration. These observations on wheat and maize clearly demonstrate that photorespiration is less in maize than in wheat, and that the primary source of CO_2 evolution from maize in the light is dark respiration.

The CO_2 compensation point is not dependent on light intensity once an illuminance above the light compensation point has been reached, suggesting that both gross photosynthesis and photorespiration change simultaneously. A constant CO_2 compensation point was obtained in Scots pine seedlings (*Pinus silvestris*) at 25° at an illuminance of 4000 lux, and this was not changed by higher light intensities although net photosynthesis was still increasing at 10,000 lux (Zelawski, 1967).

The CO_2 compensation point is not a fixed value and apparently anything that alters gross photosynthesis, such as drought (Meidner, 1967), or total respiration in the light will change this value. In severe nitrogen deficiency, the CO_2 compensation point of *Vicia faba* leaves was 40% greater with an equal number of incident quanta of blue light compared with red light (Voskresenskaya et al., 1970b), although photorespiration was not involved because the effect of the blue light did not increase at warmer temperatures. Apparently dark respiration was stimulated by blue light as was previously described for starved algal cells (Chapter 5.A). Blue light also did not stimulate photorespiration in normal tobacco leaves as determined by the post-illumination outburst (Decker, 1955). Leaves of *Kalenchoe* (possessing crassulacean acid metabolism, Chapter 4.L) gave a CO_2 compensation point of less than 5 ppm whether the leaves were in light or darkness (Tregunna et al., 1970). Supplying 10^{-3} M solutions of riboflavin phosphate to maize and wheat leaves in the light in air greatly increased their CO_2 compensation point and decreased net CO_2 assimilation (Tregunna, 1966), indicating either that respiration was stimulated or gross photosynthesis was inhibited by the treatment. On the other hand, applying inhibitors of photorespiration (Chapter 6.C) to tobacco tissue in the light has diminished the CO_2 compensation point and increased net photosynthesis thus showing that large changes in photorespiration can be detected by this assay.

J. DIFFERENCES IN LIGHT RESPIRATION WITHIN A SPECIES

Species have previously been classified in this chapter as possessing either high or low rates of photorespiration, but there is evidence that intermediate forms are also found and these may be useful in studying the inheritance as well as the biochemical control of photorespiration and its relation to net photosynthesis. Obtaining plant material with a lower rate of photorespiration than appears normal for a given species should provide increased rates of net photosynthesis if other diffusive resistance such as the stomata are not adversely affected.

Such differences in photorespiration have been found in tobacco siblings. A tobacco variety, John Williams Broadleaf (JWB), possessing yellowish leaves low in chlorophyll content was first described by Burk and Menser (1964). The yellow mutant (JWB Mutant) is dominant and is viable when the yellow character is heterozygous. When a yellow plant is selfed, one-fourth of the progeny are white and are not viable after germination, one-half are the yellow plants, and one-fourth have dark green leaves (JWB Wild) that carry the yellow trait in a homozygous recessive form. When photosynthesis was measured at concentrations from 0.45% to 5.0% CO_2 [which would inhibit photorespiration (Chapter 5.E)], JWB Mutant tobacco had a net photosynthesis superior to its normal green sibling (Schmid and Gaffron, 1967). However at normal concentrations of CO_2, net photosynthesis in the mutant is very inferior to the dark green leaf (Table 5.8). Photorespiration in these two varieties of tobacco and in a standard variety (Havana Seed) was inversely related to net photosynthesis, and low rates of photorespiration were well correlated with increased rate of growth in seedlings of these three tobacco varieties (Zelitch and Day, 1968b). The glycolate oxidase activity of JWB Mutant is also twice as great as in Wild and it is much less inhibited by blue light (Schmid, 1969). Differences between JWB Mutant and Wild involved not only photorespiration, but the type and amount of chlorophyll present as well as the fine structure of the chloroplasts themselves (Schmid and Gaffron, 1967; Homann and Schmid, 1967). Therefore our recent observations that individual plants of Havana Seed possess low rates of photorespiration and these individuals

TABLE 5.8   **Differences in Light Respiration in Varieties Within a Species and its Relation to Net Photosynthesis**[a]

Species[a]	Ratio $^{14}CO_2$ released, light/dark in ^{14}C photorespiration assay	Net photosynthesis at 300 ppm CO_2, mg CO_2 dm^{-2} hr^{-1}
Tobacco (JWB Mutant)	3.7 –6.6	10.8 ± 0.6
Tobacco (Havana Seed)	2.8 –5.2	13.6 ± 0.5
Tobacco (JWB Wild)	0.82–2.3	16.8 ± 0.4
Maize (Hybrid Penn 602A)	<0.1	26.5 ± 0.7
	CO_2 compensation point (ppm)	
Maize (W22 × NY821)	11 ± 1	75 ± 2
Maize (W22)	25 ± 1	45 ± 3
Maize (NY821)	20 ± 1	63 ± 2
Maize (Pa83)	13 ± 1	85 ± 3
Maize (Wf9)	23 ± 2	28 ± 2

[a] The experiments on photosynthesis with excised tobacco leaves (from Zelitch and Day, 1968b) were carried out in air at about 30° and an illuminance of 1500 ft-c. About twenty determinations were made on tobacco varieties and eight on maize and the standard errors of the mean are given. Photorespiration was assayed as shown in Fig. 5.6. The data on maize (from G. H. Heichel and Musgrave, 1969) were obtained with attached leaves in the field at 32° and in full sunlight in air. The leaves were at the ear-filling stage about 2 weeks following pollination. A single leaf from ten plants was assayed. The overall K_m for CO_2 assimilation varied considerably but was generally about 300 ppm.

have unusually high rates of net photosynthesis (30% greater than the highest value for tobacco in Table 5.1 and Table 5.8) are of considerable interest for future studies on the nature of the inheritance of this process.

Just as individual varieties of tobacco may possess lower than usual rates of photorespiration, so varieties of maize have been found to have higher rates of respiration in the light than is implied in Table 5.6. A comparison of the CO_2 compensation point and net photosynthesis in a number of inbred and hybrid lines of maize grown in the field has been described by G. H. Heichel and Musgrave (1969), and representative data are shown in Table 5.8. Some of the inbred leaves had CO_2 compensation points intermediate between that usually observed in maize and the less efficient photosynthetic species, and high compensation points were associated with lower rates of net photosynthesis, which varied as much as threefold in plants with different genetic backgrounds. The CO_2 compensation points of these maize varieties when grown in a chamber under artificial illumination were not as high as those shown in Table 5.8, but values exceeding 15 ppm were still observed in the high compensation point varieties. However, with maize the high CO_2 compensation point is probably caused by a high rate of dark respiration rather than by enhanced photorespiration since the CO_2 compensation point was not depressed when the O_2 concentration was decreased to 2%.

Some recent work with cattail (*Typha latifolia*) leaves shows they have high rates of net photosynthesis at 340 ppm of CO_2 (44 to 69 mg CO_2 dm^{-2} hr^{-1}) and yet phosphoglyceric acid and not malic acid is the first radioactive product formed after 2 seconds of photosynthesis in $^{14}CO_2$ (McNaughton and Fullem, 1970). Preliminary evidence based on the relatively small accumulation of glycolate in the presence of an α-hydroxysulfonate suggests that a decreased photorespiration results in these leaves because of their slow rate of glycolate synthesis.

These results provide evidence that photorespiration rates may be controlled within a species by genetic or biochemical means, and that diminishing photorespiration can bring about large increases in net photosynthesis. Biochemical aspects of the metabolism of the primary substrate of photorespiration, glycolic acid, will be examined in Chapter 6, and a more detailed evaluation will be made of the mechanism by which CO_2 is released during this process.

6   Glycolate Metabolism and the Mechanism of Photorespiration

Glycolic acid is synthesized as an early product of CO_2 assimilation during photosynthesis in many species (Chapter 4.H), and its oxidation in leaves is catalyzed by an easily extracted enzyme that carries out the reaction:

$$\overset{\circ}{C}H_2OH—\overset{*}{C}OOH + O_2 \xrightarrow[\text{Glycolate oxidase}]{} \overset{\circ}{C}HO—\overset{*}{C}OOH + H_2O_2 \qquad (6.1)$$

In the absence of catalase, the glyoxylate produced is oxidized nonenzymatically by the hydrogen peroxide to produce formate and CO_2, the CO_2 arising from the carboxyl-carbon of glycolate:

$$\overset{\circ}{C}HO—\overset{*}{C}OOH + H_2O_2 \longrightarrow H\overset{\circ}{C}OOH + \overset{*}{C}O_2 + H_2O \qquad (6.2)$$

The biosynthesis and oxidation of glycolate have many characteristics in common with those described in Chapter 5 for photorespiration, and quantitative estimates of the rates of glycolate synthesis and oxidation in photo-

173

respiring tissues support the view that glycolate is the primary source of CO_2 in photorespiration (Zelitch, 1958, 1959, 1964b). The shared characteristics include an enhanced rate of synthesis and oxidation of glycolate with increasing O_2 concentration, a greater glycolate synthesis at higher light intensities, and the relative independence of CO_2 concentrations between zero and 300 ppm in the atmosphere on glycolate synthesis and its inhibition when the CO_2 concentration is increased above about 2000 ppm. Direct evidence of the role of glycolate in photorespiration comes from the measurement of $^{14}CO_2$ output from ^{14}C-labeled glycolate in photosynthetic tissues, and the use of inhibitors of Reaction (6.1) in leaves possessing high and low rates of photorespiration.

In this chapter the biochemistry of glycolate synthesis and metabolism is examined, and several hypotheses about the mechanism of photorespiration and its connection with glycolate metabolism are evaluated in relation to the role of photorespiration in net photosynthesis.

A. BIOSYNTHESIS OF GLYCOLATE

Benson and Calvin (1950) found that barley leaves produce ^{14}C-labeled glycolate after short periods of exposure to $^{14}CO_2$ in the light, and that the

TABLE 6.1   *Percent Distribution of ^{14}C After Photosynthesis in Barley Seedlings*[a,b]

Time (seconds) in $^{14}CO_2$	Followed by time (seconds) in CO_2-free air	Percent distribution of ^{14}C				
		Glycine	Serine	Glycolic acid	Phosphate esters	Sucrose
30	—	12	4.9	8.3	49	15
30	120, light	7.4	6.7	23	10	45
30	120, dark	0.7	17	0.2	27	34
60	—	3.3	3.7	2.0	44	27
300	—	4.5	4.3	6.4	5	65

[a] From Benson and Calvin, 1950.
[b] The $^{14}CO_2$ was supplied at an initial concentration greater than 1% CO_2 in the atmosphere after a preliminary period of photosynthesis. The ^{14}C distribution in the various compounds was determined after separating the compounds in an alcoholic extract by paper chromatography and counting directly on the paper.

labeled glycolate disappears rapidly in darkness (Table 6.1). A requirement of light for the rapid biosynthesis of glycolate, and its disappearance in the dark were also observed in experiments with tobacco seedlings (Stutz and Burris, 1951). Since an active glycolate oxidase system is evidently present in all green tissues, the disappearance of glycolate in the dark must have occurred because glycolate is synthesized exclusively during photosynthesis. The light intensity required for glycolate production in tobacco leaves was similar to the light intensity curve for net photosynthesis (Zelitch, 1959; D. N. Moss, 1968). Synthesis of glycolate occurs equally well in red and white light in *Chlorella,* but not as rapidly in blue light of the same energy (Becker *et al.,* 1968). After photosynthesis in $^{14}CO_2$ for 15 to 30 seconds, both carbon atoms of glycolate were equally labeled with ^{14}C (Table 4.1), and although similar findings have been described repeatedly, there are occasional reports in which the C-2 atom of glycolate contained more ^{14}C than the carboxyl-carbon atom (Hess and Tolbert, 1967).

Many factors are known to control the biosynthesis of glycolic acid in photosynthetic tissues, but the exact biochemical reactions responsible for this synthesis are still uncertain. Several hypotheses have been suggested in attempts to explain glycolate synthesis and these will be examined. None of these mechanisms can presently be shown to function exclusively, although there is some evidence in support of each of them, and it seems likely that more than one pathway for glycolate biosynthesis occurs simultaneously even in a single tissue.

1. EFFECT OF O_2 CONCENTRATION

Increasing concentrations of O_2 in the atmosphere cause an increased rate of synthesis of glycolate in the light, but there is no completely adequate explanation for this phenomenon. In *Chlorella* cells, ^{14}C-glycolate formation from $^{14}CO_2$ was greatest in 100% O_2, less in 21% O_2, and least in an atmosphere of N_2 (Warburg and Krippahl, 1960; Bassham and Kirk, 1962). After 3 minutes in $^{14}CO_2$ the percent of total ^{14}C found in glycolate was: 5 in zero O_2, 10 in 20% O_2, and 38 in 100% O_2 (Coombs and Whittingham, 1966b; Whittingham *et al.,* 1967). The rate of glycolate synthesis could be measured in tobacco leaf disks in the light upon blocking the oxidation of glycolate *in vivo* with an α-hydroxysulfonate (Chapter 6.C). Glycolate formation was negligible in an atmosphere of nitrogen containing 300 ppm of CO_2 compared with normal air (Zelitch and Walker, 1964), and similar results were obtained with bean leaves although an apparent inhibition of synthesis was observed in 75% O_2 (Fock and Egle, 1967), perhaps because glycolate oxidase was not inhibited as effectively by the inhibitor at the higher O_2 concentrations. The O_2 requirement for glycolate synthesis is possibly related to an oxidative step in the synthesis of a two-carbon compound at the glycolaldehyde oxidation level that is derived from the photosynthetic carbon reduction cycle, as described below. Although the reason for the O_2 requirement is uncertain, the requirement for high concentrations of O_2 for rapid synthesis of glycolate is consistent with its role as a substrate in photorespiration, since photorespiration is similarly dependent on O_2 concentration.

2. DEPENDENCE ON CO_2 CONCENTRATION

Large increases in the pool size of glycolic acid in *Scenedesmus* cells were observed during steady-state photosynthesis when the CO_2 concentration was lowered from 1 to 0.003% (A. T. Wilson and Calvin, 1955). The increase in glycolate concentration was accompanied by reciprocal decreases in the pool sizes of intermediates of the Calvin cycle as would be expected if there was a precursor–product relationship, hence they proposed that glycolate is synthesized from a two-carbon fragment originating from increases in concentrations of intermediates of the cycle when CO_2 assimilation is slower. An equally plausible explanation may be that glycolate is synthesized by an independent carboxylation reaction with a greater affinity for CO_2 than the ribulose diphosphate carboxylase, and hence glycolate formation would be favored at lower CO_2 concentrations. There are considerable data showing that glycolate synthesis proceeds at rapid rates in many tissues at CO_2 concentrations between "zero" and 0.2% CO_2, and that further increases in the CO_2 concentration cause a severe inhibition of glycolate synthesis in the light.

Glycine and serine, the early products of glycolate metabolism, were heavily labeled after 30 seconds of photosynthesis by sugar beet leaves in 0.1% $^{14}CO_2$, and the radioactivity in these compounds was decreased 68% when the CO_2 concentration was increased to 0.25% although the labeling of phosphate esters was not affected (Mortimer, 1959). Likewise in experiments with beet leaves, the percent ^{14}C in glycine and serine was the same when $^{14}CO_2$ was applied at 50 ppm or 300 ppm, and was inhibited 80% at 0.3% CO_2 (Tolbert, 1963). In tobacco tissue, glycolate biosynthesis in the light was about 29% less in CO_2-free air than in normal air, about the same at 700 ppm CO_2 as in air, and was inhibited 60% in 0.5% CO_2 and 89% in 1.8% CO_2 (Zelitch and Walker, 1964; Zelitch, 1965a). Similar observations on the inhibitory effects of higher CO_2 concentrations on glycolate synthesis have been made with *Chlorella,* where maximal rates of synthesis were obtained at between 0.1% and 0.2% CO_2 (Warburg and Krippahl, 1960). Others found the maximal formation at 0.1% CO_2, and the synthesis was inhibited about 80% in 0.25% CO_2 (Pritchard *et al.,* 1962) or in 0.4% CO_2 (Coombs and Whittingham, 1966b). Thus glycolate synthesis and photo-respiration show similar properties with respect to the ambient CO_2 concentration since photorespiration is not much affected between zero and 300 ppm and is inhibited at higher CO_2 concentrations (Chapter 5.E).

3. OXIDATION OF A TWO-CARBON FRAGMENT DERIVED FROM THE CALVIN CYCLE

There is evidence, especially from work obtained with inhibitors of glyco-late oxidase (Chapter 6.C), that the rate of glycolate synthesis may be as high as 50% of the net CO_2 assimilation during photosynthesis. Hence any biochemical mechanism for glycolate biosynthesis must account for rates of at least 50 and more likely 100 μmoles of glycolate produced per mg chloro-phyll^{-1} hr^{-1} (a minimum of 13.2 mg CO_2 fixed into glycolate dm^{-2} hr^{-1}).

The kinetics of ^{14}C-labeling of glycolic acid during photosynthesis in $^{14}CO_2$ suggest that glycolic acid arises from the C-1 and C-2 of a sugar phosphate, since it has often been found that the labeling of glycolic acid becomes appreciable only after phosphate esters such as phosphoglyceric acid become heavily labeled (Orth *et al.,* 1966). Other explanations based on the sequence of labeling (Chapter 4.B) can account for such results.

Studies with model systems have shown that the "active glycolaldehyde" produced during transketolase-type reactions (Scheme 4.1) in the Calvin cycle can be oxidized to yield glycolate. Transketolase, a thiamine pyro-phosphate-containing enzyme, catalyzes the transfer of a two-carbon frag-ment from suitable donors such as xylulose-5-phosphate, ribulose-5-phos-phate, or fructose-6-phosphate to a suitable acceptor such as ribose-5-phosphate:

$$\begin{array}{ccc}
\text{CH}_2\text{OH} & \text{CHO} & \\
| & | & \\
\text{C}=\text{O} & \text{H}-\text{C}-\text{OH} & \xrightarrow{\text{Transketolase}} \quad \text{sedoheptulose-7-phosphate} \\
\text{HO}-\text{C}-\text{H} \quad + \quad \text{H}-\text{C}-\text{OH} & & + \\
| & | & \text{glyceraldehyde-3-phosphate} \\
\text{H}-\text{C}-\text{OH} & \text{H}-\text{C}-\text{OH} & \\
| & | & \\
\text{CH}_2\text{OP} & \text{CH}_2\text{OP} & \\
\\
\text{Xylulose-5-} & \text{Ribose-5-} & \\
\text{phosphate} & \text{phosphate} &
\end{array} \qquad (6.3)$$

In the absence of an acceptor, "active glycolaldehyde" is synthesized on position-2 of the thiazolium ring of thiamine:

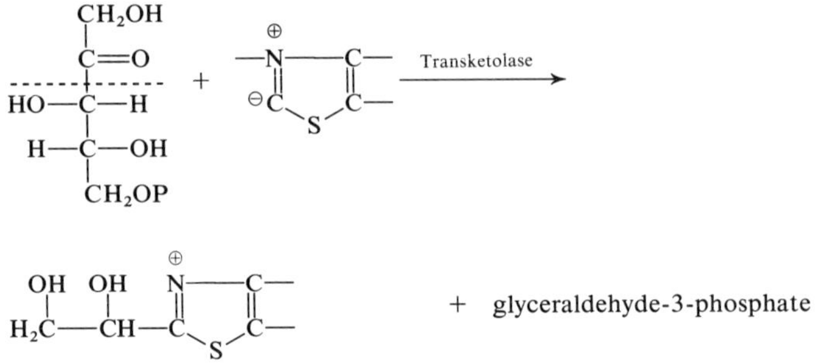

$$+ \quad \text{glyceraldehyde-3-phosphate} \qquad (6.4)$$

"Active glycolaldehyde"
2-(α,β-dihydroxyethyl
thiamine pyrophosphate)

With isolated enzymes, the "active glycolaldehyde" thus produced is oxidized to glycolate in the presence of ferricyanide, 2 moles of ferricyanide being reduced for each mole of glycolate synthesized (Bradbeer and Racker, 1961; Goldberg and Racker, 1962; Holzer and Schröter, 1962). The carboxyl-carbon of glycolate would be derived from the C-2 of the precursor sugar phosphate, and the methylene-carbon would arise from the C-1 of the sugar moiety by such a mechanism.

Although this is a plausible mechanism for glycolate biosynthesis, all attempts to obtain high rates of synthesis by such reactions with intermediates of the Calvin cycle as substrates and chloroplast preparations as a source of enzymes have thus far not been very successful (when artifacts of nonenzymatic conversion of sugar phosphate to glycolate were avoided). For example, chloroplast preparations from spinach, beet, and pea leaves with fructose-1,6-diphosphate (the best substrate) synthesized glycolate at rates of 0.4 μmole mg chlorophyll^{-1} hr^{-1} (Bradbeer and Anderson, 1967), or about 0.5% of the rate observed in intact leaves (Chapter 6.C). More recently rates of 6 μmoles glycolate mg chlorophyll^{-1} hr^{-1} were obtained with fructose-6-phosphate as substrate in a system consisting of fragmented chloroplasts from spinach supplemented with NADP$^+$, ferredoxin, and transketolase in an atmosphere of O$_2$ (Gibbs, 1969a). Since the reaction was inhibited by catalase, hydrogen peroxide may serve as the oxidant (Chapter 6.A.6).

Additional support for the view that glycolic acid may be derived from C-1 and C-2 of pentose comes from the work of Griffith and Byerrum (1959), who infiltrated young tobacco leaves with ribose-1-^{14}C. After 3 hours in light, only 0.5% of the total radioactivity was in glycolate and glycine, but 70% of the ^{14}C was found in the α-carbon of glycolate and glycine. The poor yields raise doubts about whether this is an important route of glycolate synthesis. Similar labeling results were obtained on adding glucose-1-^{14}C and -2-^{14}C (which are likely precursors of the correspondingly labeled fructose-6-phosphate) to *Chlorella* (Whittingham *et al.,* 1963). The C-1 of glucose primarily labeled the C-2 of glycolate, and the C-2 of glucose was converted to the C-1 of glycolate as if the two-carbon fragments derived from the transketolase reaction were oxidized. However, changing the O_2 concentration or even the CO_2 partial pressure did not affect the labeling results despite large changes in glycolate production, hence the quantitative significance of such a pathway is still uncertain.

A variation of this hypothesis of glycolate biosynthesis comes from the discovery of an active phosphatase in green plants that catalyzes the hydrolysis of phosphoglycolic acid to glycolic acid (Richardson and Tolbert, 1961a). A purified preparation of this enzyme obtained from extracts of tobacco leaves was highly specific for phosphoglycolate as the substrate, thus suggesting that the phosphate ester is an immediate precursor of glycolic acid. Presumably phosphoglycolate would be derived from the oxidation of a two-carbon fragment from C-1 and C-2 of fructose-1,6-diphosphate, ribulose-1,5-diphosphate, or sedoheptulose-1,7-diphosphate in the Calvin cycle (Scheme 4.1).

Phosphoglycolate was detected as one of the radioactive products excreted by isolated spinach chloroplasts during photosynthesis in $^{14}CO_2$ (Kearney and Tolbert, 1962). Almost all of the phosphoglycolate phosphatase activity in spinach leaves was found in the chloroplast fraction when isolated by nonaqueous techniques (Thompson and Whittingham, 1967), and its activity was far in excess of that needed if all of the glycolate produced in photosynthesis were synthesized by this pathway. When cell organelles were separated by sucrose gradient-density procedures (Table 6.3) phosphoglycolate phosphatase activity was located in both the mitochondrial and chloroplast fractions (Tolbert *et al.,* 1968). Some doubts have been raised about the importance of phosphoglycolate in glycolate biosynthesis because, in experiments with tobacco leaves supplied with $^{14}CO_2$ under several conditions, the specific activity of the carbon atoms of phosphoglycolic acid was only about one-tenth of the specific activity in glycolic acid (Zelitch, 1965a).

4. AN INDEPENDEDT CARBOXYLATION REACTION

Warburg and Krippahl (1960) studied glycolate synthesis in *Chlorella* suspended in acid phosphate buffer at pH 4.3 in the light and supplied with

0.1 to 0.2% CO_2 in an atmosphere of O_2. A stoichiometric carboxylation occurred in which glycolate synthesis showed the following relationship over a 60 minute period:

$$2 CO_2 + 2 H_2O \longrightarrow HOCH_2—COOH + 1.5 O_2 \tag{6.5}$$

Since the yield of glycolate excreted by the cells was 93% of that expected from the measured O_2 evolution, and this could not have occurred if O_2 uptake was involved in the oxidation of a two-carbon fragment, the observation suggested that glycolate is synthesized *de novo* from CO_2 and is not produced from intermediates of the photosynthetic carbon reduction cycle. Tolbert (1963) found similar high yields of glycolate in *Chlamydomonas*, where 75% of the $^{14}CO_2$ fixed during a 10 minute period appeared as glycolate, but not in *Chlorella*.

Kinetic evidence supporting a carboxylation pathway independent of the Calvin cycle comes from experiments on photosynthetic labeling of photoheterotrophically grown *Rhodospirillum rubrum*. During the first few seconds in $^{14}CO_2$, the ^{14}C present in glycolate was greatly in excess of that found in phosphate esters (L. Anderson and Fuller, 1967), and the radioactivity decreased in glycolate and increased in phosphate esters with time as would be expected if the glycolate were synthesized by a carboxylation independent of the Calvin cycle.

The specific activity (cpm per μmole of C) of the carboxyl-carbon atom of phosphoglyceric acid (Scheme 4.1) has been compared with that of the carbon atoms of glycolic acid in tobacco tissue supplied initially with low (0.1%) and high (1.0%) CO_2 concentrations of $^{14}CO_2$ in the light (Zelitch, 1965a). If glycolate is synthesized from the photosynthetic carbon reduction cycle, the carbon atoms of glycolate should have a lower specific activity than the carboxyl-carbon atom of phosphoglyceric acid which is first labeled in the Calvin cycle. Since the glycolic acid concentration in photosynthetic tissues is usually too low (less than 0.2 to 0.5 μmole/gm fresh weight) to be determined accurately, an inhibitor of glycolate oxidase (α-hydroxy-2-pyridinemethanesulfonic acid, Chapter 6.C) was supplied to leaf disks during $^{14}CO_2$ fixation. The disks were floated on water for 1 hour in the light before the $^{14}CO_2$ was supplied in a closed system, and the inhibitor, when added, was also supplied at zero time. Net $^{14}CO_2$ uptake was not affected by the inhibitor during the 5 minute experimental period, and less than 50% of the total $^{14}CO_2$ available in the closed system was fixed by the tissues. At the end of the experiment, the disks were rapidly killed by plunging them into boiling 20% ethyl alcohol; it required as long as 30 seconds to remove the last of the leaf disks from the vessels containing the $^{14}CO_2$.

The carbon atoms of glycolic acid possessed the same specific activity as that of the $^{14}CO_2$ available to the leaf, evidence it is an early product of photosynthesis, and the specific activities of the carboxyl-carbon of phosphoglyceric acid was one-third to one-half as great as that in glycolic acid. The specific activity in phosphoglyceric acid was more nearly equal to that in the $^{14}CO_2$ supplied to the disks when the concentration of CO_2 in the gas phase was raised to 1%, and the high concentrations of CO_2 also inhibited the rate of glycolate synthesis. These experiments suggested that, especially at CO_2 concentrations close to normal air, glycolic acid is synthesized in the light by an undefined carboxylation reaction different from that catalyzed by ribulose diphosphate carboxylase.

The interpretation of measurements of specific activities in intact tissues have certain limitations as do all other methods of deducing biochemical pathways (Chapter 4.B). For example, the specific activity of the C-1 of phosphoglyceric acid may appear to be lower than that of glycolate because there is a metabolically inactive pool of phosphoglyceric acid in the leaf cells that lowers the apparent specific activity of this compound when the two pools are mixed on isolation. This possibility has been emphasized by Hatch and Slack (1970) who believe that the rate of entry of $^{14}CO_2$ into glycolate carries more weight in interpreting the results than do data on specific activity.

In these experiments (Zelitch, 1965a), the specific activities of the C-1 of glucose-6-phosphate and the C-3 of phosphoglyceric acid (from which the hexose is derived in normal carbohydrate synthesis) were also measured, and these carbon atoms had similar values. The specific activity of the C-1 of glucose-6-phosphate was lower than the C-3 of phosphoglyceric acid. This indicates, but does not prove, that the results are not likely explained by the presence of pools of phosphoglyceric acid with differing metabolic activities. The specific activity experiments (Zelitch, 1965a) have also been criticized (Hess and Tolbert, 1966) because of the possibility that in a killing procedure which required nearly 30 seconds, normal CO_2 from air might have replaced a significant amount of the ^{14}C from the carboxyl group of phosphoglyceric acid. This is unlikely because the tissues were killed almost instantly, less than 50% of the $^{14}CO_2$ in the vessel was fixed, and if any "dark" CO_2 fixation occurred for a few seconds after the 5 minutes of photosynthesis, it would probably have been refixed preferentially from $^{14}CO_2$ available within the tissue during the post-illumination outburst (Fig. 5.7).

Comparisons of the specific activities of glycolic acid and phosphoglyceric acid in tobacco tissues supplied $^{14}CO_2$ in the light for shorter periods were also described by Hess and Tolbert (1966). However, they first allowed leaves to take up either water or the glycolate oxidase inhibitor for 1 hour in "partial shade" before the tissue was exposed to $^{14}CO_2$ at high light intensities at zero time. The tissues were killed and analyzed after 4 to 300 seconds. Unfortunately, a direct comparison of the specific activities in phosphoglyceric acid and glycolic acid with the $^{14}CO_2$ supplied cannot easily be made from the available data. In the absence of glycolate oxidase inhibitor, the specific activity of the carbon atoms of glycolic acid in normal tissue could not be determined with great accuracy by the methods described. The quantity of tissue analyzed (1.3 gm fresh weight) if normal would contain less than 0.5 μmole of glycolic acid. In the presence of an inhibitor of glycolate oxidase, the specific activity of glycolate would be expected to be less than that of phosphoglyceric acid if the inhibitor were added 1 hour before the $^{14}CO_2$, because the pool size of cold glycolate would increase, and these were the results obtained.

5. MULTIPLE PATHWAYS

Considerable evidence supports the view that glycolate may be synthesized by more than one mechanism. The reaction catalyzed by glyoxylate reductase [Reaction (6.6)] is active in photosynthetic tissues, and under some conditions glyoxylate may provide a carbon source for glycolate production. Acetate is also converted into glycolate by *Chlorella pyrenoidosa* in the light (Merrett and Goulding, 1967; Goulding and Merrett, 1967). They supplied mixtures of acetate-^3H and acetate-2-^{14}C in which the initial $^3H/^{14}C$ ratio was 4. Glycolic acid was labeled within 10 seconds with its $^3H/^{14}C$ ratio increased to 10, and the ratio was constant between 10 and 300 seconds after addition of the radioactive acetate. The increased $^3H/^{14}C$ ratio suggests that the ^3H is derived more directly from acetate while the ^{14}C is diluted with cold carbon obtained elsewhere. Both ^3H and ^{14}C were incorporated into glycolate before Calvin cycle intermediates were labeled (after 60 seconds) so that glycolate was probably not synthesized from a two-carbon fragment derived from the cycle. When $H^{14}CO_3{}^-$ was added together with the labeled acetate, the $^3H/^{14}C$ ratio decreased with time indicating that at least two independent pathways were operating in glycolate biosynthesis, one from CO_2 and another from acetate. The relative importance of these pathways in various photosynthetic tissues is not known.

Glycolic acid labeling during photosynthesis has also been investigated during steady-state conditions at high light intensities in *Chlorella* suspended in ^3H_2O in an atmosphere of 0.1% $^{14}CO_2$ at high O_2 concentrations (Plamondon and Bassham, 1966). If glycolate synthesis from glyoxylate is significant [Reaction (6.6)] compared to the biosynthesis of glycolate from

CO_2, one might expect a higher $^3H/^{14}C$ ratio in glycolate than in intermediates of the Calvin cycle such as phosphoglyceric acid after short periods in the presence of the tracers. They observed that in the interval between 50 and 250 seconds the $^3H/^{14}C$ ratio increased from 2.0 to 2.5-fold greater in glycolate than in phosphoglyceric acid, showing that glycolate was continuously synthesized by the reductive formation of an unlabeled carbon compound (such as glyoxylate) as well as from $^{14}CO_2$. This provides additional evidence that multiple pathways for glycolate biosynthesis exist *in vivo.*

Glycolate is always present in leaves at low concentrations, and in long periods of time very much greater quantities of CO_2 are released by photorespiration in CO_2-free air than are available from the catalytic concentration of glycolate (Fig. 5.7). D. N. Moss (1966) has pointed out that if glycolate is the primary source of the CO_2 released under these conditions, it must also be synthesized in part from endogenous sources of carbon independent of the photosynthetic reduction of CO_2 because more CO_2 cannot be produced from glycolate than is fixed into it from CO_2. These tissues store large quantities of carbohydrates and organic acids, and presumably these provide sources of endogenous carbon that are readily converted into glycolate under photosynthetic conditions by reactions that may be still unknown.

6. GLYCOLATE EXCRETION IN ALGAE AND STUDIES WITH ISOLATED CHLOROPLASTS

During the assimilation of $^{14}CO_2$ by photosynthesizing *Chlorella* or *Chlamydomonas,* at least 10 to 20% of the assimilated CO_2 may be excreted into the medium as glycolate and its metabolic products (Tolbert and Zill, 1956; Tolbert, 1963; Whittingham *et al.,* 1963). This excretion of glycolic acid by algae requires light as for photosynthesis, the presence of CO_2, and aerobic conditions; i.e., the conditions for optimal glycolate synthesis.

Isolated spinach chloroplasts also produce large quantities of glycolate-^{14}C in the supernatant fluid from $^{14}CO_2$ (Kearney and Tolbert, 1962), and glycolate and its metabolic products account for about 50% of the ^{14}C in the supernatant liquid. In contrast to $^{14}CO_2$ fixation in the light by intact tissues where a large number of labeled compounds are produced, few compounds are formed during fixation of CO_2 by isolated chloroplasts, and glycolic acid is one of the main products from pea leaves (Walker, 1967) and spinach leaves (Jensen and Bassham, 1966; Bassham *et al.,* 1968; Gibbs, 1969a). This occurs partly because, unlike the conditions in intact leaves, the glycolate oxidase activity is largely missing from isolated chloroplasts (Chapter 6.B). A comparison of the kinetics of labeling of glycolic acid with intermediates of the Calvin cycle has been made in experiments with isolated spinach chloroplasts (Bassham and Jensen, 1967; Chan and Bassham, 1967). The ^{14}C was found in phosphoglyceric acid, sedoheptulose-7-phosphate, fructose-6-phosphate, and hexose-phosphates under the conditions of these experiments, which included high concentrations of CO_2, before it could be detected in glycolic acid (Chan and Bassham, 1967). However, their data show that glycolic acid was in fact labeled with ^{14}C before the total "pentose monophosphates" from which it is allegedly derived (Fig. 2 in Chan and Bassham, 1967).

Rates of synthesis of glycolate by isolated chloroplasts thus far represent only a small fraction of $^{14}CO_2$ fixation, which is itself still low compared to rates of CO_2 uptake by intact leaves. Isolated spinach chloroplasts assimilated $^{14}CO_2$ supplied at low concentrations at rates of 3.5 μmoles CO_2 mg chlorophyll^{-1} hr^{-1}, and 2 to 3 μmoles of the CO_2 were fixed into glycolate in air (Plaut and Gibbs, 1970). The CO_2 assimilation and glycolate synthesis by the chloroplasts were stimulated in the presence of ascorbate and inhibited by DCMU (Fig. 3.5), while the addition of both ascorbate and DCMU partly restored CO_2 fixation and glycolate synthesis. This suggested that the O_2 requirement in glycolate formation results because an oxidant produced in Photosystem II reacts with a reductant such as ferredoxin to produce hydrogen peroxide (Telfer *et al.,* 1970), and the peroxide then oxidizes the "active glycolaldehyde", a proposal made earlier by Coombs and Whittingham (1966a). There is other evidence that inhibitors of Photosystem II strongly block the synthesis of glycolate, as shown below, but because of the low rates of glycolate production obtained with isolated chloroplasts it is still difficult to evaluate the importance of this proposed pathway.

7. INHIBITORS OF GLYCOLATE SYNTHESIS

Biochemical inhibitors that specifically diminish the rate of glycolate synthesis without affecting CO_2 assimilation would be desirable to study the mechanism of biosynthesis of glycolate as well as to evaluate its role in photorespiration. There are no published reports of such inhibitory substances, although α-hydroxysulfonates, inhibitors of glycolate oxidase, also inhibit glycolate synthesis in intact tissues to some extent and may increase net CO_2 assimilation under some conditions because of this activity.

The herbicide atrazine (2-chloro-4-ethylamino-6-isopropylamino-*s*-triazine) suppressed photorespiration in barley leaves, and preliminary evidence suggests it did not inhibit glycolate oxidation, but perhaps prevented glycolate synthesis (Imbamba and Moss, 1969). Inhibitors of Photosystem II, such as CMU, blocked glycolate synthesis during $^{14}CO_2$ fixation by

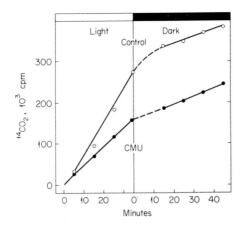

FIG. 6.1. $^{14}CO_2$ released by tobacco leaf disks in the presence of CMU at 35° (from Zelitch, 1968). At zero time (2.95×10^6 cpm previously fixed) CO_2-free air was passed over the leaf disks as in Fig. 5.6. The CMU (0.1 mM final concentration) was added at zero time. Control experiments showed that stomatal widths were unaffected by CMU in the light under these experimental conditions.

Ankistrodesmus, although the synthesis of glycolate and its products was diminished proportionately to the inhibition of total CO_2 assimilation (Tolbert, 1963). The inhibition of glycolate formation by CMU in tobacco leaf disks has been observed by supplying this inhibitor together with an inhibitor of the oxidation of glycolate and measuring the decrease in accumulation of glycolate-^{14}C when $^{14}CO_2$ was supplied in the light (Zelitch, 1968). The inhibition of glycolate-^{14}C accumulation by CMU and of the net yield of glycolate synthesized was generally greater than the inhibition of net CO_2 assimilation caused by the CMU. The specific activity of the glycolate-^{14}C that accumulated in the presence of the α-hydroxysulfonate alone was similar to that of the $^{14}CO_2$ supplied to the disks (Zelitch, 1965a). When CMU was added together with the glycolate oxidase inhibitor, the specific activity of the glycolate-^{14}C was always lower. Thus CMU not only interfered with glycolate synthesis, but the glycolate produced was synthesized from carbon sources with a lower specific activity, again indicating that there are alternative precursors of glycolate synthesis in photosynthesizing tissues.

When CMU was added to tobacco leaf disks under conditions where net photosynthesis was inhibited about 65%, photorespiration was inhibited 43% without decreasing dark respiration (Table 5.5) although stomatal widths were not affected by the inhibitor. Figure 6.1 shows the time course of the $^{14}CO_2$ release in a photorespiration assay similar to Fig. 5.6, in which

CMU was added to one vessel at zero time. The $^{14}CO_2$ released in the presence of CMU in the light was 46% less than the control, while the rate of dark respiration was about the same in both vessels. Thus CMU inhibited the release of $^{14}CO_2$ in the light by photorespiration to a similar but somewhat lesser extent (since dark respiration is unaffected) than it does net photosynthesis, indicating that CO_2 assimilation, glycolate metabolism, and photorespiration are closely linked processes.

If the metabolism of glycolate is primarily responsible for CO_2 release in photorespiration, those species with a low rate of photorespiration may have a lower rate of glycolate synthesis, since their rates of glycolate oxidation do not appear to differ greatly at high substrate concentrations (Chapter 6.B, Table 6.2). Glycolate synthesis in maize is about one-third as rapid as in tobacco as estimated from the rate of accumulation of glycolic acid in tissues in which the oxidation of glycolate has been blocked with an α-hydroxysulfonate (Table 6.4; Zelitch, 1958).

B. GLYCOLATE OXIDASE

1. THE ENZYMATIC REACTION

There has been considerable work aimed at assessing the quantitative role of glycolate oxidase of higher plants [Reactions (6.1) and (6.2)] in the process of O_2 uptake and CO_2 release in the light. The addition of glycolic acid to extracts of barley leaves was first observed to cause an increased O_2 uptake, and glyoxylic acid was produced (Kolesnikov, 1948). Clagett and co-workers (1949) found that extracts of a number of leaves catalyze the oxidation of glycolate and L-lactate, and they partially purified this enzyme from tobacco leaf extracts. Experiments with ^{14}C-labeled glycolic acid as substrate showed that all of the formic acid produced in the reaction arose from the methylene-carbon while the CO_2 was produced from the carboxyl-carbon atom of glycolic acid (Tolbert *et al.*, 1949). The dye 2,6-dichloro-phenolindophenol can substitute for O_2 as the electron acceptor for the enzyme from higher plants, and with use of this sensitive assay the enzyme from spinach leaves was purified considerably and shown to contain riboflavin phosphate (FMN) as the prosthetic group (Zelitch and Ochoa, 1953). The stoichiometry shown in Reactions (6.1) and (6.2) was established. Similar results on the role of hydrogen peroxide were obtained independently by Kenten and Mann (1952).

Glycolate oxidase was further purified and crystallized from extracts of spinach leaves, and the properties of the purified protein have been investigated (Frigerio and Harbury, 1958). The glycolate oxidase flavoprotein can utilize quinones, such as *p*-benzoquinone, as electron acceptors in addition to O_2 and certain dyes, and the corresponding hydroquinone is then produced in the reaction (Kolesnikov *et al.*, 1958, 1959). Other substrates oxidized by this enzyme besides L-lactate (from which pyruvate is the product) include phenyllactate and *p*-hydroxyphenyllactate (Gamborg *et al.*, 1962), but the activity with glycolate is much greater than with other substrates.

The rate of glycolate oxidase activity with O_2 as acceptor increases greatly with increasing O_2 concentration in the gas phase as with all flavoprotein oxidases, and this property is consistent with the stimulatory effect of O_2 concentration on photorespiration. Since the rate of photorespiration may equal 50% of the rate of gross CO_2 assimilation, glycolate oxidase activity should be sufficient in species with a high rate of photorespiration to account for rapid rates of CO_2 evolution greatly in excess of dark respiration. If only the carboxyl-carbon atom of glycolate is involved in the CO_2 released in photorespiration, then rates of about 100 μmoles of glycolate oxidized mg chlorophyll^{-1} hr^{-1} would be needed to account for known rates of photorespiration.

Table 6.2 shows that a number of species known to have high rates of photorespiration have glycolate oxidase activity well within the expected range and greatly in excess of the rates of dark respiration. Glycolate oxidase activity (in 21% O_2) and gross CO_2 uptake (determined in 1% O_2 to avoid photorespiration) were compared in leaves of red kidney bean and sunflower at different ages of development (Fock and Krotkov, 1969), and in all instances the glycolate oxidase activity was at least 50% as great as gross photosynthesis. Some papers state that the glycolate oxidase activity in species known to have low rates of photorespiration is considerably lower, but when efficient methods of grinding and extracting leaves are used several species which show low rates of photorespiration also have considerable glycolate oxidase activity (Rehfeld *et al.*, 1970). Glycolate oxidase activity is generally about three times greater when measured with O_2 as the acceptor (in 21% O_2) compared with rates of dye reduction, but the results based on dye reduction given by Rehfeld *et al.* were adjusted by extrapolating to saturating dye concentrations. Thus species possessing high rates of photorespiration have adequate glycolate oxidase activity to account for the CO_2 released by this process, and species with low photorespiration rates also have considerable enzyme activity when assayed with high concentrations of glycolate.

Evidence of the high rate of glycolate metabolism in leaves has also been

TABLE 6.2   *Glycolate Oxidase Activity in Leaf Extracts of Various Species*

Species	Glycolate Oxidase Activity[a] μmoles glycolate oxidized mg chlorophyll^{-1} hr^{-1}		Method of assay	Reference
	glycolate + $\frac{1}{2}O_2 \rightarrow$ glyoxylate + H_2O	glycolate + dye$_{ox} \rightarrow$ dye$_{red}$ + glyoxylate		
Tobacco	82–113		O_2 uptake	Zelitch (1959)
Tobacco	112		O_2 uptake	Osmond (1969)
Spinach	171		O_2 uptake	Osmond (1969)
Spinach	167		Dye reduction	Rehfeld *et al.* (1970)
Wheat	43		Ketoacid formation	Gamborg *et al.* (1962)
Wheat	119		O_2 uptake	Osmond (1969)
Wheat	74		Dye reduction	Rehfeld *et al.* (1970)
Red kidney bean	104		O_2 uptake	Fock and Krotkov (1969)
Lima bean	16		O_2 uptake	Noll and Burris (1954)
Sunflower	145		O_2 uptake	Fock and Krotkov (1969)
Swiss chard	43		O_2 uptake	Noll and Burris (1954)
Barley	53		O_2 uptake	Noll and Burris (1954)
Atriplex hastata	163		O_2 uptake	Osmond (1969)
Maize	16		O_2 uptake	Osmond (1969)
Maize	38		Dye reduction	Rehfeld *et al.* (1970)
Maize	20		O_2 uptake	Noll and Burris (1954)
Sugarcane	23		Dye reduction	Rehfeld *et al.* (1970)
Sorghum	41		O_2 uptake	Osmond (1969)
Atriplex spongiosa	29		O_2 uptake	Osmond (1969)
Atriplex rosea	26		Dye reduction	Rehfeld *et al.* (1970)

[a] Excess catalase is generally present in crude leaf extracts. It was assumed that 1.5 mg chlorophyll = 1 gm fresh wt = 0.5 dm^2 = 3 mg N = 20 mg protein.

obtained by feeding glycolate to intact leaf tissues. Vickery and Palmer (1956) supplied 0.2 M potassium glycolate at pH 4, 5, and 6 (pK = 3.8) to tobacco leaves through their bases in darkness and observed that within 24 hours between 69 and 89% of the glycolic acid was converted into other products. Supplying 0.1 M glycolate through the petiole to tobacco leaves in the dark increased CO_2 evolution fourfold, while acetate at the same concentration had a negligible effect by comparison, and feeding glycolate to maize leaves in the dark also apparently failed to stimulate CO_2 evolution greatly (D. N. Moss, 1967). However, supplying 0.05 M glycolate to wheat leaves doubled the CO_2 released in an atmosphere of 21% O_2 compared with 2% O_2 in darkness, consistent with glycolate oxidase, while there was a negligible stimulation in CO_2 output when acetate was added to the tissues (Downton and Tregunna, 1968). Similar experiments with maize leaves gave essentially the same results showing that the oxidation of glycolate could be carried out by maize in the dark in a reaction whose rate was dependent on the ambient O_2 concentration, as is glycolate oxidase. Maize leaf disks also assimilated ^{14}C into other products from 0.04 M glycolate-1-^{14}C as well as tobacco tissue did in the light, but the quantity of $^{14}CO_2$ released was only 2% as large from maize as from tobacco (Chapter 6.D).

2. IN ALGAE

Although glycolate oxidase activity was readily detected in a wide variety of higher plants (Noll and Burris, 1954), attempts to demonstrate this activity in several species of algae by manometric, isotopic, or spectrophotometric assays were unsuccessful (Hess and Tolbert, 1967). Glycolate-1-^{14}C and -2-^{14}C were assimilated in the light and in darkness by *Scenedesmus* suggesting that glycolate oxidation, the first step in glycolate metabolism, occurred in these cells (Schou *et al.,* 1950). The oxidation of glycolate to glyoxylate was established in cell-free preparations of *Chlamydomonas* and *Chlorella* cells grown in air by means of the assay based on the reduction of 2,6-dichlorophenolindophenol (Zelitch and Day, 1968a). Extracts obtained from a strain of *Chlorella* also converted glycolate to glyoxylate in the absence of added acceptor (Lord and Merrett, 1968).

These apparent discrepancies were partly clarified by Nelson and Tolbert (1969), who confirmed that extracts of *Chlamydomonas* could reduce the dye in the presence of glycolate, but found that this enzymatic reaction did not couple to O_2 as does the enzyme from higher plants. The nature of the acceptor that replaces the dye *in vivo* is unknown. They also made the interesting observation that, when the cells were grown in air, the dye-reducing activity was about four times greater than when the cells were grown in

$1\% \ CO_2$. When cells grown at high concentrations of CO_2 were transferred to air, there was a two- to fourfold increase in the dye-reducing activity. Thus high concentrations of CO_2 in the gas phase repressed glycolate oxidizing activity.

A 16-fold increase in dye-reducing activity was also observed in *Euglena* grown in air compared with cells grown in $5\% \ CO_2$, and the failure of these extracts to couple with O_2 as acceptor was also observed (Codd *et al.*, 1969). Exogenous glycolate was rapidly metabolized by *Chlorella* suspensions in the light and in darkness, and the oxidation was inhibited by α-hydroxysulfonate and isonicotinylhydrazide (Chapter 6.C) (Lord and Merrett, 1970). Dye reduction, but not O_2 uptake, has also been detected in cell-free preparations of the blue-green algae *Anabena* and *Oscillatoria*, and this activity was likewise inhibited by α-hydroxysulfonates (Grodzinski and Colman, 1970).

In spite of the uncertainties about the nature of the enzyme responsible for the oxidation of glycolate in algae, low rates of photorespiration (CO_2 output) were detected in *Chlorella* with the ^{14}C assay of photorespiration, and high rates of photorespiration were found with *Chlamydomonas* (Zelitch and Day, 1968b). The quantity of CO_2 released by chloroplast preparations obtained from *Acetabularia* cells was greater in the light than in darkness, and CO_2 evolution was enhanced at high concentrations of O_2. Hence these chloroplast preparations displayed characteristics typical of photorespiration (Bidwell *et al.*, 1969).

3. GLYOXYLATE REDUCTASES

An enzyme which catalyzes the reduction of glyoxylate to glycolate in the presence of NADH according to the following reaction was found in extracts of spinach leaves (Zelitch, 1953).

$$\begin{matrix} CHO \\ | \\ COOH \end{matrix} + NADH + H^+ \xrightarrow[\text{reductase}]{\text{Glyoxylate}} \begin{matrix} CH_2OH \\ | \\ COOH \end{matrix} + NAD^+ \qquad (6.6)$$

Glyoxylate   glycolate

Glyoxylate reductase combined with the glycolate oxidase reaction and a NAD^+-linked dehydrogenase into a system which served as an electron carrier to oxidize substrates with the uptake of O_2. In the presence of an excess of catalase, to remove the hydrogen peroxide produced by glycolate oxidase, the following glycolate-glyoxylate cycle was demonstrated with partially purified enzymes:

$$\text{Substrate} + NAD^+ \xrightarrow[\text{Dehydrogenase}]{} \text{oxidized substrate} + NADH + H^+$$

$$NADH + H^+ + \text{glyoxylate} \xrightarrow[\substack{\text{Glyoxylate}\\\text{reductase}}]{} \text{glycolate} + NAD^+$$

$$\text{Glycolate} + O_2 \xrightarrow[\substack{\text{Glycolate}\\\text{oxidase}}]{} \text{glyoxylate} + H_2O_2$$

$$H_2O_2 \xrightarrow[\text{catalase}]{} H_2O + 0.5\ O_2$$

$$\text{Net:}\quad \text{Substrate} + 0.5\ O_2 \longrightarrow \text{oxidized substrate} \qquad (6.7)$$

If such a sequence of reactions functions *in vivo* during photosynthesis, one may obtain O_2 uptake and lesser quantities of CO_2 evolution, depending on the portion of glyoxylate that is oxidized further to produce CO_2. A sequence similar to the one shown could account for the observations that the photosynthetic quotient is frequently not 1.0 because O_2 uptake probably exceeds CO_2 evolution (Chapter 5.A).

The NADH-linked glyoxylate reductase from tobacco leaves has been purified and crystallized (Zelitch, 1955), and an enzyme of similar specific activity has been obtained from spinach leaves (Holzer and Holldorf, 1957). The purified enzyme is specific for NADH, does not react with pyruvate as a substrate, and the equilibrium of the reaction is very far in the direction of glyoxylate reduction. Hydroxypyruvate is also an excellent substrate for this enzyme (Stafford *et al.,* 1954), and the following reaction is catalyzed:

$$\begin{array}{c} COOH \\ | \\ C{=}O \\ | \\ CH_2OH \end{array} + NADH + H^+ \xrightarrow[\substack{\text{Glyoxylate}\\\text{reductase}}]{} \begin{array}{c} COOH \\ | \\ H{-}C{-}OH \\ | \\ CH_2OH \end{array} + NAD^+ \qquad (6.8)$$

$$\text{Hydroxypyruvate} \qquad\qquad\qquad\qquad \text{D-glycerate}$$

Further details of the chemistry of the purified glyoxylate reductase from spinach leaves have recently been described (Kohn *et al.,* 1970; Kohn and Warren, 1970).

A NADPH-linked glyoxylate reductase with different kinetic properties has been partly separated from the NADH enzyme in extracts of spinach and tobacco leaves (Zelitch and Gotto, 1962), and it has been found to be associated with chloroplast fractions (Tolbert *et al.,* 1970). Particulate preparations from leaves that effectively carried out oxidative phosphorylation with substrates of the tricarboxylic acid cycle failed to synthesize ATP

TABLE 6.3   *Distribution of Peroxisomal Enzyme Activities in a Crude Chloroplast Preparation from Spinach Leaves after Centrifugation on a Sucrose Density-Gradient*[a,b]

Fraction No.	Glycolate Oxidase		Catalase		NADH-glyoxylate reductase	
	Sp. Act.	% Distribution	Sp. Act.	% Distribution	Sp. Act.	% Distribution
9 (1.3 M sucrose, Top)	0.014	24	110	24	0.015	21
8 (Chlorophyll)	0.007	5	49	5	0.008	5
7 (Mitochondria)	0.013	3	102	3	0.029	6
6	0.021	2	107	1	0.029	2
5	0.035	6	184	4	0.036	5
4	0.190	13	1190	10	0.259	14
3 (Peroxisomes)	0.920	36	6780	34	1.190	37
2	0.370	11	4690	17	0.478	11
1 (2.5 M sucrose, Bottom)	0.030	1	68	17	0	0

[a] From Tolbert *et al.*, 1968.

[b] A "chloroplast" fraction (containing many broken chloroplasts) was obtained by centrifugation at 3000 g for 20 minutes. The pellet was suspended in 0.5 M sucrose containing 0.02 M glycylglycine, pH 7.5, and the resulting suspension was layered in a sucrose density-gradient prepared in a centrifuge tube. The specific activities of several enzymes in the suspension (in μmoles min^{-1} mg protein^{-1}) were: glycolate oxidase, 0.047; catalase, 299; and NADH-glyoxylate reductase, 0.033. After centrifugation in the density gradient for 3 hours at 45,000 to 107,000 g, the fractions shown were collected by withdrawal from the bottom of the centrifuge tube and assayed for enzymatic activity. Fraction 3 was rich in peroxisomes as determined by electron microscopy.

when glycolate was oxidized (Zelitch and Barber, 1960), but experiments on the anaerobic uptake of glucose by *Chlorella* in the light suggested a role of glycolate metabolism in ATP synthesis by photophosphorylation (Butt and Peel, 1963). The uptake of glucose was inhibited by α-hydroxysulfonates, which are inhibitors of glycolate oxidase, and the inhibition was reversed upon addition of glycolate. Butt and Peel suggested that the glycolate-glyoxylate cycle [Reaction (6.7)] participates in the generation of ATP and this promotes the uptake of glucose as follows:

$$NADP^+ + ADP + P_i + H_2O \xrightarrow{\text{Light}} NADPH + H^+ + ATP + 0.5\ O_2$$

$$NADPH + H^+ + glyoxylate \xrightarrow[\substack{\text{Glyoxylate} \\ \text{reductase}}]{} NADP^+ + glycolate$$

$$Glycolate + 0.5\ O_2 \xrightarrow[\substack{\text{Glycolate} \\ \text{oxidase, catalase}}]{} Glyoxylate + H_2O$$

$$\text{Net:}\quad ADP + P_i \xrightarrow{\text{Light}} ATP \qquad\qquad (6.9)$$

The first reaction in the sequence depicts noncyclic photophosphorylation (Chapter 3.E). The synthesis and metabolism of glycolate brings about the reoxidation of NADPH to $NADP^+$ that is necessary to permit photophosphorylation to proceed at rapid rates. Such a sequence of reactions would be especially favored at low concentrations of CO_2, when less NADPH might otherwise be oxidized in the process of CO_2 reduction (Chapter 4.C), and the synthesis of glycolate would then also be enhanced. Such a glycolate-glyoxylate system for the generation of ATP has been demonstrated in isolated illuminated chloroplasts (Asada *et al.*, 1965b). A similar mechanism *in vivo* is suggested by experiments in which the addition of glycolic acid stimulated the uptake of ^{32}P-phosphate and the formation of organic phosphate by cells of *Ankistrodesmus* in the light (Jacobi, 1959).

NADH-glyoxylate reductase activity is concentrated in the peroxisome fraction (Table 6.3) where its specific activity is about the same as that of glycolate oxidase. The rate of its activity in crude extracts of tobacco leaves is also about the same as that of glycolate oxidase (Zelitch, 1955); therefore, this reaction could participate in the biosynthesis of glycolate. However, effective precursors of glyoxylate in leaves are presently not known. Glyoxylate-^{14}C was rapidly taken up by isolated spinach chloroplasts, and in the light a large fraction of ^{14}C was found in glycolate suggesting that glyoxylate reductase may function in this manner (Kearney and Tolbert, 1962). There is evidence that glycolate is synthesized from carbon sources other than CO_2 (Chapter 6.A), and perhaps glyoxylate is one of these precursors.

4. ENHANCEMENT OF GLYCOLATE OXIDASE ACTIVITY BY LIGHT

Glycolate oxidase activity is usually low in etiolated leaves compared with their normal green counterparts (Tolbert and Burris, 1950). Spraying glycolic acid on leaves of wheat plants stored in the dark caused an increased glycolate oxidase activity in extracts on the next day, suggesting that the light activation of glycolic acid resulted from the formation of the substrate during photosynthesis (Tolbert and Cohan, 1953). This cannot account for the stimulation by light since the enzyme is activated without photosynthesis. The activity was low in bean seedlings grown in the dark, and after an exposure to light for only 10 minutes the activity increased more than sixfold in the seedlings receiving the brief illumination, while it increased slowly in plants kept in continuous darkness (Filner and Klein, 1968). Glycolate oxidase activity is associated with the peroxisomes of photosynthetic tissue, and such organelles are not restricted to photosynthetic tissues. Hence although the nature of the light activation of this enzyme is still not understood, it is probably not related to peroxisome development. In dark-grown mustard seedlings (*Sinapis alba*), the application of far-red light induced both glycolate oxidase and NADH-glyoxylate reductase simultaneously, indicating that the development of these enzymes is under control of phytochrome, the photomorphogenic pigment, and that these enzymes function together *in vivo* (Van Poucke *et al.*, 1970).

5. ASSOCIATION WITH PEROXISOMES

The discovery and characteristics of peroxisomes in photosynthetic tissue have been described in Chapter 1.F. There are at present five known enzyme activities that are localized in these organelles in relatively large concentrations: glycolate oxidase, catalase, NADH-glyoxylate reductase, NAD^+-malate dehydrogenase, and transaminases (Tolbert and Yamazaki, 1969). The separation of peroxisomes from a crude chloroplast preparation of spinach leaves on a sucrose density-gradient and the distribution of some enzymes on the gradient is shown in Table 6.3. The peroxisomes contain a large excess of catalase compared with glycolate oxidase activity, and the glyoxylate reductase and glycolate oxidase activities are similar. Cytochrome oxidase (associated with mitochondria) was concentrated in about Fraction 6 on the gradient while chlorophyll (representing intact chloroplasts) was largely present in Fraction 7. Phosphoglycolate phosphatase activity was located in the same area as cytochrome oxidase as well as in a small band of whole chloroplasts.

A survey of ten species for peroxisome enzyme activity has been made (Tolbert *et al.*, 1969), and significant yields of peroxisomes were obtained in five species known to have high rates of photorespiration including spinach, sunflower, tobacco, pea, and wheat leaves while yields of peroxisomes in species such as maize and sugarcane with low photorespiration were poor although they can be clearly seen in Fig. 2.1 B. It is uncertain whether this low yield is due to the fragility of the peroxisomes in these species during isolation or to their relatively fewer numbers. Some observations indicate there are fewer peroxisomes in maize and these are small and not so closely associated with chloroplasts as in tobacco (Frederick and Newcomb, 1969a; Fig. 1.5). These microbodies may vary in their enzyme complement according to cell type and the state of differentiation in the cells. Frederick and Newcomb (1969b) have shown that all leaf peroxisomes contain catalase based on a cytochemical analysis with the electron microscope, and in some peroxisomes the catalase is visibly present as a crystalline structure. If glycolate is the primary substrate of photorespiration and its oxidation occurs in an organelle different from the site of glycolate synthesis, in the chloroplasts, this would readily account for the diffusive resistances implied in Fig. 5.1 that permit a portion of the evolved CO_2 to be released into the atmosphere in CO_2-free air as well as allowing a portion to be refixed within the leaf tissue.

Intact isolated spinach chloroplasts do not appear to oxidize glycolate (Everson and Gibbs, 1967), and chloroplasts isolated by nonaqueous techniques as well as in aqueous media also did not show glycolate oxidase activity (Thompson and Whittingham, 1968). Crude chloroplast preparations from *Tetragonia expansa* oxidized glycolate-1-^{14}C to $^{14}CO_2$ especially during illumination, thus suggesting that the process of photorespiration was involved. Isolated spinach chloroplasts capable of relatively high rates of CO_2 assimilation converted glycolic acid-2-^{14}C to glyoxylic acid and glycine, thus indicating that some glycolate oxidase activity was associated with these chloroplast preparations perhaps because they contain peroxisomes (Chan and Bassham, 1967). Tolbert and Yamazaki (1969) have calculated that the peroxisomal glycolate oxidase is active enough to account for the metabolism of glycolate at rates greater than 50% of the total CO_2 assimilation in photosynthesis, although some of the later reactions in the glycolate pathway are not yet active enough according to present assays to account for photorespiration (Chapter 6.D).

C. INHIBITION OF GLYCOLATE OXIDASE

1. ISOLATED ENZYMES

Bisulfite addition compounds of aldehydes are salts of α-hydroxysulfonates possessing the general structure $R-CHOH-SO_3Na$, and are thus analogs of glycolic acid. At a glycolic acid concentration of 10^{-2} M, sodium hydroxymethanesulfonate (formaldehyde-bisulfite) at 5×10^{-5} M (less than 1% of the substrate concentration) inhibited a partially purified preparation of glycolate oxidase of spinach leaves 50%, and the inhibitor was competitive with the substrate as suggested by their similarity in structure (Zelitch, 1957). Sodium bisulfite also acted as an inhibitor because it combined with the glyoxylate produced in the enzymatic reaction and formed the corresponding α-hydroxysulfonate, disodium sulfoglycolate. By a similar exchange reaction, glyoxylate bisulfite may always be the active inhibitor no matter which α-hydroxysulfonate is supplied (Corbett and Wright, 1970). These inhibitors were relatively ineffective against the oxidation of malate by malate-NAD^+ dehydrogenase or against the reduction of glyoxylate by NADH-glyoxylate reductase. One of the most effective α-hydroxysulfonates for studies of the inhibition of glycolate oxidase *in vivo* is the 2-pyridine aldehyde analog (Zelitch, 1958),

α-Hydroxy-2-pyridinemethanesulfonic acid (α-HPMS)

Evidence of the relatively high specificity of this α-hydroxysulfonate against glycolate oxidase was obtained on studies of O_2 uptake and oxidative phosphorylation with particulate preparations from spinach leaves (Zelitch and Barber, 1960). With substrate concentrations of 10^{-2} M and α-HPMS concentrations of 5×10^{-4} M (less than 1% of the substrate), these preparations were unaffected in their succinoxidase activity or accompanying oxidative phosphorylation, while glycolate oxidation was inhibited about 70%. The specificity of this compound has recently been questioned because it also inhibited the activity of phosphoenolpyruvate carboxylase (Chapter 4.F) in a competitive manner and also slowed malate-NAD^+ dehydrogenase (Osmond and Avadhani, 1970). However, the inhibitor concentration required was ten times greater than the substrate concentration in order to obtain 50% inhibition of the carboxylase activity, and the effect on malate dehydrogenase was barely detectable.

Isonicotinylhydrazide acts somewhat like α-hydroxysulfonates and causes glycolic acid to accumulate in photosynthetic tissues (as shown below), although this substance does not inhibit isolated glycolate oxidase. Blue light at low intensities (150 to 200 erg sec^{-1} cm^{-2}) inhibits glycolate oxidase activity 50% in extracts whereas higher intensities of red light (20,000 erg

sec^{-1} cm^{-2}) have no inhibitory effect (Schmid, 1969). This suggests that the enhancement of respiration by blue light in some tissues (Chapter 5.B) is not related to photorespiration, although the extracted glycolate oxidase of some tobacco varieties is more sensitive to inhibition by blue light, and this may have some effect on controlling photorespiration if the inhibition by blue light also functions *in vivo*.

2. IN INTACT TISSUES

The inhibitory action of α-hydroxysulfonates provided a means of evaluating the extent of participation of glycolate oxidase in the respiration of leaves under various physiological conditions. In the early experiments the inhibitors were supplied to excised leaves through their bases, and the glycolate that accumulated in the light was determined quantitatively after separation of the glycolic acid by ion exchange chromatography (Zelitch, 1958). When a tobacco leaf was placed in a 10^{-2} M solution of α-HPMS in sunlight for 30 minutes, glycolic acid accumulated at a rate of 20 μmoles gm fresh weight^{-1} hr^{-1} (greater than 40-fold increase in concentration). There was little effect on net CO_2 uptake when $^{14}CO_2$ was supplied in the atmosphere at the same time as the inhibitor at 25° for 7.5 or 15 minutes, and the radioactivity in glycolate in the presence of the inhibitor accounted for 50% of the ^{14}C in the tissue compared with only about 5% in the absence of inhibitor (Zelitch, 1959). Thus at least one-half of the carbon fixed in photosynthesis in the normal leaf may be metabolized by the glycolate oxidase reaction, consistent with its role in photorespiration.

Net CO_2 assimilation was not inhibited and glycolate accumulated at rates of 40 μmoles gm fresh weight $^{-1}$ hr^{-1} in leaf disks floated on a solution of 10^{-2} M α-HPMS at 30° for 5 to 10 minutes in the light at concentrations of $^{14}CO_2$ not greater than 0.1% (Zelitch, 1965a). Using shorter times of exposure to the inhibitor, rates of 60 to 70 μmoles have subsequently been obtained.

A 10^{-2} M solution of α-HPMS does not inhibit CO_2 assimilation by leaf disks at 35° if they are floated on inhibitor solution less than 15 to 20 minutes. Perhaps this toxicity on longer exposures is caused by the release of a pyridine residue when the α-hydroxysulfonate exchanges with glyoxylate (Corbett and Wright, 1970), and in part because α-hydroxysulfonates close stomata under many conditions (Zelitch, 1961). In contrast, solutions of 10^{-2} M α-HPMS when taken in through the base of leaves inhibited net photosynthetic CO_2 uptake in tobacco 40% in 5 minutes, before stomata were affected (D. N. Moss, 1968). One hour after solutions of this inhibitor were taken up through the petioles, net $^{14}CO_2$ uptake in tobacco leaves (during 15 minutes) was inhibited about 70% (Asada and Kasai, 1962). When the inhibitor was fed to tobacco leaves in subdued light for 1 hour prior to supplying $^{14}CO_2$ for 5 minutes in strong light, 48% of the ^{14}C was found in glycolic acid and there was no inhibition of $^{14}CO_2$ uptake for the first 30 seconds (Hess and Tolbert, 1966). Thus large accumulations of glycolate in the presence of the inhibitor have been confirmed, and even as much as 76% of the total ^{14}C was in glycolic acid during photosynthesis after an α-hydroxysulfonate was supplied to spinach leaves for 3 hours before providing $^{14}CO_2$ (Asada *et al.,* 1965a).

In *Ankistrodesmus* cells the addition of 9×10^{-3} M α-HPMS for 15 minutes had no effect on net $^{14}CO_2$ fixation while the radioactivity in excreted glycolate accounted for 6% of the total fixation and 30% of the ^{14}C in the supernatant fluid (Urbach and Gimmler, 1968). Supplying this inhibitor through the petioles of *Atriplex hastata* leaves (a high photorespiration species) for 30 minutes at 28° increased net $^{14}CO_2$ uptake in the light by 50% during a 10-minute period compared with the rates of leaves in water, while in *Atriplex spongiosa* leaves (low photorespiration) CO_2 uptake was inhibited 80% under the same conditions (Osmond and Avadhani, 1970). These results, together with those carried out with leaf disks (indicated above), emphasize that the specificity of an inhibitor supplied to a tissue depends on the temperature and time it is available, its concentration, the tissue itself, and the manner in which it is administered.

Isonicotinylhydrazide added to *Chlorella* inhibited the metabolic conversion of glycine to serine (Pritchard *et al.,* 1962, 1963), a reaction related to the further metabolism of glycolate (Chapter 6.D). At a concentration of 10^{-2} M the inhibitor increased CO_2 assimilation slightly, and about 10% of the carbon fixed was excreted as glycolic acid and approximately an equal amount accumulated in the supernatant fluid as glycine. This inhibitor does not affect isolated glycolate oxidase activity directly, hence the blocking of glycine metabolism must control glycolate oxidase activity in some indirect manner, presumably by a feedback mechanism operating at a later step in the metabolism of glycolate.

3. EFFECT ON NET PHOTOSYNTHESIS AND PHOTORESPIRATION

Blocking Reactions (6.1) or (6.2) should inhibit the losses of fixed carbon and thereby increase net photosynthesis by the extent to which respiratory CO_2 is normally refixed (Fig. 5.1) if glycolate is the main source of the CO_2 evolved in photorespiration. This effect would be magnified at warmer temperatures if photorespiration increases greatly (Chapter 5.E and 5.H) and thus one might be able to convert a plant with a high photorespiration (and

TABLE 6.4   *Effect of α-Hydroxysulfonate on $^{14}CO_2$ Uptake by Tobacco and Maize Leaf Disks in Light at 25° and 35°* [a,b]

| | | | μmoles $^{14}CO_2$ Uptake gm fr wt^{-1} hr^{-1} | | | |
| | | | 25° | | 35° | |
Expt. No.	Species	Min in $^{14}CO_2$	Disks in water	Disks in inhibitor	Disks in water	Disks in inhibitor
1	Tobacco	5.0	56	67	61	181
2	Tobacco	2.5	136	62	58	128
3[c]	Tobacco	2.5	39	34	37	188
4	Tobacco	3.0			47	203
5	Tobacco	5.0			31	66
6	Tobacco	2.5			45	268
7[c]	Maize	4.0			78	77
8[c]	Maize	5.0	40	11	57	28
9[c]	Maize	5.0			51	10

[a] From Zelitch, 1966.

[b] After a preliminary period at 2000 ft-c of illuminance at the temperature shown, the fluid in Warburg vessels was replaced with 0.01 M α-hydroxy-2-pyridinemethanesulfonic acid (the inhibitor) and after 2 min 3 μmoles of $^{14}CO_2$ (2.5 μmoles with maize, initial concentrations about 0.1%) were liberated for the time shown. The glycolic acid formed in the presence of inhibitor in μmoles gm fr wt^{-1} hr^{-1} was: Expt. No. 1, 15.7 at 25° and 29.9 at 35°; Expt. No. 2, 26.9 at 25° and 25.8 at 35°; Expt. No. 3, 5.3 at 25° and 9.0 at 35°; Expt. No. 8, 3.5 at 25° and 4.9 at 35°; Expt. No. 9, 3.6 at 35°.

[c] The concentration of inhibitor in this experiment was 0.005 M.

a lower rate of net CO_2 assimilation) into one with a low photorespiration (and a high net CO_2 assimilation rate).

In earler experiments with tobacco leaf disks in the light at 30° (Zelitch, 1965a), addition of α-HPMS had no effect on the rate of $^{14}CO_2$ uptake during experimental periods of 5 to 10 minutes. The rate of CO_2 assimilation by leaf disks floating on water is usually about one-half the rate observed with leaves in air because the disks have only one surface freely exposed to the CO_2 in the atmosphere. The rate of $^{14}CO_2$ uptake by tobacco and maize disks was compared in the presence and absence of the glycolate oxidase inhibitor at 25° and 35° (Table 6.4). With tobacco, the presence of the inhibitor for as long as 7 minutes stimulated $^{14}CO_2$ uptake at 35° from 2.1 to 6.0 times, with a mean 3.8-fold increase. In experiments at 25° the inhibitor changed the $^{14}CO_2$ uptake only slightly, results similar to those previously observed at 30°. With maize the inhibitor caused no increase in $^{14}CO_2$ uptake and was in fact inhibitory and more variable in its effect at both tem-

TABLE 6.5   **Model of CO_2 Budget for a Tobacco Leaf at 25° and 35° in High Illuminance and 300 ppm CO_2 in Air**

	mg CO_2 dm^{-2} hr^{-1}	
	25°	35°
1. Gross photosynthetic CO_2 uptake	− 25.0	− 47.0
2. CO_2 output from dark respiration	+ 3.0	+ 6.0
3. CO_2 output from photorespiration	+ 7.0	+26.0
4. Net CO_2 uptake observed	− 15.0[a]	− 15.0[a]
5. When photorespiration inhibited, net CO_2 uptake observed	− 22.0	− 41.0

[a] 15.0 mg CO_2 dm^{-2} hr^{-1} is equivalent to about 114 μmoles CO_2 mg chlorophyll^{-1} hr^{-1} (3 mg chlorophyll dm^{-2}).

These estimations are based on the results in Tables 5.1 and 6.4, estimates of the relative magnitude of dark respiration and photorespiration (Chapter 5.I), and observed rates of net photosynthesis in maize at 25° and 35° (Chapter 8.A).

peratures, and considerably less glycolate accumulated in the presence of the inhibitor in maize than in tobacco. Thus when the oxidation of glycolate (and perhaps its synthesis to some extent) was inhibited in tobacco leaf disks at 35°, there was about a threefold increase in the rate of photosynthetic CO_2 uptake while the inhibitor did not stimulate photosynthesis greatly in tobacco at 25° or in maize at either temperature. Gross photosynthesis must also increase at higher temperatures in species like tobacco, but it is ordinarily masked by the high rate of photorespiration since 50 to 100% increases in net CO_2 uptake usually occur in maize at 35° compared with 25° and not in tobacco (Chapter 8.D). Slowing photorespiration may cause photosynthetic tissues like tobacco to have net rates of CO_2 assimilation similar to those of maize.

To account for these results of enhanced photosynthesis at 35°, a balance sheet was prepared (Table 6.5) that provides the basis of further experiment. This CO_2 budget shows that photorespiration alone accounts for about 30% of the gross photosynthesis and 50% of the net photosynthesis at 25°. When assayed as the post-illumination outburst (Table 5.1), the CO_2 produced in photorespiration (and dark respiration) accounted for 45% of the net photosynthesis at 25° in tobacco leaves, in reasonable agreement with the model. The budget also indicates that at 35° photorespiration would account for about 55% of the gross CO_2 uptake and be 1.7 times greater than the net photosynthetic CO_2 assimilation. The data in Table 5.1 show photorespiration at 34° to be 66% of net photosynthesis, far less than that predicted by the budget, but at warmer temperatures photorespiration is greatly underestimated because of the increased rate of internal refixation of CO_2 (as shown below).

The budget also shows that total CO_2 evolution in the light (from photorespiration plus dark respiration) should be 3.3 times greater than dark respiration at 25° and 5.3 times greater at 35°, in general agreement with the results obtained by the extrapolation method (Table 5.2) and the ^{14}C assay of photorespiration (Table 5.5). The model (Table 6.5) was tested further by comparing rates of [14]CO_2 release by leaf disks supplied acetate-[14]C (a substrate for dark respiration) and glycolate-[14]C (the primary substrate for photorespiration) in the light. The evolution of [14]CO_2 from glycolate-1-^{14}C, but not -2-^{14}C, was greatly enhanced in tobacco at 35° compared with 25°, and the increased [14]CO_2 release in the light arose specifically from the carboxyl-carbon atom of glycolate. This forms the basis of an independent assay for photorespiration (Chapter 6.D). Maize released [14]CO_2 poorly from glycolate-1-^{14}C in comparison with tobacco, while acetate-1-^{14}C produced similar amounts of [14]CO_2 in the light in both tobacco and maize, as suggested by the model which shows that dark respiration changes little relative to photosynthesis with increasing temperature.

Goldsworthy (1966) found that 10^{-2} M α-HPMS or isonicotinylhydrazide at pH 5 had little effect on [14]CO_2 uptake by tobacco leaf segments during a pretreatment period in the light. The [14]CO_2 released during a subsequent period when CO_2-free air was swept over the tissue had a lower specific activity in the presence of either of these inhibitors and was similar to that obtained in the [14]CO_2 released in darkness. This also indicates that a considerable portion of the CO_2 produced by a tobacco leaf in the light is derived from the metabolism of glycolate since blocking its oxidation produced CO_2 with a specific activity characteristic of dark respiration.

A test of whether glycolate fulfilled the requirement of a substrate for photorespiration was also made by placing tobacco leaves in a stream of CO_2-free air while their bases were in 10^{-2} M α-HPMS for 30 minutes in the dark and 10 minutes in the light (D. N. Moss, 1968). The inhibitor markedly diminished CO_2 evolution only from the illuminated leaves and had no effect on dark respiration. In five of six experiments the number of moles of glycolate that accumulated when the inhibitor was present was sufficient to account for the inhibition in the CO_2 released in the light.

The α-HPMS inhibits [14]CO_2 released in photorespiration but not dark respiration in the ^{14}C-assay (Chapter 5.H) under conditions where net CO_2 uptake is unaffected (Fig. 6.2), further showing that the substrates for photorespiration and dark respiration differ. A quantitative evaluation of

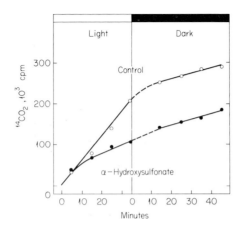

FIG. 6.2. $^{14}CO_2$ released by tobacco leaf disks in the presence of α-hydroxysulfonate at 35° (from Zelitch, 1968). At zero time, the fluid was replaced with water or 10 mM α-hydroxy-2-pyridinemethanesulfonic acid, and CO_2-free air at 7 flask volumes per minute was passed through the vessels as in Fig. 5.6. Control experiments showed stomatal widths were not affected by the inhibitor under these experimental conditions.

the fraction of CO_2 evolved by dark respiration plus photorespiration compared with the fraction of CO_2 refixed within the tissues when CO_2-free air is passed over the leaf can be made from experiments like those in Fig. 6.2. Assume that glycolate is the ultimate source of CO_2 evolution in photorespiration, 50% of it is synthesized directly from CO_2, and the evolution of CO_2 from glycolate is blocked by the inhibitor without slowing CO_2 assimilation (if the experimental time does not exceed 15 minutes). Therefore the ^{14}C present in the glycolate that accumulates with inhibitor should be at least equal to the decrease in quantity of $^{14}CO_2$ released in CO_2-free air. However, if more glycolate-^{14}C accumulates than the decrease of $^{14}CO_2$ released from the tissue, it indicates that a part of the evolved $^{14}CO_2$ is refixed and only some fraction of the evolved $^{14}CO_2$ was released to the atmosphere.

The biochemical and algebraic basis of such an analysis has already been described in detail (Zelitch, 1968, 1969b). At 35° the quantity of ^{14}C accumulating in glycolate was as much as twice as great as could have arisen from the decreased $^{14}CO_2$ output, indicating that even with open stomata and fast flow rates of CO_2-free air the $^{14}CO_2$ released represented only about 33 to 50% of the gross photorespiration. The remainder is being recycled within the leaf. At 25° and 30° relatively more of the gross photorespiratory CO_2 was released to the atmosphere than at 35°, and this explains the apparently small temperature effect on photorespiration observed in the ^{14}C-assay (Table 5.5). These results further emphasize that photo-respiration is always greatly underestimated even when it is measured by the sensitive ^{14}C assay of photorespiration (Chapter 5.H).

D. THE GLYCOLATE PATHWAY OF CARBOHYDRATE SYNTHESIS AND RELATED REACTIONS

1. ^{14}C-LABELED INTERMEDIATES

The sequence of reactions shown in Scheme 6.1 has been demonstrated in photosynthetic tissues by use of ^{14}C-labeled precursors, and the necessary enzymes for these steps have also been detected.

This pathway may provide an important route of carbohydrate syn-thesis in which losses of CO_2 by photorespiration are merely a byproduct, although quantitative estimates of the proportion of carbohydrate synthe-sized by these reactions relative to total carbohydrate synthesis have still not been made. Nevertheless, increasing the O_2 concentration in the atmo-sphere, which increases the rate of glycolate synthesis and oxidation, has often been shown to increase the labeling of glycolate, glycine, and serine in tissues supplied with [14]CO_2 in the light (Zak, 1965; Coombs and Whitting-ham, 1966a; Whittingham *et al.,* 1967; Fock *et al.,* 1969a; Voskresenskaya *et al.,* 1970a).

Inhibitors of glycolate oxidation block this pathway (Zelitch, 1958; Mifflin *et al.,* 1966), and it is inhibited by high concentrations of CO_2 in the gas phase (Tolbert, 1963). Serine and glycine are the only amino acids containing ^{14}C after short exposures (15 seconds) to [14]CO_2 in tobacco leaves (Ongun and Stocking, 1965a). The existence of this pathway was shown when glycolate-2-^{14}C supplied to wheat leaves in the light was incorporated equally into C-2 and C-3 in serine and glycerate and there was little ^{14}C in the carboxyl group, while serine-3-^{14}C produced glycerate-3-[14]C (Rabson *et al.,* 1962). The major products of serine metabolism by tobacco leaves in the light were sucrose and starch (Ongun and Stocking, 1965a), and serine-3-^{14}C labeled primarily the C-1 and C-6 of the glucose moiety of sucrose, while glycolate-2-^{14}C, glycine-2-^{14}C and glyoxylate-1,2-^{14}C labeled carbons 1,2,5, and 6 of glucose almost equally (Jimenez *et al.,* 1962; Wang and Waygood, 1962). Addition of glycine or serine together with glyoxylate-^{14}C diminished the radioactivity in sugars presumably by diluting the radio-active pools in this pathway.

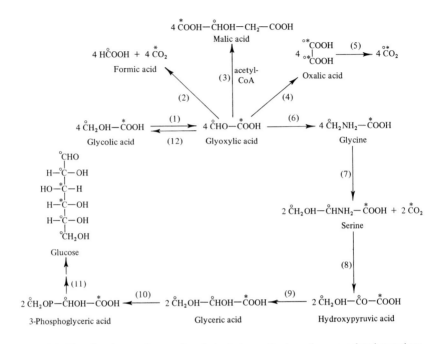

SCHEME 6.1. The glycolate pathway of carbohydrate synthesis and some related reactions. The symbols represent the fate of the carbon atoms of glycolate. (1) Glycolate is oxidized by the flavoprotein glycolate oxidase. (2) Glyoxylate is oxidized to formate and CO_2 by hydrogen peroxide or perhaps by an enzymatic reaction. (3) Glyoxylate condenses with acetyl coenzyme A in the presence of malate synthetase to produce malate. The carbonyl-carbon of glyoxylate becomes the hydroxyl-carbon of malate. (4) Glyoxylate is oxidized to oxalate by an enzyme activity that appears to be associated with glycolate oxidase. (5) Oxalate oxidase, probably a flavoprotein, catalyzes the oxidation to CO_2. (6) Glyoxylate is transaminated to glycine in the presence of glutamate. (7) Two glycine molecules react to produce one serine and CO_2, a reaction catalyzed by serine hydroxymethyltransferase. (8) Serine is transaminated to hydroxypyruvate in the presence of pyruvate. (9) Hydroxypyruvate is reduced by NADH or NADPH to D-glycerate by glyoxylate reductase. (10) Glycerate is phosphorylated by ATP and requires glycerate kinase. (11) 3-Phosphoglycerate is reduced and converted to carbohydrate by a reversal of the steps in glycolysis (Scheme 5.1). (12) Glyoxylate is reduced to glycolate by a NADH- and NADPH-specific glyoxylate reductase.

The carboxyl-carbon of glycolate would be expected to produce CO_2 readily in the light (Chapter 6.C), and most of this CO_2 would be refixed within the tissues to produce equally labeled glycolate. This would tend to obscure the labeling pattern in carbohydrate, and when glycolate-1-^{14}C (Jimenez *et al.*, 1962) or glycine-1-^{14}C (Wang and Waygood, 1962) were supplied to wheat leaves or glycolate-1-^{14}C to pea leaves (Marker and Whittingham, 1967), an even distribution of the ^{14}C within glucose was in fact observed.

2. ENZYME ACTIVITIES

Some of the enzymes shown in Scheme 6.1 have already been discussed in this chapter. Purified chloroplast preparations oxidize glyoxylate to CO_2 slowly [reaction (2) in Scheme 6.1] and purified peroxisomes did not oxidize glycolate beyond glyoxylate because of the large excess of catalase present in these particles (Kisaki and Tolbert, 1969). Whether glycolate oxidation to CO_2 occurs rapidly elsewhere in the cell or whether an enzymatic reaction occurs catalyzing the rapid oxidation of glyoxylate to CO_2 is not known.

The condensation of glyoxylate with acetyl coenzyme A to form L-malate by malate synthetase (3) is an enzymatic reaction whose equilibrium is far in the direction of malate formation. This enzyme is very active in fatty tissues, and it has been detected in mature leaves of tomato, tobacco, and barley (Yamamoto and Beevers, 1960).

The enzymatic oxidation of glyoxylate to oxalate (4) is associated with glycolate oxidase activity in leaf extracts (Kenten and Mann, 1952; Richardson and Tolbert, 1961b), but the rate of glyoxylate oxidation is much slower than the rate of glycolate oxidation. Oxalate is generally believed to be metabolically inert in leaves, but particulate preparations from leaves of sugar beet and *Bougainvillea* rapidly oxidize oxalate (5) to CO_2 with the production of hydrogen peroxide (Finkle and Arnon, 1954; Srivastava and Krishnan, 1962). Since the $^{14}CO_2$ arising from glycolate by this reaction would be equally radioactive starting with either glycolate-1-^{14}C or -2-^{14}C, and this does not occur, oxalate oxidase probably does not contribute much CO_2 during respiration in the light.

A transaminase (6) which catalyzes the conversion of glyoxylate to glycine in the presence of glutamate is widespread in plant tissues (D. G. Wilson *et al.*, 1954), and it is located in peroxisomes, as is glycolate oxidase, but although the latter activity is adequate to account for known rates of photorespiration the rate of reaction of this transaminase was found to be only 2% of the glycolate oxidase (Kisaki and Tolbert, 1969). Recently, however, it has been described as being 30% as great (Tolbert and Yamazaki, 1969). Serine hydroxymethyltransferase activity (7) has been found in extracts of pea and wheat leaves (Cossins and Sinha, 1966). The reaction involves a decarboxylation of glycine and the use of the activated one-carbon unit to react with a second glycine molecule to yield serine,

$$\text{NH}_2\overset{\circ}{\text{C}}\text{H}_2\text{—}\overset{*}{\text{C}}\text{OOH} + \text{tetrahydrofolate} \rightarrow N^5N^{10}\text{-methylenetetrahydrofolate} + \text{NH}_3 + \overset{*}{\text{C}}\text{O}_2 \quad (6.10)$$
Glycine

$$\text{NH}_2\overset{\circ}{\text{C}}\text{H}_2\text{—}\overset{*}{\text{C}}\text{OOH} + N^5N^{10}\text{-methylenetetrahydrofolate} \rightarrow$$
Glycine

$$\text{HO}\overset{\circ}{\text{C}}\text{H}_2\text{—}\overset{\circ}{\text{C}}\text{HNH}_2\text{—}\overset{*}{\text{C}}\text{OOH} + \text{tetrahydrofolate} \quad (6.11)$$
Serine

In wheat leaf extracts, they showed that Reaction (6.10) and (6.11) occur and that glycine-1-^{14}C is incorporated into carboxyl-labeled serine and glycine-2-^{14}C into the C-2 and C-3 of serine. This reaction may be responsible for the CO_2 evolution in photorespiration rather than the direct oxidation of glyoxylate, as discussed below, but its enzymatic rate in leaves has not yet been determined, and in wheat leaf extracts glycine-1-^{14}C yielded considerably more ^{14}C in serine than in the $^{14}CO_2$ produced. Perhaps this is because formaldehyde is a product of Reaction (6.10) as well as CO_2.

In *Rhodospirillum spheroides* cells the glycine decarboxylase activity was activated by added glyoxylate, and the activity increased markedly when the cells were grown in the presence of intermediates of the glycolate pathway (Tait, 1970). This enzyme is not found in isolated peroxisomes, but is apparently located in mitochondria (Tolbert and Yamazaki, 1969), hence if this is the source of photorespiratory CO_2 it would arise in the mitochondria as does dark respiration.

A transaminase in the glycolate pathway (8) converts serine to hydroxypyruvate, and the reaction occurs in wheat leaves (Kretovich and Stepanovich, 1963) and in leaf extracts of many species (Willis and Sallach, 1963). This activity (with pyruvate as the amino acceptor) is associated with peroxisomes (Tolbert and Yamazaki, 1969). The reduction of hydroxypyruvate to D-glycerate (9) is catalyzed by glyoxylate reductase in the presence of NADH and it also occurs at rapid rates in the peroxisomes (Table 6.3).

The phosphorylation of D-glycerate (10) by ATP is catalyzed by a glycerate kinase which is present in leaf extracts of a number of species, and its activity is comparable to photosynthetic rates especially in species with a rapid photorespiration (Hatch and Slack, 1969c). The enzyme is located in the chloroplast in fractions from maize leaves isolated by non-aqueous methods.

3. THE ORIGIN OF CO_2 PRODUCED IN PHOTORESPIRATION

Measurement of the rate of the $^{14}CO_2$ released into a stream of moistened air from tobacco leaves floating in 0.04 M glycolate-1-^{14}C in the light has

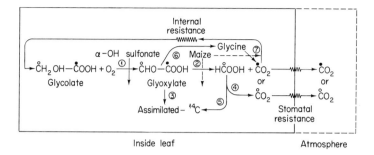

FIG. 6.3. Biochemical reactions of glycolic acid related to an assay of photorespiration based on the ratio of the $^{14}CO_2$ released from glycolic acid-1-^{14}C relative to the assimilated-^{14}C. Glycolate-1-^{14}C (●) or -2-^{14}C (○) was added to leaf disks in the light and the $^{14}CO_2$ in the atmosphere was swept out by a rapid stream of air and trapped in alkali. Reaction (1) is the glycolate oxidase reaction that is inhibited by α-hydroxysulfonate. Reaction (2) is nonenzymatic, requiring H_2O_2, or perhaps enzymatic, and is presumably slow in maize relative to (3). Reaction (3) represents all of the biochemical reactions whereby glyoxylate is converted to other products. Assimilated-^{14}C was determined from the total ^{14}C in the tissue minus that in glycolate-^{14}C at the end of the experiment. Reaction (4) represents the formic dehydrogenase reaction, known to be present in leaves (Mazelis, 1960). Reaction (5) indicates the conversion of formate into products other than $^{14}CO_2$. Reaction (6) is the transamination of glyoxylate to produce glycine. Reaction (7) represents the decarboxylation of glycine, another likely source of $^{14}CO_2$, which is also presumably slow in maize. Stomata must be open in these experiments to minimize the stomatal diffusive resistance and thus aid the release of $^{14}CO_2$.

provided an independent assay of photorespiration (Zelitch, 1966). The ratio of the $^{14}CO_2$ released to the assimilated -^{14}C increased three- to fourfold at 35° compared with 25° in tobacco, did not change appreciably with temperature in maize tissue, and the quantity of $^{14}CO_2$ evolved was only about 2% as great in maize as in tobacco. This is consistent with the low rate of photorespiration by maize in the ^{14}C-assay of photorespiration (Chapter 5.H). With the glycolate-1-^{14}C assay, a high rate of photorespiration was found in *Chlamydomonas* and in the high photorespiration JWB Mutant variety of tobacco, and a lower photorespiration was observed in the JWB Wild variety (Zelitch and Day, 1968b) which also shows low rates by other assay methods (Chapter 5.J). Under the same conditions, the $^{14}CO_2$ released by tobacco tissue from the C-2 of glycolate (or from ^{14}C-formate) did not increase at the warmer temperatures as it did from C-1, showing that the carboxyl-carbon atom is the primary source of CO_2 in photorespiration (Fig. 6.3).

If glycolate oxidase activity is found exclusively in the peroxisomes and these organelles possess a large excess of catalase (Table 6.3), the CO_2 evolved in photorespiration could not arise from the nonenzymatic oxidation of glyoxylate by hydrogen peroxide [Reaction (6.2)], although an enzymatic oxidation of glyoxylate is possible. The photorespiratory CO_2 may come from the decarboxylation of glycine [Reaction (7) in Scheme 6.1] as suggested by Tolbert and Yamazaki (1969), although some of the rates of the enzymatic reactions are at present not adequate to account for known rates of such a pathway, particularly the step involving the transamination of glyoxylate (6). The rate of the serine hydroxymethyltransferase reaction (7) has not yet been determined and only its existence in leaf extracts has been established.

Comparisons of the rate of $^{14}CO_2$ released into a stream of flowing air in the light in pea leaves supplied glycolate-1-^{14}C or glycine-1-^{14}C have been made, and both substrates were equally effective precursors of CO_2 and far superior to glycine-2-^{14}C (Marker and Whittingham, 1967). A similar finding of the equal effectiveness of the carboxyl-carbon atoms of glycine and glycolate as a source of CO_2 evolution in the light was observed with tobacco leaf disks by the present author. In some experiments glycine-1-^{14}C appeared to be a superior substrate for CO_2 release in tobacco leaf disks (Kisaki and Tolbert, 1970), but in these instances the substrates were supplied in low concentrations (about 1 mM), a closed system rather than a flowing air stream was used, and the results were expressed on the basis of the ^{14}C absorbed rather than the quantity actually metabolized. It is not yet known whether glycolate can be synthesized directly from glycine, hence experiments with radioactive glycolate and glycine have thus far failed to provide an unequivocal result indicating whether glycolate or glycine is the immediate precursor of photorespiratory CO_2. In either case, the CO_2 produced in photorespiration would ultimately be derived from the carboxyl-carbon atom of glycolate since it is synthesized first during CO_2 assimilation.

E. THE MECHANISM AND CONTROL OF PHOTORESPIRATION

Since high rates of photorespiration are well established in many species, and must surely decrease net photosynthesis greatly, the following important questions arise: (*1*) Does photorespiration have a useful function, and if not, why does it occur? (*2*) Do species which appear to have low rates of photorespiration merely have more efficient ways of refixing evolved CO_2 so that the photorespiration cannot be detected? (*3*) What are the consequences of regulating photorespiration in species with high rates and how can this be accomplished?

Photorespiration appears to be a largely wasteful process from an energetic standpoint, since CO_2 is lost and ATP may not be synthesized directly. But glycolate metabolism stimulates ATP synthesis [Reaction (6.9)] under conditions where there is an excess of reducing power (NADPH) during photosynthetic electron transport, and may also produce carbohydrate (Scheme 6.1). The reactions indicate that a penalty of at least two CO_2's (25%) must be paid for every four molecules of glycolate needed to produce one molecule of glucose. Conceivably under some circumstances all of the carbon of glycolate could be respired leaving none available for carbohydrate synthesis by this pathway. With respect to the efficiency of coupling of ATP synthesis during electron transport (Chapters 3.E and 5.A), it should be recognized that not all respiration need be tightly coupled to ATP formation, and the relative efficiency of respiration may influence plant productivity (Chapter 9.E). For example, the respiration by brown adipose tissue of mammals produces considerable heat and little ATP (R. E. Smith and Horowitz, 1969). Moreover, respiration not used to synthesize ATP may function to produce carbon skeletons essential for metabolism as do the tricarboxylic acid cycle (Chapter 5.A) and the glycolate pathway. Another function of photorespiration may be related to the consumption of O_2 during the oxidation of glycolate in higher plants. This would tend to lower the O_2 concentration in the vicinity of the chloroplasts and perhaps protect them from photooxidative damage and thereby enhance carboxylation reactions (Chapter 8.E).

The restricted glycolate synthesis and oxidation found in certain grasses and dicotyledons with low rates of photorespiration probably occurs because these species originated in tropical regions where photorespiration is more important in limiting net photosynthesis than in cooler climates. Thus selection pressures may have resulted in certain tropical species evolving systems for controlling photorespiration, including the regulation of glycolate synthesis or oxidation. We must not forget that evolution is concerned with species survival and not plant productivity or yield, and that photorespiration may have some unknown functions advantageous for the survival of species, especially at colder temperatures.

The question whether low rates of photorespiration result from rapid refixation of evolved CO_2 or low rates of CO_2 evolution would be answered differently by two schools of investigators. If low photorespiration is an artifact it must be brought about by a more efficient leaf morphology or a carboxylation system which enables refixation to occur so efficiently that CO_2 is not released rapidly. Either hypothesis has to account for the fact that photorespiration in maize is about 2% of that in tobacco tissue (Fig. 5.6).

The morphology of species showing extremes in photorespiration does differ. Species with low rates of photorespiration have a close proximity of the mesophyll and bundle-sheath cells (and have interconnections by means of plasmodesmata), and many chloroplasts are close to the vascular system (Fig. 1.1). The distribution of enzymes between the two types of chloroplasts in these efficient species (Chapter 4.F) suggests that the transfer of CO_2 from C-4 of malate to C-1 of 3-phosphoglyceric acid occurs without loss of CO_2 into the intercellular spaces (Johnson and Hatch, 1968; Slack *et al.*, 1969). The efficient tropical grasses also have a large ratio of their internal cell surface exposed to air compared to their cell volume (El-Sharkawy and Hesketh, 1965), and this would tend to favor refixation of CO_2.

However, the dimorphism exhibited by the two types of chloroplasts in maize (Fig. 2.1A, 2.1B) is not characteristic of all species with low photo-respiration, although the chloroplasts in mesophyll cells are always smaller than those in the bundle-sheath. Species possessing both types of chloro-plasts have considerable variation in their internal structures suggesting that the dimorphism has little relationship to the efficiency of photosynthetic CO_2 fixation. A characteristic peripheral reticulum or membrane is found next to the chloroplast envelope in species with low rates of photorespiration (Osmond *et al.*, 1969; Laetsch, 1970), and this structure could play a role in enhancing diffusion of CO_2.

If the low level of photorespiration in maize comes about because maize refixes its evolved CO_2, then the carboxylation reaction in maize must be about 50 times more efficient than in tobacco in 21% O_2. The overall K_m for photosynthetic CO_2 uptake in tobacco and maize in low O_2 concentra-tions (to eliminate photorespiration) is about the same, 300 ppm CO_2 in the atmosphere (about 9×10^{-6} M free CO_2 in solution), and at high illu-minance and saturating CO_2 concentrations net photosynthesis is also similar in both species even in 21% O_2 (Chapter 8.C, Table 8.3). Thus differ-ences in anatomy and carboxylation efficiency alone cannot account for the divergence in net photosynthesis in normal air by these species whereas differences in photorespiration largely do so (Table 6.5).

Experiments showing increased net photosynthesis in tobacco in air by blocking photorespiration at warmer temperatures suggest that gross photo-synthesis does not differ greatly between the efficient and the less efficient photosynthetic species, and this view is also supported by the enhancement of net photosynthesis by low O_2 in species with high levels of photorespira-tion (Chapter 8.E). Since glycolate synthesis is depressed in the more efficient species and it is known to be inhibited by high CO_2 levels (Chapter 6.A), perhaps the higher CO_2 concentration resulting from the decarboxylation of malate in the bundle-sheath in species possessing these structures inhibits the synthesis of glycolate in the chloroplasts. The efficient species may also possess a glycolate metabolism involving minimal losses of CO_2 in the glycolate pathway and a somewhat less active glycolate oxidase (Table 6.2), and this would also tend to diminish photorespiration.

Goldsworthy and Day (1970) tried another approach to this question of refixation with results which further challenge the hypothesis that it need explain apparent low photorespiration. Leaves of maize and *Phragmites* (rapid photorespiration) were supplied 0.05 M NaH^{14}CO$_3$ through the vas-

TABLE 6.6   *Some Differences between Leaves of Species with High and Low Rates of Photorespiration*[a]

	High Photorespiration	Low Photorespiration
Representative species	Most crop plants, cereals, spinach, tobacco, bean, sugar beet	Maize, sugarcane, sorghum and some weeds in *Amaranthus* and *Atriplex*
Rate of Photorespiration in various assays, CO_2 evolved	Greater than net photo-synthesis under many conditions, 3 to 5 times dark respiration	Less than 10% of dark respiration
Net Photosynthesis, mg dm^{-2} hr^{-1}, at 25°, bright light, 300 ppm CO_2, 21% O_2	10–30 (50% increase in 2% O_2)	50–70 (no change in 2% O_2)
Light intensity for maximal rate of photosynthesis	1000 to 3000 ft-c	Not saturated at 10,000 ft-c
Effect of temperature on net photosynthesis, 25° vs. 35°	No change, or less at warmer temperature	50% to 100% greater at warmer temperature
"First product" of $^{14}CO_2$ fixation	3-Phosphoglyceric acid	Oxaloacetic acid
Anatomy and Chloroplasts	No bundle-sheath with specialized chloroplasts, only mesophyll chloroplasts	Higher proportion cell surface exposed to air; larger chloroplasts in the bundle-sheath around the vascular tissue
Rate of translocation of photosynthate	Slow	Rapid

[a]It should be emphasized that some forms between these extremes are already known (Chapter 5.J), and additional ones will undoubtedly be discovered.

cular system, and the rate of $^{14}CO_2$ released in the light into the atmosphere surrounding the leaves was then measured in a gas stream. In an atmosphere of 21% O_2 and 350 ppm of CO_2, four times as much $^{14}CO_2$ leaked from *Phragmites* as from maize, but in an atmosphere of less than 2% O_2 and 350 ppm of CO_2 (to inhibit photorespiration) the $^{14}CO_2$ output was unchanged in maize and it was decreased to the same rate in *Phragmites*. These results may be explained if nonradioactive CO_2 (derived largely from photorespiration) were preferentially refixed by *Phragmites* in 21% O_2, whereas in low O_2 the $^{14}CO_2$ available within the leaf from the vascular system was fixed as well as in maize and hence less diffused to the atmosphere. It seems unlikely that more efficient refixation alone could have accounted for these experimental results in maize, since refixation must have been similar in both species at low levels of O_2. The rate of leakage of $^{14}CO_2$ supplied to both types of leaves through the vascular system in light is thus similar when photorespiration is eliminated, and maize probably has an inherently low rate of photorespiration.

In answer to the final question of the outcome if photorespiration is diminished, although such work is just beginning, there are already some plant forms known that have rates of photorespiration between the high and low rates typical of a species (Chapter 5.J). Lowering photorespiration will greatly increase net photosynthesis in many species and increase plant growth (Chapters 8.A and 9.E), provided no adverse effects are introduced by breeding or the use of biochemical inhibitors. A summary of some of the characteristics of species with extremes of high and low rates of photorespiration is given in Table 6.6.

Section III

*Photosynthesis and Plant
Productivity in Single Leaves
and in Stands*

7   Photosynthesis as a Diffusion
Process in Single Leaves

A. FICK'S LAW AND OHM'S LAW

During photosynthesis a concentration gradient exists between the CO_2 in the ambient air and the chloroplasts. The flux of CO_2 (net photosynthesis) is determined by the size of this gradient and by a series of resistances to the diffusion of CO_2 (Fig. 5.1). The rate of exchange of matter between leaves and the atmosphere surrounding them can be expressed by the equation, flux = potential difference/total resistance, where the numerator is the difference between the interior of the leaf and the ambient air. The resistances to diffusion are both within and outside the leaf. The concept of diffusive resistance was first applied to leaves by H. T. Brown and F. Escombe in 1900, and has helped in recent years to specify the resistances that are most limiting to net photosynthesis. In a system of successive or several resistances such as occurs in photosynthesis, the total resistance equals the sum of the individual parts, and these can be evaluated separately (Gaastra, 1959).

The flux of CO_2 may be analyzed by Fick's first law of diffusion as if CO_2 diffused through a series of tubes of similar diameter but differing in length,

$$P = D(X_a - X_{chl})/(L + S + M) \tag{7.1}$$

P (in gm cm^{-2}) is the rate of net photosynthesis per unit of leaf surface, and D (in cm^2 sec^{-1}) is the coefficient of diffusion of CO_2 in air (about 0.14). The X_a and X_{chl} are the concentrations of CO_2 (in gm cm^{-3}) in the ambient air and at the site of an acceptor within the chloroplasts. The L, S, and M (in cm) are the apparent path lengths for diffusion in the surrounding atmosphere, through the stomata, and through the mesophyll cells around the substomatal cavities to the sites of CO_2 assimilation. Since these path lengths are imaginary, it has become common in recent years to express the diffusion equation directly in terms of resistances and potentials and in the familiar form of Ohm's law. Thus L/D, S/D, and M/D represent the corresponding resistances (in sec cm^{-1}) in the air, through the stomata, and in the mesophyll, therefore

$$P = (X_a - X_{chl})/(R_a + R_s + R_m) \tag{7.2}$$

The "mesophyll resistance" has ceased to be a useful concept because it includes such parameters as the physical resistances, biochemical and photochemical resistances, and even dark respiration plus photorespiration. The latter are probably the largest part of the "mesophyll resistance" and are not even considered in the conceptual arrangement of Equation (7.2). Additional resistances and the total respiration I have since been identified (Fig. 5.1, from Waggoner, 1969a).

In this chapter the magnitude of the various resistances and the manner in which they may affect net photosynthesis in single leaves is discussed. The function and manipulation of the various resistances in a stand of plants is considered in Chapter 9.D. These internal resistances to diffusion of CO_2 are most important because together they exceed R_a, the resistance of the boundary layer of air, and R_s, the stomatal resistance when stomata are wide open.

B. THE LEAF BOUNDARY LAYER RESISTANCE

To visualize the diffusion of CO_2 in terms of an electrical analog, it is helpful to recall that when a potential of 1 V is maintained across an electrical resistance, then the value of the resistance (in ohms) is the time (in seconds) for the passage of unit charge. Similarly, the diffusive resistance from the surrounding air to the leaf surface (R_a in Fig. 5.1) in sec cm^{-1} equals the time in seconds for unit mass of CO_2 to diffuse from a cm^2 when the concentration difference is unit mass per cm^3. The time depends partly on the wind speed and partly on the geometry of the surface, particularly the size and shape of the leaf.

At all wind speeds, when a gas stream passes over a surface there is a thin layer of nonturbulent air adjacent to the surface where the air stream adheres and the flow is said to be laminar. This region of laminar flow constitutes the boundary layer. When the layer is thick, diffusion (of water vapor or CO_2) is slow and R_a is large because diffusion through the boundary layer is much slower than transfer through the turbulent air above the boundary layer. The boundary layer concept was taken from engineering and introduced into plant physiology by Raschke (1956, 1958), who calculated the thickness of the boundary layer of leaves and confirmed these calculations by measurement of humidity gradients close to the leaf and the temperature around the leaf. An alternative geometric method of calculating the boundary layer resistance based on a physical treatment of laminar flow has recently been described (N. C. Turner and Parlange, 1970), and results similar to Raschke's were obtained (see below).

R_a is affected by both wind speed and leaf size. Faster wind diminishes the boundary layer and decreases R_a, but the effect decreases in high wind. Also the average boundary layer and hence R_a are smaller for a narrow leaf than a wide one. This happens because the boundary layer grows from zero at the leading edge to greater and greater depths as the air moves across the leaf. Thus R_a, sec cm^{-1}, can be estimated from:

$$R_a = K \sqrt{\frac{\text{leaf width, cm}}{\text{wind speed, cm/sec}}} \qquad (7.3)$$

A survey of wind tunnel observations indicated that K is about 1.3 sec$^{1/2}$ cm^{-1} for leaves (Monteith, 1964). This K is somewhat less than is estimated for flat surfaces by engineers, presumably because of turbulence over the leaves. Some workers believe that K and hence R_a is even less for flapping leaves and the turbulent air found in a natural canopy.

To illustrate the use of Equation (7.3), consider a wind speed of 100 cm sec^{-1} traversing a leaf width of 1 cm, then R_a would equal 0.13 sec cm^{-1}; for a 4 cm leaf, R_a would be 0.26 sec cm^{-1}. [100 cm sec^{-1} = 2 miles per hour, and wind speeds less than this are seldom encountered in nature.] These values, as shown below, are reasonably close to those obtained experimentally in a number of laboratories. Pubescence or hairs on leaves have only a slight effect on the thickness of the boundary layer (Raschke, 1960).

There are primarily two different methods of experimentally estimating R_a. One is based on a calculation made from the Ohm's law equation after determining the rate of water loss by evaporation [Equation (7.4)], and the other relies on measurement of the half-life of the rate of cooling of a leaf when a heat source is suddenly removed. The evaporation method depends on the fact that during transpiration water evaporates from the moist mesophyll cells in the leaf interior, and through the stomatal openings into the atmosphere. By analogy with Equation (7.2), if there is no internal diffusive resistance in the mesophyll cells and R_m' is equal to zero, the rate of transpiration (gm cm^{-2} sec^{-1}) may be expressed by,

$$T = (X_i' - X_a')/(R_a' + R_s') \qquad (7.4)$$

where X_i' (gm cm^{-3}) is the saturation water vapor concentration at the leaf temperature in the intercellular spaces of the leaf and X_a' is the concentration of vapor in the air surrounding the leaf. These values can be determined from measurements of the leaf temperature and the relative humidity in the air. If the leaf is simulated by a piece of wet filter paper of the same shape, there will be zero stomatal diffusive resistance (R_s'). Thus by measuring the rate of water evaporation, the temperature of the paper, and the relative humidity of the surrounding air, R_a' can be calculated from Equation (7.4). Since the diffusion coefficients of CO_2 [D in Equation (7.1)] and H_2O vapor in air differ, the correction is $R_a (CO_2) = 1.7 R_a' (H_2O)$ (Gaastra, 1959).

The second method of estimating R_a is based on the principle that the boundary layer serves as a resistance to the transfer of heat as it does for the diffusion of matter. This resistance can then be derived from measurement of the rate of change of the leaf temperature after a sudden alteration of the intensity of irradiation (Linacre, 1967). With this method R_a was about 1.3 sec cm^{-1} if the half-life of the rate of cooling was 25 seconds. Measurements of leaves made outdoors generally gave R_a values below this figure, while leaves or blotting paper in still air had values of R_a as high as 3.8 sec cm^{-1}.

Even in moderate wind speed, the R_a is generally between 0.5 and 2.0 sec cm^{-1}, a small fraction of the total resistance (Table 7.2), hence lack of wind would seldom limit net photosynthesis in the field (Chapter 9.D). Thus Raschke (1958) calculated the boundary layer of *Canna* and sunflower leaves from heat transfer coefficients to vary from 0.1 to 1.7 sec cm^{-1}. The observed R_a varied from 0.5 to 0.8 sec cm^{-1} in leaves of several woody and herbaceous species (Holmgren *et al.,* 1965), was 0.5 sec cm $^{-1}$ for bean leaves (Chartier *et al.,* 1970), 0.26 sec cm^{-1} for lucerne (Begg and Jarvis, 1968), and 1.6 sec cm^{-1} for cotton leaves in slow wind indoors (Slatyer and Bierhuizen, 1964).

When surface length is constant, the aerodynamic resistance R_a is inversely proportional to the square root of the wind speed [Equation (7.3)], and the effect of wind on fixation of CO_2 is curvilinear as expected from Equation (7.3) (Decker, 1947). An estimate of the relatively small influence of wind velocity on net photosynthesis can be made from data prepared by Gaastra (1963). If a wind speed of 100 cm sec^{-1} is taken as a reference point, at lower wind velocities the effect of wind is relatively large, and net photosynthesis is reduced 20% when the wind speed is lowered to 16 cm sec^{-1}. Even at such low wind velocities R_a accounted for only about 17% of the total diffusive resistance. At a wind velocity of 100 cm sec^{-1}, R_a provided only about 9% of the total diffusive resistance, and if the wind speed were increased ten times to 1000 cm sec^{-1}, net photosynthesis would increase by only 8%.

The effect of constant wind velocities of 30, 70, 170, and 400 cm sec^{-1} has been compared on the growth of young plants of rape (*Brassica napus*), barley, and peas (Wadsworth, 1960). Since no significant difference was found in the relative growth rate, the boundary layer resistance must be small compared with the other diffusive resistances even when lower than normal wind speeds are used. In an extreme example, however, net photosynthesis in brightly illuminated sugarcane leaves at 200 ppm of CO_2 was increased greatly when the air in a cylinder was stirred vigorously compared with still air (Waggoner *et al.,* 1963); in these abnormal conditions turbulence was equivalent to increasing the external CO_2 concentration by 50% (Chapter 8.C). In the field, it is not likely that the air is ever as still as it can be made in a closed cylinder.

The diffusive resistance encountered by CO_2 entering the leaf is exerted by the stomata once CO_2 passes the boundary layer. Before the magnitude of the stomatal resistance R_s can adequately be considered, these unusual pores and the control of their widths will be examined.

C. CONTROL OF STOMATAL OPENING

1. MORPHOLOGY OF STOMATA

The outside covering of leaves consists of a waxy cuticle composed of a complex mixture of long-chain compounds (Chapter 1.A) that makes the

Fɪɢ. 7.1. Photomicrograph of the surface of a maize leaf showing the linear arrangement of open stomata. The photograph was made from a replica of a silicone rubber impression prepared as described in the text. The longitudinal distance between stomata is about 120 μ, and the stomatal openings are 25 μ long and 5 μ in width. (Photo by Fritz Goro, *Life Magazine.*)

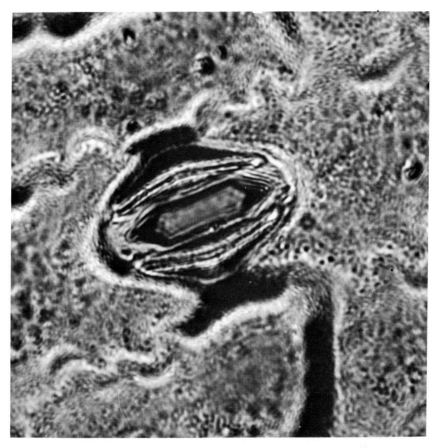

Fig. 7.2. Photomicrograph of an open stoma of a maize leaf. This represents a higher magnification made from a replica similar to Fig. 7.1. (Photo by Fritz Goro, *Life Magazine*)

leaf nearly impervious to the diffusion of CO_2 and water vapor. Therefore gases must primarily diffuse through the stomata, the microscopic pores in the leaf surface encompassed by the guard cells of the epidermis (Figs. 1.1, 1.2). The essential role of stomata in CO_2 diffusion was observed by F. F. Blackman in 1894 and was confirmed by H. T. Brown and F. Escombe in 1895. Recent measurements showed that net photosynthesis through the surface of leaves lacking stomata varied from zero to 0.1 mg CO_2 dm^{-2} hr^{-1} while net photosynthesis on the surface possessing stomata varied in different species from 4 to 14 mg CO_2 dm^{-2} hr^{-1} at normal concentrations of CO_2 (Holmgren *et al.*, 1965). Gaastra (1963) estimated the cuticular resistance in many species to be at least 35 sec cm^{-1} and Linacre (1967)

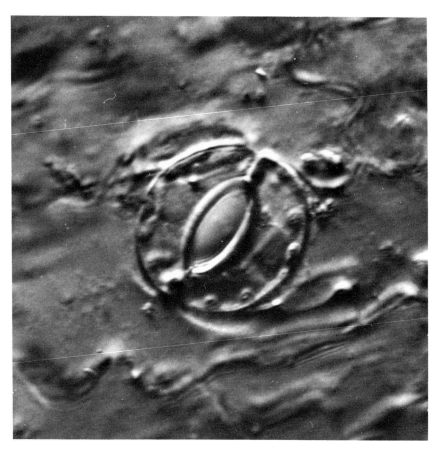

FIG. 7.3. Photomicrograph of an open stoma of a tobacco leaf in a strip of living epidermal tissue. The elliptical pore encompassed by the pair of guard cells is 20 μ long and 8 μ in width. (Photo by Fritz Goro, *Life Magazine*.)

cites various references that give values from 20 to 80 sec cm^{-1}. Since these values include leaking stomata, the cuticular resistance must be greater than 40 sec cm^{-1}. Thus the cuticular diffusive resistance is so great that leaves are dependent on stomatal apertures to facilitate CO_2 diffusion into the leaf interior. Factors that control stomatal opening will be considered first, and in the next section the consequences of altering stomatal dimensions and numbers on the uptake of CO_2 and the loss of water vapor in single leaves will be examined.

Esthetic photographs are shown in Figs. 7.1 and 7.2 of the open stomata of a grass, maize, and in Fig. 7.3 of a dicotyledon, tobacco, as they would appear to entering molecules of CO_2. These unusual structures and their mechanism of action have fascinated plant biologists for the more than 100 years since 1856 when H. von Mohl first proposed clear hypotheses to account for their movement.

The upper and lower surfaces of maize and many other grasses are similar (Fig. 1.1), and are perforated by almost equal numbers of stomata arranged linearly. In maize this density is about 10,000 stomata per square centimeter. In the light the guard cells at each side of the pore in a well-watered leaf (Chapter 8.F) separate as the regions at each end take up water and increase in turgor relative to adjacent epidermal cells. The water uptake occurs largely because of an increase that had occurred in the osmotic potential within the guard cells (Stålfelt, 1966). In most species stomata close in the dark. Thus whatever the mechanism of stomatal movement it must account for the uptake and exit of water in and out of these highly specialized guard cells. This movement in maize opens a pore 1 to 10 μ wide, 25 μ long, and about 10 μ deep and exposes the moist leaf interior to the drier atmosphere outside.

In the upper epidermis of many herbaceous dicotyledonous plants, however, pores are quite densely distributed, although generally stomata on the upper surface are fewer and they do not open as widely as those on the lower side. Thus tobacco has about 4000 pores per square centimeter of upper and 8000 per square centimeter of lower epidermis, and in dicotyledons these are arranged in a more random fashion than in grasses (Fig. 7.1). Leaves of trees have practically no stomata on their upper surface.

2. MEASUREMENT OF STOMATAL WIDTH

Because the width of the pore changes greatly while the length varies only slightly during stomatal movement, the change in width greatly affects the diffusive resistance. Several rapid and reliable methods for the assay of stomatal widths have become available in recent years. A rapid and simple method of casting leaf impressions in silicone rubber, followed by the preparation of replicas of the cast with a cellulose acetate film was described by Sampson (1961) and has been used to assay the effect of inhibitors on stomatal opening and to measure stomatal dimensions accurately (Zelitch, 1961; Zelitch and Waggoner, 1962a). This method permits numerous leaf samples to be measured easily under the microscope, but it cannot be used with accuracy when apertures are much narrower than 1 μ or with species whose stomata are recessed in pits rather than being located on the leaf surface.

Beginning with F. G. Gregory and his colleagues in 1934, physiologists have measured the porosity of leaves by forcing air through the leaves. Air is allowed to escape from a chamber clamped to a leaf, and the rate at which the pressure decreases in the chamber indicates the porosity in the leaf. A convenient viscous-flow porometer was developed by Alvim (1965); it can be easily used in the field, but the utility of this and other viscous-flow porometers that force air through the leaf is limited. They measure viscous, not diffusive, resistance and the porosity of two leaf surfaces are measured in series, not in parallel, as when vapor is diffusing in or out of the leaf. Nevertheless, such porometers have been calibrated to give diffusive resistance by applying the relation between viscous and diffusive resistance, and the results obtained were well correlated with resistances predicted from stomatal dimensions determined from silicone rubber impressions on barley leaves (Waggoner, 1965).

The most generally useful method now available is probably a diffusion porometer that rapidly senses the increase in humidity next to a leaf as water diffuses out (Wallihan, 1964). Such porometers have recently been calibrated to give absolute resistances in the conventional units of seconds per centimeter from equations derived from physical principles (N. L. Turner and Parlange, 1970). The porometer consists of two parts: a chamber containing a humidity sensor that is clamped onto a leaf (the enclosed air is dried so that water vapor diffuses from the leaf and increases the humidity in the chamber), and a meter to record the increases in humidity. Since a large resistance of the air within the chamber would obscure the resistance of the stomata, the unwanted air resistance is decreased in some instruments by a fan or by placing the sensor near the leaf. In practice, the few seconds required for the humidity to increase a fixed amount is measured at a known temperature and this allows a wide range of stomatal diffusive resistances to be measured quickly, accurately, and directly. But discussing stomatal diffusive resistance must be postponed until more is explained about the behavior of guard cells.

3. ENVIRONMENTAL FACTORS REGULATING OPENING

Hypotheses about the mechanism that enables guard cells to increase and decrease in turgor have stimulated considerable controversy in the literature. Part of the difficulty in the interpretation of experiments with guard cells comes from the dependence of guard cell function to some degree on the activities of other cells. Confusion also results because indirect effects have often been observed, since most of the factors that influence photosynthesis and water relations in leaves (such as light, O_2 concentration, temperature, leaf hydration, biochemical inhibitors, and CO_2 concentrations) also influence stomatal movement. Guard cells occupy only a small portion of the epidermis, and either intact tissues or, at best, epidermal strips must be used as experimental material.

A number of articles concerned with the mechanism and control of stomata and a concise book (Meidner and Mansfield, 1968) have been published in recent years, and some of these proposed mechanisms are mentioned briefly. Zelitch (1965b) summarized work with chemical inhibitors and reviewed the evidence for the role of glycolic acid metabolism in stomatal opening in the light; Meidner and Mansfield (1965) concluded that the CO_2 concentration in the intercellular spaces near the guard cells largely controls stomatal activity; Pallas (1966) doubted that a universal mechanism could account for stomatal movement; Levitt (1967) rejected all current interpretations and suggested that a variation of the classical explanation that an increase in soluble carbohydrate in guard cells in the light and its conversion to starch in the dark best fits the facts. Many experiments indicate that the classical hypothesis has more pedagogic than scientific merit. Zelitch (1969a) reviewed the earlier evidence obtained from several laboratories that a large increase in potassium ion concentration by means of a potassium "pump" largely accounts for the increase in solute concentration in guard cells in light and suggested that this may be the primary mechanism for opening, a hypothesis that has since received strong support (Fig. 7.4).

Irradiance is perhaps the most important factor controlling stomatal opening in most species, the exception being plants exhibiting crassulacean-type metabolism whose stomata open in the dark (Chapter 4.L), but maximal stomatal widths are obtained at light intensities considerably below those required for maximal rates of net photosynthesis (Zelitch, 1961), although photosynthesis is needed for stomatal opening. Presumably this occurs because guard cell chloroplasts are located on the outside of leaves and are immediately exposed to light, although stomata in the upper and lower epidermis do not always respond equally to irradiance (Chapter 9.D).

The chloroplasts of guard cells are smaller, contain fewer lamellae, and contain less chlorophyll than chloroplasts of the mesophyll (W. V. Brown and Johnson, 1962; Thomson and De Journett, 1970). The quantitative importance of chlorophyll was demonstrated with etiolated wheat leaves, when a close relationship was found between the stomatal light response and the chlorophyll a content of the tissue after a short exposure to light, during which time chlorophyll synthesis was initiated (Virgin, 1956). Further evidence for the importance of photosynthesis in increasing the turgor of the guard cells comes from the action spectrum for maintaining opening of stomata in isolated epidermis (Kuiper, 1964). The spectrum

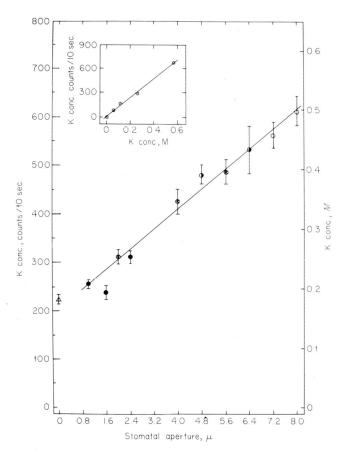

Fig. 7.4. Relation of potassium concentration in guard cells and stomatal aperture in tobacco leaves (from Sawhney and Zelitch, 1969). The determinations were carried out with an electron microprobe. Leaf disks were treated variously and then epidermal tissue was stripped from disks and rapidly frozen in isopentane and then freeze-dried. Each point shows the mean of six to eight determinations on a 1 to 2 μ diameter region of the guard cells and on a typical epidermal cell. $\triangle$, Epidermal cell; $\bullet$, light (zero hour); $\bigcirc$, light (1.5 hours); $\bigcirc\!\!\!\!\bullet$, dark (0.5 hour); $\otimes$, dark (1 hour). The insert illustrates the relation between counts per 10 seconds detected by the electron microprobe and the potassium content of a KCl-impregnated membrane filter used to calibrate the instrument.

for maintaining opening was the same as the absorption spectrum of a dilute chloroplast suspension. Stomata were closed by low concentrations of DCMU so that the effect of light is very likely on the production of ATP by photophosphorylation (Chapter 3.E) that is needed to drive the potassium "pump."

Higher temperatures have sometimes been accompanied by stomatal closure in the light, as in "midday closure," but this was probably caused by the drying of the warm leaf rather than by the higher temperature alone. Stomata of leaves of several species floated on water widened as they warmed from 5 to 35° (Stålfelt, 1962). Also when tobacco leaf disks were floated on water for 4.5 hours their stomatal widths were 2.3 μ at 10° and 7.7 μ at 30°, and when the temperatures were reversed they quickly changed to a new aperture dictated by the new temperature (Walker and Zelitch, 1963). Stomata of intact well-watered maize increased from 2 μ at 14° to about 6.5 μ at 40°, diffusive resistance decreased, and photosynthesis and transpiration increased (D. N. Moss, 1963). In other experiments, plants of a number of species were grown at day temperatures ranging from 15 to 36° in a humid atmosphere, and the stomatal widths on each epidermis were measured separately (Hofstra and Hesketh, 1969b). In bean, maize, wheat, soybean, and eucalyptus the stomata were progressively wider as the temperature increased, while there was no regular pattern in cotton leaves. For most species the stomata on the lower surface were more widely open than in the upper surface, particularly at lower temperatures.

The night temperature at which plants are grown also has an effect on the stomatal apertures. Tobacco plants kept for five nights at 16° had much narrower stomata during the day than plants at 27°, although the day temperatures were the same (Zelitch, 1963). Inhibitions of photosynthesis that have been noted after cold nights may be caused by reduced stomatal apertures that follow such low night temperatures. Although stomata respond readily to increasing temperature when leaves are well hydrated, this effect may not be observed if water deficits coincide with the increased leaf temperature (Chapter 8.F).

High concentrations of CO_2 in the atmosphere (1000 to 3000 ppm) cause stomata to close in the light (Pallas, 1965). Similar earlier observations led O. V. S. Heath in 1948 to suggest that the opening of stomata in light may be entirely due to removal of CO_2 inside the leaf by photosynthesis. More recent versions of this CO_2 hypothesis have been described by Heath *et al.* (1965) and Meidner and Mansfield (1968). A number of findings suggest that the CO_2 hypothesis does not fully account for the opening of stomata in the light in normal air or for the variations in stomatal opening that can be induced (Zelitch, 1965b, 1969a).

The effect of the CO_2 concentration in air on stomatal opening in light is small in the normal range for most species, and CO_2 concentration in the intercellular space is probably within 50 ppm of the ambient 300 ppm of CO_2 (Fig. 5.1). Changes in the viscous-flow resistance of wheat leaves, a function of stomatal width, at ambient CO_2 concentrations from 0 to

300 ppm in the light have frequently been cited to show that the internal CO_2 concentration contributes to changes in stomatal aperture (Heath and Russel, 1954; Mansfield, 1969). If viscous resistance changes as the third power of diffusive resistance (Waggoner, 1965), Heath and Russell's observations show that increasing the CO_2 level from 0 to 300 ppm increased the stomatal diffusive resistance by about 50%. Since stomatal width varies approximately with the reciprocal of the diffusive resistance (R_s) (Waggoner, 1965), at 300 ppm of CO_2 the stomatal widths may have been decreased by one-third by the 300 ppm change. There was no change in R_s at high light intensities in turnip leaves between 0 and 300 ppm of CO_2, and increasing the CO_2 level to 1500 ppm increased R_s from 2.5 sec cm^{-1} to 5.0 sec cm^{-1} (Gaastra, 1959). Since the CO_2 concentration within a crop canopy undergoing photosynthesis does not usually decrease by more than 50 ppm from the normal 300 p$_r$/m in air (Chapter 9.D), the effect of ambient CO_2 concentration on stomatal width may be considered to be unimportant in nature.

4. GENETIC AND BIOCHEMICAL REGULATION

There are varieties of grasses known in which significant stomatal opening cannot be demonstrated, and such plants are efficient during drought in their production of dry matter per unit of water transpired (Zelitch, 1965b). On the other hand, abnormal stomatal behavior has been described in which stomata remain open even in wilted leaves in potato (Waggoner and Simmonds, 1966) and in tomato (Tal, 1966). The open stomata in potato were closed by spraying the leaves with an inhibitor of opening, phenylmercuric acetate (Waggoner and Simmonds, 1966), while stomata of one of the wilty mutants of tomato were closed by spraying with a naturally-occurring growth regulator, abscisic acid (Imber and Tal, 1970). Dilute solutions of abscisic acid applied to the leaf surfaces or introduced through the base of cut leaves of normal plants closed stomata for up to 9 days (Mittelheuser and Van Steveninck, 1969; R. Jones and Mansfield, 1970). The rates at which stomata open in the light may also be under genetic control, since stomata of a yellow mutant of tobacco opened in the light twice as rapidly as leaves of the wild type (Zelitch and Day, 1968b).

Scattered observations of stomatal closure after chemical treatment have been made in the past, but these never seem to have been exploited. Perhaps this was because the observations were made casually, because they were not reproducible, or—more likely—because the inhibitors were extremely toxic. The first observations that stomata could be closed without causing a generalized toxicity to the plant appears to have been made by Mateus

TABLE 7.1   Some Inhibitors of Stomatal Opening in the Tobacco Leaf Disk Assay[a,b]

Compound	Concentration to inhibit opening 50% ($M \times 10^{-5}$)
Decenylsuccinic acid (esters)	3
Phenylmercuric acetate	5
3-(4-Chlorophenyl)-1,1-dimethylurea (CMU)	5
2-Chloro-4-ethylamino-6-isopropylamino-s-triazine (Atrazine)	10
Carbonyl cyanide m-chlorophenylhydrazone (CCCP)	10
Sodium azide (pH 4.5)	20
Sodium α-hydroxydecanesulfonate	80
Sodium 1-naphthaleneacetate	100
8-Hydroxyquinoline sulfate	100
Di-n-butyloxalate	100
Potassium iodoacetate	100
Sodium 2,4-dichlorophenoxyacetate	200
α-Hydroxy-2-pyridinemethanesulfonic acid (α-HPMS)	300
Sodium fluoride	1000
Sucrose	70,000

[a] From Zelitch, 1961, 1967; Walker and Zelitch, 1963.

[b] Leaf disks 1.6 cm in diameter were cut from tobacco leaves kept in the dark for at least one hour. They were floated right side up on water and the inhibitor solution was tested at different concentrations in 2000 ft-c of irradiance at 30° for 90 minutes. Silicone rubber impressions were then made of the lower epidermis, and stomatal widths were measured in replicas of the cast under a microscope fitted with an ocular micrometer at a magnification of × 1000.

Ventura (1954). He placed excised leaves of *Stizolobium* in dilute solutions of sodium arsenite, 2,4-dinitrophenol, or Janus green and found that all solutions decreased stomatal opening and transpiration.

The observation that α-hydroxysulfonates, effective competitive inhibitors of glycolate oxidase (Chapter 6.C), prevent stomatal opening in the light led to the development of a standard leaf disk assay for investigating stomatal movement (Zelitch, 1961). Leaf disks of tobacco were used because they have large stomata that open and close relatively uniformly. In the assay, leaf disks with closed stomata are floated on water, or the solution to be tested, for 90 minutes at 30° in the light. Stomatal apertures are then measured on replicas prepared from silicone rubber impressions of the lower epidermis. The standard assay permits a rapid determination of whether a given compound effectively prevents stomatal opening, and the large diversity of compounds that inhibit stomatal opening suggests that the process can be interrupted at a number of points (Table 7.1). With three different inhibitors, a good correlation was found between closure in the disk assay and the closure observed several hours after tobacco leaves were sprayed with solutions of these reagents (Zelitch and Waggoner, 1962a).

Although the disk assay has proven useful and sensitive to effective compounds, it has limitations: the assay does not reveal the duration of closure after a single spraying, or whether the compound will be toxic or translocated. Nor does the assay indicate whether CO_2 uptake will be inhibited in mesophyll cells. Some compounds effective in the standard tobacco disk assay have been listed together with the concentrations of each which brings about a 50% mean decrease in width (Table 7.1). Phenylmercuric acetate is effective at low concentrations and has been useful in demonstrating the role of stomatal resistance in photosynthesis and transpiration (see below).

Several compounds, including herbicides such as CMU and atrazine that are inhibitors of Photosystem II (Chapter 3.D), also effectively inhibit stomatal opening when sprayed on plants (D. Smith and Buchholtz, 1964). The phenylhydrazones of carbonyl cyanide at 10^{-4} M inhibit stomatal opening in the disk assay. Since these compounds are uncouplers of both oxidative and photosynthetic phosphorylation (Heytler, 1963), the inhibition of formation of ATP may be responsible for their activity on guard cells. Alkenylsuccinic acids and their derivatives, $CH_3—(CH_2)_n—CH = CH—CH_2—CH(COOH)—CH_2—COOH$, effectively close stomata at low concentrations in the disk assay (Zelitch, 1964a). These surface-active agents probably damage membranes through the C_8-C_{12} long hydrocarbon chain acting on the lipid portion of membranes. The monomethylester of decenylsuccinic acid, at 10^{-3} M and 10^{-4} M, inhibited electron transport and photophosphorylation as well as changes in the volume of chloroplast membranes in isolated spinach chloroplasts in the light (Siegenthaler and Packer, 1965).

A number of correlations exist between factors that affect the metabolism of glycolate and those that influence stomatal opening (Zelitch, 1969a). These include the responses of stomata to CO_2 concentration, light, O_2 levels, and the biochemically active inhibitors, α-hydroxysulfonates. Perhaps the most dramatic demonstration is that closure of stomata in the light induced by 1.8% CO_2 (an inhibitory concentration for glycolate synthesis, Chapter 6.A), was specifically reversed by adding glycolate to the tissues and not by any of a large number of other metabolites. The effect of glycolate on stomatal opening is probably attributable to the activity of the glycolate-glyoxylate cycle which reoxidizes NADPH, especially at low CO_2 concentrations, and hence stimulates the noncyclic photophosphorylation needed for a "pump" [Reaction (6.9)].

Biochemical substances are also known which either stimulate stomatal opening in the dark or prevent the normal closing in the dark. For example, sodium azide at appropriate concentrations completely inhibits stomatal closing (Walker and Zelitch, 1963), and when excised barley, wheat, or oat leaves were placed in the light in a 10^{-5} M solution of the growth regulator kinetin, stomatal opening was greatly enhanced within 2 hours (Livne and Vaadia, 1965). Fusicoccin a, a sterol-like glycoside produced by a pathogenic fungus *Fusicoccum amygdali* increased stomatal opening in the light or dark when painted on bean or tobacco leaves at 10^{-5} M (N. C. Turner and Graniti, 1969). The mechanism by which these substances enhance stomatal opening is not known, but there is some evidence that the stimulation of opening by kinetin results from a decrease in the osmotic potential and turgor induced in the epidermal cells adjacent to the guard cells rather than directly affecting the guard cells themselves (Pallas and Box, 1970).

5. ROLE OF POTASSIUM

The possible role of metal ions on stomatal opening has been examined by many investigators, but serious consideration of the possibility that potassium ions may be the active solute responsible for the increased turgor in guard cells has occurred only in recent years. Studies on the effect of a slight potassium deficiency on stomata in maize showed that while stomata in normal leaves opened about 6.5 μ, the widths in the deficient stomata were less than 1 μ (Peaslee and Moss, 1968). The resulting lower net photosynthesis was similar when normal stomata were closed by spraying a solution of phenylmercuric acetate, hence it appeared that guard cells could not compete efficiently with mesophyll cells in obtaining limiting quantities of potassium ions.

Leaves generally contain about 4% by dry weight of potassium, or a concentration of about 0.1 M (Altman and Dittmer, 1968). Guard cells of several species contained large concentrations of potassium as determined by cytochemical methods when they opened in the light, and relatively small quantities when they closed in the dark (Fujino, 1967). When epidermal strips of *Commelina communis* were floated on phosphate buffer containing 5×10^{-2} M potassium chloride, the stomata opened 7 μ, and when 10^{-2} M ATP was added to the medium, the stomata opened 14 μ and a still greater quantity of potassium accumulated in the guard cells. Fischer (1968) independently described experiments on the uptake of potassium ions (or $^{86}Rb^{+}$ as a substitute) in epidermal strips of *Vicia faba* floated on a medium containing buffer and 10^{-2} M potassium chloride. Added potassium ion stimulated stomatal opening in the light, and from a determination of the surrogate $^{86}Rb^+$ uptake, on the assumption that it occurred in the guard cells, calculations showed the increase in potassium concentration to be 0.3 M. Since the increase in solute concentration in the guard cells needed to accommodate opening has been reported to range from 0.1 M to 0.5 M in various experiments (Meidner and Mansfield, 1968), Fischer's observations suggested that the potassium concentration may account for a large part of the increase in solute in the light. The specific nature of the stimulation of stomatal opening in epidermal strips by potassium ions has been studied in further detail (Humble and Hsaio, 1969).

A quantitative test of this hypothesis was carried out on individual cells of intact leaf tissue by taking advantage of the sensitive electron microprobe technique to measure the concentration of potassium and other elements (Sawhney and Zelitch, 1969). Essentially, a narrow beam of electrons of desired diameter is focused on the specimen surface. The chemical elements irradiated by the beam emit their characteristic X-ray spectra, and the wavelength and intensity of the emitted X-rays are then analyzed, permitting quantitative analysis of volumes smaller than guard cells. The stomatal width and mean potassium concentration in the guard cells were linearly related in the quickly frozen epidermal tissue (Fig. 7.4). The concentration of potassium in guard cells of stomata about 1 μ wide was 0.21 M; in guard cells with pores 8 μ wide was 0.50 M; and in adjacent epidermal cells the concentration was 0.19 M. Hence an increase in concentration of 0.29 M potassium was observed in the guard cells during stomatal opening. Assuming that there is an accompanying anion, the increase in solute concentration would be 0.58 M, which is sufficient to account for the largest accumulation of solute estimated to be required in the literature (Meidner and Mansfield, 1968).

This movement of potassium ions into the guard cells during opening and its loss during closing must involve an exchange between the guard cells and the adjacent epidermal cells that surround the stomata. Protoplasmic connections (plasmodesmata) occur between guard cells and adjacent cells (Litz and Kimmins, 1968), and the exchange of potassium may be facilitated by these connections. Thus a mechanism of active transport of ions can reasonably account for stomatal opening in the light, and a potassium pump probably functions in guard cells of intact leaves to transport and concentrate this ion. Hence factors that affect ATP synthesis by noncyclic photophosphorylation in the guard cells (including the oxidation of glycolate) will influence stomatal opening by affecting the potassium ion pump. Studies on the nature of the anion transported to or synthesized within the guard cells during photosynthesis and on the mechanism of action of the ion pump are now urgently needed.

6. RELATIVE EFFECT ON PHOTOSYNTHESIS AND TRANSPIRATION

The diffusion of CO_2 has two resistances in common with the diffusion of water vapor during transpiration, namely the air boundary layer and the stomatal resistance [compare Equations (7.2) and (7.4)]. Photosynthesis has additional resistances beyond the substomatal cavities and the intercellular spaces of the leaf, hence the pathway for diffusion of CO_2 must have a greater total resistance than that for water vapor. Therefore decreasing stomatal width, and increasing R_s and R_s', should have a greater relative effect on transpiration than on net photosynthesis. Consequently if a compound is sprayed on the leaf to close the stomata and it acts on little else but guard cells, one would expect photosynthesis to be diminished relatively less than transpiration. This result was obtained in a number of examples indicated below.

Experiments with biochemical inhibitors of stomatal movement have provided crucial evidence of the relationship between stomatal aperture and rates of net photosynthesis and transpiration. Since it was no longer necessary to close stomata in darkness or by wilting leaves, these experiments have provided some insight into the biochemical mechanism responsible for this movement, and clearly demonstrated the regulation of diffusion of CO_2 and H_2O.

Guard cells are located on the outer surfaces of leaves; therefore the greatest chance of achieving specificity and controlling stomata with inhibitors would likely be by supplying such compounds as a spray on leaves or by floating leaf tissue in solutions. If a substance acts only on the guard cells of the leaf to close stomata, the transpiration rate should be inhibited more than net photosynthesis. On the other hand, if photosynthesis in the mesophyll cells is also seriously impaired by an inhibitor that closes stomata, the decrease in photosynthesis could be greater than in transpiration. Thus a comparison of the effect of an inhibitor on these two processes provides a simple test of the specificity of the substance on guard cell metabolism.

Phenylmercuric acetate is outstanding in that it produces stomatal closure at low concentrations (Table 7.1), and the closure induced by spraying phenylmercuric acetate on tobacco leaves lasted 2 weeks and did not greatly affect growth (Zelitch and Waggoner, 1962b). Results similar to those with tobacco were obtained when cotton leaves in the greenhouse were wet with 10^{-4} *M* phenylmercuric acetate (Slatyer and Bierhuizen, 1964). Transpiration was decreased about 40% while photosynthesis was lowered about 10%. Spraying with phenylmercuric acetate similarly affected intact maize plants (Shimshi, 1963) and grasses (Davenport, 1967), so there is little reason to doubt that stomata affect transpiration from a single leaf or an isolated plant, and closure of stomata lowers transpiration more than photosynthesis.

The role of stomatal diffusive resistance in transpiration under field conditions was first demonstrated with barley plants that were sprayed with the monomethyl ester of nonenylsuccinic acid (10^{-3} *M*) (Waggoner *et al.*, 1964). A reduction in transpiration of 20 to 30% was observed that persisted for several days, but only as long as the stomata were closed by the inhibitor.

Species differ greatly in their efficiency of net photosynthesis relative to their use of water, because of differences between plants in the magnitude of their resistances to diffusion and respiration. Thus net photosynthesis in maize leaves at $26°$ and high illuminance was 46 mg CO_2 dm^{-2} hr^{-1} and transpiration was 1058 mg H_2O dm^{-2} hr^{-1} (23 times greater than the CO_2 uptake), while for spinach these values were 21 mg CO_2 dm^{-2} hr^{-1} and 735 mg H_2O dm^{-2} hr^{-1} (35 times greater) (Bull, 1969). Maize is therefore more efficient in its CO_2 absorption per unit of water transpired, and J. D. Hesketh (1963) has observed that maize was twice as efficient in its photosynthesis per water use compared with tobacco at high irradiance.

The effectiveness of stomatal change outdoors and of increasing stream flow in watersheds has been shown by closing stomata in a forest with phenylmercuric acetate (Waggoner and Bravdo, 1967; N. C. Turner and Waggoner, 1968). They sprayed the foliage of red pine trees 16 meters tall at the beginning of the summer, observed increased stomatal diffusive resistance in the trees during daylight hours, and found a significantly higher soil moisture content around the treated trees at the end of the summer. This showed that stomatal closure is significant in controlling the hydrologic cycle and conserving soil water.

D. STOMATAL DIFFUSIVE RESISTANCE

The air boundary layer resistance (R_a) is very small compared with the total diffusive resistance to CO_2 in photosynthesis. Since stomatal widths can be controlled by environmental and especially by biochemical manipulations, the relative role of stomata in CO_2 uptake during photosynthesis deserves evaluation. In general, the stomatal resistance (R_s) is about an order of magnitude greater than the boundary layer resistance at wind speeds usually encountered outdoors, and stomata can act as an effective variable resistor to the diffusion of CO_2 (and water vapor) as has already been illustrated in Figs. 5.1 and 5.8.

There are two methods of determining R_s in addition to use of calibrated porometers (Chapter 7.C). One can assume that R_s is the same resistance for transpiration as for photosynthesis and then subtract R_a after substituting the water vapor concentration within the leaf and in the air in Equation (7.4). A constant must then be applied because of differences in the diffusivity of water and CO_2 in air. Alternately, the R_s can be calculated from stomatal dimensions and numbers per unit leaf area.

The history of studies of diffusive resistance through stomata has been summarized by Waggoner and Zelitch (1965). Earlier workers regarded the stomatal pore as a cylinder, and H. T. Brown and F. Escombe in 1900 visualized three parts to the resistance to diffusion of water vapor through stomata: the resistance inside and near the pore where lines of flow converge, the resistance inside the stomatal tube where the lines are close and parallel, and finally, outside where the lines diverge. In 1910 O. Renner added a resistance similar to R_a, which corrected for the slower diffusion from a large rather than a small area perforated by stomata. During some years of confusion, perhaps because Renner's work was overlooked, the opinion became prevalent that stomata affected diffusion only when they were closed or nearly closed.

In a landmark paper, however, Penman and Schofield (1951) explained the misunderstanding. This was reinforced by geometric analyses by Bange (1953), Lee and Gates (1964), and recently by Parlange and Waggoner (1970). Penman and Schofield, assuming that elliptical stomata function like circular stomata of equal cross-sectional area, wrote

$$R_s = \frac{1}{nD}\left(\frac{d}{\pi ab} + \frac{1}{2\sqrt{ab}}\right) \tag{7.5}$$

where n is the number of stomata per cm² (12,000 per cm², the mean of both surfaces for tobacco), d is the depth of stomatal opening (10 μ), a is the semiwidth (equivalent to the radius of a circle), and b is the semilength of the aperture (10 μ). D is the coefficient of diffusion of CO_2 (or water vapor) in air. Equation (7.5) of course shows that decreasing stomatal width over the full range of widths, or decreasing stomatal numbers will increase R_s.

Theoretical analyses of R_s were given an experimental foundation by observing transpiration from leaves whose stomatal widths were varied by

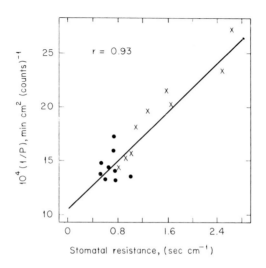

FIG. 7.5. Relation between the reciprocal of net photosynthesis (*P*) and stomatal diffusive resistance in unsprayed (●) tobacco leaves and in leaves sprayed with phenylmercuric acetate (×) (data from Zelitch and Waggoner, 1962a). Nine pairs of excised tobacco leaves were observed during a 1-hour exposure in a brightly illuminated and well-ventilated chamber. Photosynthesis was determined as $^{14}CO_2$ uptake (cpm cm^{-2} min^{-1}), and stomatal dimensions were measured (from silicone rubber replicas of both epidermes). The mean stomatal widths varied from 1 to 8 μ.

sprays of phenylmercuric acetate (Zelitch and Waggoner, 1962a). To permit a conventional regression analysis, Equation (7.2) was rewritten as

$$\frac{1}{P} = \frac{R_a + R_m}{(X_a - X_{chl})} + \frac{R_s}{(X_a - X_{chl})} \tag{7.6}$$

$$= \text{constant} + \text{constant}' \, (R_s) \tag{7.7}$$

R_s was calculated from Equation (7.5) and microscopic observations of *a*, *b*, *d*, and *n*. The reciprocal of transpiration was also written as a linear function of $R_s{}'$ [Equation (7.4)].

   The effect of stomatal width on net photosynthesis was first tested in an experiment carried out in a chamber with nine pairs of excised tobacco leaves, one member of each pair having previously been sprayed with 0.33×10^{-4} or 10^{-4} *M* phenylmercuric acetate to close the stomata. As expected from Equation (7.7), 90% of the variability of the reciprocal of net photosynthesis for these eighteen leaves was explained by the relation between the reciprocals of photosynthesis or transpiration and stomatal resistance (Fig. 7.5). A decrease in the mean stomatal width from about 8μ to 1μ increased R_s from 0.6 to 2.7 sec cm^{-1} and resulted in about a 40% inhibition of net photosynthesis and a greater inhibition of transpiration. Since similar results were obtained with intact maize plants (Shimshi, 1963), little doubt remains that throughout the range of stomatal width stomata affect the diffusion of CO_2 from a single leaf or an intact plant.

Even though the relative change in net photosynthesis per 1μ change in stomatal width will never be zero, it will not always be the same. From Equation (7.5) it can be seen that the larger the width, a, the smaller the change in either stomatal or total diffusive resistance per 1μ change. The shape of the stomatal walls may also affect the percentage change considerably (Waggoner and Zelitch, 1965), and the importance of stomatal diffusive resistance will be greater in a brisk wind with a narrow leaf when R_a is small [Equation (7.2)].

Recent refinements in evaluating stomatal geometry reveal that when the stomatal pore is more nearly a slit (Fig. 7.2) than a circle (Fig. 7.3), as is usual, the $1/(2\sqrt{ab})$ in Equation (7.5) should be replaced by $\log_e(4a/b)/\pi a$

TABLE 7.2   **Comparison of Stomatal Diffusive Resistance (R_s) for CO_2 with the Total Diffusive Resistance in Single Leaves of Various Species**

Species	R_s in bright light (sec cm^{-1})		Total diffusive resistance with open stomata sec cm^{-1}	Reference
	Stomata open	Stomata closed		
Sugar beet, turnip	2.7– 3.1	22[a]	10	Gaastra (1959, 1963)
Tobacco	0.6	2.7[b]		Zelitch and Waggoner (1962a)
Cotton	1.0– 1.5	4.5[b]	11	Slatyer and Bierhuizen (1964)
Various	1.0– 3.0			Monteith (1964)
Various	0.7–12.4[c]		4–20	Holmgren et al. (1965)
Rumex acetosa	1.7– 2.1		13	Holmgren and Jarvis (1967)
Various	0.3– 2.0			Linacre (1967)
Lucerne	2.8	35[a]	4.2	Begg and Jarvis (1968)
Bean	0.6– 2.2		5–13	Chartier et al. (1970)

[a] In darkness or during wilting. Hence this value represents primarily the cuticular diffusive resistance.

[b] Stomatal widths were narrowed (but not completely closed) by spraying or dipping leaves in a dilute solution of phenylmercuric acetate. Net photosynthesis was inhibited 25 to 40% by the treatment and the transpiration rate was inhibited more than photosynthesis.

[c] *Circaea lutetiana* had an R_s of 12.4 sec cm^{-1} and a net CO_2 uptake of only 4.4 mg CO_2 dm^{-2} hr^{-1}. In contrast, sunflower had an R_s of 0.7 sec cm^{-1} and a net CO_2 assimilation rate of 35 mg CO_2 dm^{-2} hr^{-1} under similar conditions.

(Parlange and Waggoner, 1970). These workers also showed that the outer resistance R_a is independent of individual stomatal shape and size. R_a is a function of leaf shape and size and ventilation. R_s, on the other hand, is independent of ventilation, and they also found that if interstomatal spacing is at least three times stomatal length, interstomatal interference does not affect R_s significantly. Finally, if the depth d is at least 2.5 times stomatal width, $2a$, there is negligible error in adding the three independent resistances for a stoma: one outside, and through, and one beneath the pore. For instance, for a slit one can calculate R_s with little error by adding $\frac{1}{2} \log_e (4a/b)/\pi a$ plus $d/(\pi ab)$ and again $\frac{1}{2} \log_e (4a/b)/\pi a$.

Estimates of R_s obtained from stomatal geometry or by measuring transpiration and subtracting R_a in Equation (7.4) give relatively similar results (Table 7.2). These determinations of R_s show that stomatal opening and numbers can play an important role in controlling net photosynthesis. Therefore in comparing net photosynthesis between species or in varieties within a species, it is essential to determine whether any differences encountered are brought about by alterations in stomatal aperture or numbers before seeking other explanations for such variability (Chapter 8.G).

E. INTERNAL PHYSICAL RESISTANCES AND CARBOXYLATION RESISTANCE (R_o, R_i, R_c)

1. DEFINITION AND MAGNITUDE

After passing through the stoma (Fig. 5.1), the incoming CO_2 must diffuse through two liquid-phase resistances during assimilation in photosynthesis. The R_o is located between the substomatal cavities and the junction of the incoming dark respiration and photorespiration, and it may be visualized as occurring at the cell wall and the plasma membrane (Fig. 1.4). The diffusive resistance to CO_2 through a 1 μ thickness of air is only 7.1 $\times$ 10^{-4} sec cm^{-1} and that of water about 10,000-fold greater, 6.7 sec cm^{-1} (Gaastra, 1959; Waggoner, 1967). Hence the physical diffusion of CO_2 through the liquid phase would provide an infinitely large diffusive resistance unless it were somehow diminished. The R_i is pictured as being inside the junction where respiratory CO_2 joins the main flux and represents the physical barrier that separates the CO_2 coming in from the atmosphere and from respiration from the carboxylation system. Lastly, there is a resistance produced by limitations of the enzymatic carboxylation reaction itself, R_c. The R_c will increase infinitely in darkness, and will be larger with increasing CO_2 concentration in the light when the increased flux overburdens the enzymatic carboxylation system; it will become smaller as a function of increasing temperature and increasing illuminance. These three resistances together and the effects of respiration constitute the "mesophyll" resistance calculated by Gaastra (1959) and others.

These resistances and the respiration taken together have been estimated to total from 2.3 to 14.3 sec cm^{-1} (Holmgren *et al.*, 1965), and they constitute a large proportion of the total diffusive resistance (Table 7.2) in six different species. The equivalent of R_o plus R_i was calculated to be between 4.5 and 9.1 sec cm^{-1} for bean leaves at saturating illuminance, while R_c was estimated between 0.1 and 0.6 sec cm^{-1} (Chartier *et al.*, 1970). Respiratory CO_2 was said to have little effect on the internal physical resistance, but the conceptual model used by these workers did not distinguish between R_o and R_i, and presumably mainly R_i may be affected by a large flux of photorespiratory CO_2, which could alter the diffusive properties of internal membranes.

The conception of photosynthesis in Fig. 5.1 has been incorporated into a simulator of photosynthesis by a single leaf (Waggoner, 1969a). The system incorporates the effects of light, temperature, CO_2 concentration, photorespiration and dark respiration, as well as the diffusive resistances of the boundary layer, the stomata, and the internal physical and carboxylation resistances. The simulator mimics net photosynthesis in a normal leaf in a number of ways when different parameters are altered.

For a standard case, a simulated tobacco leaf, the following values were used at 30°, 300 ppm of CO_2, and saturating irradiance: $R_a = 0.3$ sec cm^{-1}, $R_s = 3.2$ sec cm^{-1}, $R_o = 2.0$ sec cm^{-1}, $R_i = 2.0$ sec cm^{-1}, $R_c = 1.4$ sec cm^{-1} (total diffusive resistance = 8.9 sec cm^{-1}); photorespiration = 11 mg CO_2 dm^{-2} hr^{-1}, dark respiration = 4 mg CO_2 dm^{-2} hr^{-1}, maximal gross photosynthesis = 240 mg dm^{-2} hr^{-1}. In the standard example, net photosynthesis was a reasonable 16 mg dm^{-2} hr^{-1}, and the simulated leaf responded to irradiance and CO_2 concentration as would be expected (Chapter 8.B, 8.C). Decreasing the physical resistance, R_o, from the standard 2.0 to 0.5 sec cm^{-1} in the model increased net photosynthesis by 20%, while decreasing the internal resistance, R_i, from the standard 2.0 to 1.0 increased net photosynthesis 24%. Hence R_o and especially R_i, both seem likely areas where increases in net photosynthesis might be achieved. If total respiration in the light is large, decreasing R_i and R_o is very effective, but if respiration is small (as in maize) diminishing any of the resistances in Fig. 5.1 will have a large effect on increasing net photosynthesis. The carboxylation resistance, R_c, is affected by the maximum gross photosynthesis, the response of carboxylation to light and CO_2 concentration, and leaf temperature. Because of the many other resistances, however, even doubling the maximal gross photosynthesis in a plant like tobacco would only increase net photosynthesis by 15%; hence this does not seem to be a very effective factor. Moreover, there is ample confirmatory evidence from nature that although net photosynthesis may vary more than 10-fold between species, there is relatively little difference in their maximum gross photosynthesis (Table 8.3).

2. REGULATION BY LEAF ANATOMY AND MORPHOLOGY

Just as leaf morphology and the stomatal arrangement will influence the boundary layer and stomatal resistance, they may also affect the internal physical resistance, with possible large consequences upon net photosynthesis as discussed above. Having a larger proportion of the internal cell surface exposed to air would tend to diminish the physical resistances, and leaves of the photosynthetically efficient tropical grasses apparently have such an adaptation (El-Sharkawy and Hesketh, 1965). The compact arrangement of small cells containing chloroplasts radiating from the bundle-sheath in certain plants would tend to decrease the diffusion resistance, and this is also characteristic of the more efficient species (Table 6.6). Leaf thickness may affect the diffusion resistance although it is not clear what the outcome might be. As thickness increases, CO_2 assimilation would increase if more chloroplasts and a larger cell surface were exposed to CO_2 without accompanying increases in respiration and path lengths, but J. D. Hesketh (1963) could find no correlation between net photosynthesis and thickness of the lamina in a comparison of nine species with greatly different rates of photosynthesis.

The physical diffusive resistance of the chloroplasts themselves would affect the R_i, and differences in chloroplast morphology, the chemistry of the outer membrane, and especially the presence of an additional peripheral membrane system (Chapter 6.E, Fig. 2.1B) could affect CO_2 diffusion. The chloroplasts of aquatic plants such as *Elodea* and *Lemna* have the ability to rearrange themselves (Virgin, 1964) and become oriented in such a way as to increase the efficiency of light absorption, and this may enhance carboxylation at low light intensities in these plants. Since CO_2 must be dissolved in the cytoplasm (where it is presumably present largely in the form of bicarbonate), protoplasmic streaming may play a role in minimizing the internal physical diffusive resistance.

3. EFFECT OF CARBONIC ANHYDRASE

Since free CO_2 is the form in which carboxylation occurs in at least two of the enzyme systems known to function in photosynthesis (Chapter 4.E and 4.F), the solubility of CO_2 itself presents some problems. For example, at equilibrium at 25° and 760 mm atmospheric pressure and 300 ppm of CO_2 in the atmosphere, the concentration of free CO_2 in water is only 9×10^{-6} M. According to the Henderson–Hasselbach equation (Chapter 4.E), if the cytoplasm had a pH of only 6.3, at equilibrium there would be equal concentrations of free CO_2 and bicarbonate ion in solution, at pH 7.3 there would be 10 times as much bicarbonate, and at pH 8.3 100 times as much. The pH of the cytoplasm of leaf cells is not known, but it is presumably at pH 7 or higher. Thus it would be a great advantage for the plant cell to store CO_2 as bicarbonate (which is vastly more soluble than free CO_2), and the enzyme carbonic anhydrase (Chapter 4.E) if present at or near the plasma membrane (Fig. 1.4) would catalyze the hydration of free CO_2 to bicarbonate. Carbonic anhydrase would therefore act as a concentrating mechanism and diminish R_o. Since free CO_2 must be fixed, however, it would also be advantageous to have carbonic anhydrase at or near the outer chloroplast envelope membrane to hasten the production of free CO_2 for carboxylation, and thus decrease R_i. Although such a mechanism seems reasonable, present knowledge does not allow us to evaluate its capacity to function in living tissues.

Model studies have demonstrated the role of carbonic anhydrase on the diffusion of CO_2 through membranes composed of silicone rubber in which $NaH^{14}CO_3$ solution was on the donor side of the membrane at pH 7.4 (Broun *et al.*, 1970). Neither water nor ions (including bicarbonate) could cross the hydrophobic membrane, but CO_2 could diffuse quite easily. Carbonic anhydrase was fixed to the membrane with glutaraldehyde on one or both sides, and when the enzyme was on the donor side the rate of CO_2 transport increased 1.5 times, and the rate doubled when the enzyme was on both sides.

G. O. Burr in 1936 had calculated that the rate of CO_2 uptake in leaves was more than 200 times greater than the rate of uncatalyzed hydration of CO_2 to bicarbonate, and concluded that carbonic anhydrase was essential although he could not detect it. The enzyme has since been found in many species, and carbonic anhydrase from parsley leaves has recently been purified to homogeneity and found to account for 1% of the soluble leaf protein (Tobin, 1970). It has a molecular weight near 180,000 and there is 1 gm-atom of zinc per 30,000 gm of protein.

Chlamydomonas has 20 times as much carbonic anhydrase activity when the cells are grown in 300 ppm rather than in 10,000 ppm CO_2 (Nelson *et al.*, 1969). Other results suggesting a role of the enzyme in facilitating CO_2 transport come from experiments with *Chlorella* (Reed and Graham, 1968). Low photosynthetic rates were observed initially at low CO_2 concentrations when the cells were grown on 5% CO_2, and this subnormal rate was correlated with low activities of carbonic anhydrase which increased greatly during the period of adaptation to low CO_2 concentrations. The role of carbonic anhydrase in higher plants is less clear, although a comparison has been made of the activity in some photosynthetically efficient (maize, sugarcane, sorghum, *Amaranthus*) species and some less-efficient species (spinach, beet, wheat, pea) (Everson and Slack, 1968). Surprisingly, much less activity was found in the efficient species, but extracting enzymes from the efficient species is not always easily accomplished (Chapters 4.F and 6.B). Hence these results as well as the finding that carbonic anhydrase is located mostly in the chloroplast fraction of the less-efficient species and predominantly in the cytoplasm of the efficient ones must presently be regarded as tentative. The exact location and possible function of carbonic anhydrase in facilitating CO_2 diffusion therefore still awaits further experiment.

8 Environmental and Physiological Control of Net Photosynthesis in Single Leaves

A. VARIATION IN NET PHOTOSYNTHESIS BETWEEN AND WITHIN SPECIES

A constant rate of CO_2 uptake is achieved by leaves in a constant environment after about 30 minutes when the stomata open, and this rate will remain unchanged for many hours. If an environmental factor such as irradiance is altered, the net photosynthetic rate may also be changed, but on return to the original conditions the initial CO_2 uptake will be restored. However, large differences exist in the rate of CO_2 assimilation in air at saturating light intensities among common species (Table 8.1).

The most photosynthetically efficient species are monocotyledons and dicotyledons with low rates of photorespiration (Chapter 5), such as maize, sugarcane, sorghum, Bermuda grass, and *Amaranthus,* and these commonly have CO_2 uptake rates of 40 to 60 mg CO_2 dm^{-2} hr^{-1} although exceptions within these species have been found (Table 8.2). There are also some species

TABLE 8.1   *Typical Rates of Net Photosynthesis in Single Leaves of Various Species at High Illuminance and 300 ppm of CO$_2$ in Air*

Species	Net photosynthesis, mg CO$_2$ dm^{-2} hr^{-1}	Reference
Maize	46–63	J. D. Hesketh (1963); J. D. Hesketh and Moss (1963); Waggoner *et al.* (1963); El-Sharkawy and Hesketh (1965)
Sugarcane	42–49	J. D. Hesketh (1963); J. D. Hesketh and Moss (1963)
Sorghum	55	El-Sharkawy *et al.* (1967)
Bermuda grass, *Cyandon dactylon*	35–43	Murata and Iyama (1963)
Pigweed, *Amaranthus edulis*	58	El-Sharkawy *et al.* (1967)
Sunflower	37–44	J. D. Hesketh (1963); J. D. Hesketh and Moss (1963); Waggoner *et al.* (1963); J. Hesketh (1967)
Cattail, *Typha latifolia*	44–69	McNaughton and Fullem (1970)
Tobacco	16–21	J. D. Hesketh and Moss (1963); El-Sharkawy and Hesketh (1965); J. Hesketh (1967); Table 5.1.
Sugar beet	24–28	Gaastra (1959); Hofstra and Hesketh (1969a)
Orchard grass	13–24	J. D. Hesketh (1963); Murata and Iyama (1963)
Wheat	17–31	Murata and Iyama (1963); J. Hesketh (1967)
Bean	12–17	Hesketh and Moss (1963); J. Hesketh (1967)
Oak	10	J. D. Hesketh (1963)
Maple	6	J. D. Hesketh (1963)
Dogwood, *Cornus florida*	7	Waggoner *et al.* (1963)

with high rates of photosynthesis which may be efficient for some reason not directly associated with low photorespiration according to present knowledge, and two examples, sunflower and cattail, are shown in Table 8.1. Sunflower leaves presumably have a high photorespiration since they show the Warburg effect (Table 8.5) and release CO$_2$ in the light (Fig. 5.4). Wolf (1969) listed seventy-seven species having rates in excess of 20 mg CO$_2$ dm^{-2} hr^{-1}, and about half of these had rates of 40 or more. These in-

TABLE 8.2   *Variation of Net Photosynthesis of Single Leaves Within a Species at High Illuminance and 300 ppm of CO_2 in Air*

Species	Plants compared	Range of net photosynthesis (mg CO_2 dm^{-2} hr^{-1})	Reference
Maize	15 Inbreds and hybrids	28–85	G. H. Heichel and Musgrave (1969); Table 5.8
Maize	20 Races	48–59	Duncan and Hesketh (1968)
Rice	3 Varieties	12–30	Tanaka *et al.* (1966)
Rice	50 Varieties	34.5–62	Chandler (1969a)
Barley	7 Varieties	17.7–21	Apel and Lehmann (1969)
Bean	5 Varieties	13.7–18.5	Izhar and Wallace (1967)
Soybean	36 Varieties	12–24	Curtis *et al.* (1969)
Soybean	9 Varieties	18.7–23	Dreger *et al.* (1969)
Soybean	20 Varieties	29–43	Dornhoff and Shibles (1970)
Gossypium	26 Species	24–40	El-Sharkawy *et al.* (1965)
Tobacco	3 Varieties	11–17	Zelitch and Day (1968b); Table 5.8
Tobacco[a]	5 Varieties	9.0–13	Avratovscukova (1968)

[a] Leaf disks floated on water and exposed to 1% CO_2.

cluded plants known to have low rates of photorespiration and many others in which this process has not yet been investigated.

Species with lower photosynthetic efficiency on a leaf area basis, with rates of CO_2 uptake between 12 and 28 mg CO_2 dm^{-2} hr^{-1}, include those with high rates of photorespiration and encompass many common agricultural crops such as tobacco, sugar beet, orchard grass, wheat, and bean. Finally, leaves of woody plants often have rates between 7 and 10 mg CO_2 dm^{-2} hr^{-1}, and maple leaves also have an unusually high CO_2 compensation point (Chapter 5.I) of 145 ppm (D. N. Moss, 1962a), indicating that they have a high internal diffusive resistance to CO_2 assimilation or an unusually rapid photorespiration, or both.

Differences in photosynthetic rate as large as threefold also occur among varieties of maize (Table 8.2), with a range from 28 to 85 mg CO_2 dm^{-2} hr^{-1} at high illuminance in air. Some varieties of rice and soybean have twice the CO_2 absorbing efficiency in their leaves as others, while 50% differences have been observed between varieties in tobacco. Variations in the diffusive resistance or biochemistry between such varieties that may account for enhanced photosynthetic rates have generally not been described except for one example in tobacco where low rates of photorespiration were associated with greater net photosynthesis and growth (Table 5.8).

B. EFFECT OF IRRADIANCE

The photochemistry of photosynthesis (Chapter 3.A) requires visible light of wavelengths between 400 and 700 nm, which is absorbed primarily by chlorophyll (Figs. 2.2 and 2.4). The absorption spectrum of leaves (Fig. 3.1) is thus similar to the absorption spectrum of chlorophyll in solution or more nearly like that of the chlorophyll–protein complex. The photosynthetic action spectrum for higher plants has recently been determined for radish leaves (*Raphanus sativus*) (Chapter 8.E) (Bulley *et al.,* 1969) and for bean leaves (Balegh and Biddulph, 1970). The latter examined CO_2 uptake at steps of 12.5 nm with changing light intensity and measured CO_2 exchange during 10-minute intervals of light and darkness. Two peaks of photosynthetic activity were found in the red at 670 and 630 nm and a blue peak at 437 nm. The CO_2 absorption declined rapidly at wavelengths longer than 680 down to 700 nm, and the rate decreased progressively from 437 to 400 nm. There was thus great similarity between the action spectrum and the absorption spectrum shown in Fig. 3.1, especially at wavelengths longer than 500 nm, and the similarities are the more remarkable because, under the experimental conditions used, the stomata were probably not well opened and there could have been an effect of wavelength on stomatal opening that would have affected the rates of photosynthesis observed.

The visible region of the solar spectrum provides the energy source for photosynthesis outdoors. The solar constant is the flux density of total solar radiation at the boundary of the atmosphere at the earth's mean distance from the sun, and it equals 1.94 cal cm^{-2} min^{-1} (Sanderson and Hulburt, 1955). In passing through the atmosphere, much of the radiation is absorbed and scattered by atmospheric water vapor, CO_2, ozone, and dust so that about 1.3 cal cm^{-2} min^{-1} of radiation reaches the earth's surface (Gates, 1965). Most (52%) of this radiation has a complex spectral distribution in the infrared region at wavelengths greater than 700 nm. Since the atmosphere does not greatly affect the solar spectrum between 400 and 700 nm, a surface perpendicular to the sun's rays at the earth's surface will receive 44% of the solar radiation on a clear day in this region, while 4% is in the ultraviolet region. Therefore about 0.6 cal cm^{-2} min^{-1} of incident radiation is available for photosynthesis. In the visible region, 1 watt m^{-2} = 10^3 erg sec^{-1} cm^{-2} = 1.43×10^{-3} cal cm^{-2}; it follows that full sunlight containing 0.6 cal cm^{-2} min^{-1} of visible radiation is equivalent to 42×10^4 erg sec^{-1} cm^{-2}, which is approximately 10,000 ft-c or 108,000 lux.

In single maize leaves less than 10% of the solar radiation (400 to 700 nm) is transmitted through the leaves in the blue and red regions and 15%

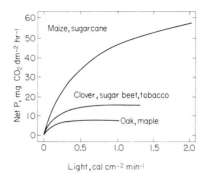

Fɪɢ. 8.1. The effect of light intensity on net photosynthesis by various species at 30° and 300 ppm CO_2 in air [adapted from J. D. Hesketh (1963); J. D. Hesketh and Moss (1963)]. About 1.2 cal cm^{-2} min^{-1} of the artificial illumination used is equivalent to full sunlight at wavelengths between 400 and 700 nm (0.6 cal cm^{-2} min^{-1}).

in the green, while 10 to 20% of the incident light is reflected (C. S. Yocum *et al.*, 1964). About 50% of the near-infrared radiation is also transmitted.

Kinetic models have been presented by Rabinowitch (1951) to account for the rate of photosynthesis as a function of incident radiation, and these various models lead to quadratic equations. At low irradiances (below 10^4 erg sec^{-1} cm^{-2}) photosynthesis is linear with irradiance, hence the photochemical processes are limiting. This is confirmed in practice since rates of CO_2 assimilation are about the same for all species and environments in this region of irradiance (Gaastra, 1962). At these low irradiances the maximum efficiency of light energy conversion (moles of CO_2 absorbed/quanta absorbed, or calories carbohydrate formed/calories incident visible irradiation) of about 12% is achieved.

With irradiance greater than 10^4 erg sec^{-1} cm^{-2}, a less rapid phase of photosynthesis occurs until saturation is reached (Fig. 8.1). Light saturation varies greatly with species as does the rate of net CO_2 absorption, and the higher the light intensity for saturation the greater the net CO_2 uptake. Thus single leaves of deciduous trees are saturated at less than one-fifth and many crop species at one-fourth of full sunlight, while light saturation has not yet been achieved at intensities twice that of sunlight in species like maize, sugarcane, and sorghum. At light saturation, sun-adapted species have considerably higher rates of photosynthesis than shade-adapted species, and these differences may exceed 15-fold in *Plantago lanceolata* compared with *Trillium oratum* (Björkman, 1968). These differences in light-saturated CO_2 uptake by clones from high and low irradiance in *Solidago virgaurea* may be caused by differences in stomatal as well as internal diffusive resistances

(Holmgren, 1968). In a pea mutant deficient in chlorophyll that is not saturated even at 113,000 lux, there is some evidence that the photosynthetic unit (Chapter 3.A) is more efficient because it contains more cytochromes on a chlorophyll basis than normal (Highkin *et al.*, 1969), but aside from this suggestion of a faster electron transport there has been little work attempting to explain the lack of light saturation that is commonly observed in the highly efficient CO_2 absorbing species.

C. INFLUENCE OF CO_2 CONCENTRATION

"Light saturation" is removed by increasing the ambient CO_2 concentration, suggesting that diffusive resistance limits photosynthetic CO_2 uptake under these conditions (Fig. 8.2). At saturating light, net photosynthesis increases as an approximately linear function of CO_2 concentration up to about 300 ppm (Fig. 5.3). Increasing the CO_2 concentration from 300 ppm to a saturating level of 1000 to 1500 ppm doubles or triples net photosynthesis in leaves of most species (Fig. 8.3), except in those cases where the internal diffusive resistances are exceedingly large, as might be expected from Equation (7.2). Even in leaves of maize in full sunlight, net photosynthesis was 50% greater at 500 ppm of CO_2 than at 300 ppm (J. D. Hesketh and Moss, 1963). Increasing the CO_2 concentration to saturating levels tends to close stomata and therefore increase the stomatal diffusive resistance (Chapter 7.C), but this handicap is easily overcome when the CO_2 concentration gradient becomes large enough.

Species differ greatly in their net photosynthesis in normal air in high illumination (Table 8.1), but at saturating CO_2 concentration at high irradiation they differ only slightly (Table 8.3). Thus at 150 ppm of CO_2, maize fixed CO_2 at 2.7 times the rate of tobacco, but at 1000 ppm of CO_2 maize

TABLE 8.3   *Effect of CO_2 Concentration on Rate of Net Photosynthesis in Single Leaves of Maize and Tobacco[a,b]*

CO_2 Concentration (ppm)	150	300	600	1000
CO_2 Uptake maize (mg dm^{-2} hr^{-1})	27	47	74	92
CO_2 Uptake tobacco (mg dm^{-2} hr^{-1})	10	24	48	69
Ratio: maize/tobacco	2.7	2.0	1.5	1.3

[a] Adapted from J. D. Hesketh, 1963.

[b] The experiments were conducted at a light intensity equivalent to twice sunlight (400 to 700 nm) at leaf temperatures between 25° and 31°.

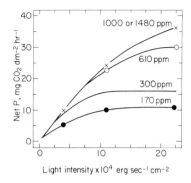

FIG. 8.2. Net photosynthesis of intact young spinach plants in relation to light intensity at different CO_2 concentrations (from Gaastra, 1959). Leaf temperature was 21 to 24° and irradiance is given for the range from 400 to 700 nm.

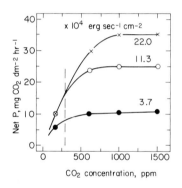

FIG. 8.3. Net photosynthesis of intact young spinach plants in relation to CO_2 concentration at different light intensities (from Gaastra, 1959). Conditions as in Fig. 8.2, with light intensities of 3.7, 11.3, and 22.0 × 10^4 erg sec^{-1} cm^{-2} as shown. The normal CO_2 concentration of air is represented by the dashed line.

leaves were only 1.3 times better. Therefore the total capacity for CO_2 assimilation (the maximal velocity, V_{max}, in terms of enzyme kinetics) differs relatively little in spite of differences that may exist in the types of carboxylation reactions that occur in different plants (Chapter 4). The more photosynthetically efficient species become even more superior as the CO_2 concentration in the air approaches the CO_2 compensation point (D. N. Moss, 1962a), as would be expected if the CO_2 concentration gradient [Equation (7.2)] remained relatively high in the efficient species because of a lower rate of CO_2 evolution by photorespiration.

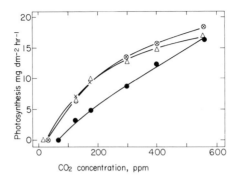

FIG. 8.4. Effect of O_2 concentration on net photosynthesis in tobacco and maize at various CO_2 concentrations (from Zelitch, 1969b). Leaf disks (3.2 cm diameter) were floated upside down on water in light for at least 30 minutes in air and net photosynthesis was measured in a closed system at about 30° and 3,000 ft-c of illuminance. After measurements were made in air at 1800 ppm, the CO_2 concentration was lowered with an absorbent to 600 ppm. After several 5-minute measurements near 600 ppm, the leaf disks were allowed to deplete the CO_2 to the CO_2 compensation point. The system was then flushed with nitrogen (final concentration of O_2 was less than 2%), and the measurements were then repeated. Each point on the curve is the mean of three experiments. Stomatal opening was similar in both species under the conditions used. The overall K_m was determined graphically. (●), Tobacco in air ($K_m = 488$); (○), tobacco in 2% O_2 ($K_m = 336$); (×), maize in air ($K_m = 307$); (△), maize in 2% O_2 ($K_m = 284$). Only the data from 550 ppm to the compensation point are shown since the rates at 1800 ppm were all about the same for all tissues and conditions, 22 mg CO_2 dm^{-2} hr^{-1}. (Reprinted from *Physiological Aspects of Crop Yield*, page 213, 1969, and reproduced with permission of the publisher.)

Since great differences in their carboxylation efficiency do not exist between species that differ greatly in CO_2 assimilation in normal air at high light intensities, it is of interest to determine if their affinity for CO_2, the K_m or CO_2 concentration to give half-maximal rate, differs. Goldsworthy (1968) found that if net photosynthesis was measured at about 2% O_2 in the ambient atmosphere, there was little difference in the overall K_m, about 300 ppm of CO_2, between maize or sugarcane and tobacco. These experiments were repeated and extended and the results were confirmed as shown in Fig. 8.4. The overall K_m for maize in 2% O_2 or in air and for tobacco in 2% O_2 was about the same and was close to 300 ppm of CO_2. The overall K_m for tobacco in air was considerably higher, 488 ppm CO_2. Since net photosynthesis of tobacco in low O_2 is similar to maize at all concentrations of CO_2, differences in the internal diffusive resistances or the affinity for CO_2 between these species cannot account for most of their differences in net photosynthesis in normal air. At higher illumination than was used in the experiments in Fig. 8.4, even in 2% O_2 maize is superior to species such as tobacco (Chapter 8.E), but probably more than half of the differences result from a lower photorespiration in maize. The overall K_m for CO_2 uptake in a number of varieties of maize has been determined (from experiments carried out at CO_2 concentrations not exceeding 300 ppm) at an illumination equivalent to full sunlight in air (G. H. Heichel and Musgrave, 1969). Although there was some variation, most of the K_m values were still close to 300 ppm of CO_2, and the effect of CO_2 concentration on stomatal width was probably slight in the range of CO_2 levels used by Heichel and Musgrave (D. N. Moss *et al.,* 1961; D. N. Moss, 1963).

A concentration of 300 ppm of CO_2 in air is equivalent to about 9×10^{-6} M free CO_2 in solution (Chapter 7.E), thus the overall K_m for CO_2 by intact leaves is probably lower than that for chloroplasts (1 to 3×10^{-5} M, Chapter 4.D), for isolated ribulose diphosphate carboxylase (4.5×10^{-4} M, Chapter 4.E), or phosphoenolpyruvate carboxylase (1.5×10^{-5} M, Chapter 4.F). This lower overall K_m for leaves is observed in spite of the greater diffusive resistance to carboxylation (through stomata and in the cytoplasm) encountered by leaves in comparison with isolated chloroplasts or enzymes. Therefore at the present time isolated preparations appear inferior to intact leaves in their affinity for CO_2 and in their total activity compared with photosynthesis in intact tissues.

D. DEPENDENCE ON TEMPERATURE

Increasing the temperature $10°$ in the range between 20 and $35°$ approximately doubled dark respiration (Chapter 5.A) and more than doubles photorespiration (Chapters 5.E, 5.I, 6.C). Since temperature will likewise affect enzymatic carboxylation reactions and stomatal opening (Chapter 7.C), it is not surprising that temperature controls net photosynthesis. Only the manner in which this control differs in various species is somewhat unexpected.

The energy budget of a leaf is the sum of the incoming radiation and the losses by convection and transpiration. Because the energy budget is regulated by diffusive resistances, as is net photosynthesis (Chapter 7), the temperature of a leaf in sunlight depends on the air temperature, the wind speed, the dimensions of the leaf surface, and the stomatal aperture. Calculations given in Table 8.4 show that with moderate wind speeds and at least partly open stomata, leaf temperatures in sunlight will be similar and slightly warmer than the air temperature. Observations in nature agree

TABLE 8.4   *Influence of Wind Speed on the Leaf to Air Temperature Difference for Various Stomatal Widths*[a,b]

Leaf length (cm)	Wind speed (cm sec⁻¹)	Temperature difference between leaf and air (°C)		
		0μ Width	1μ Width	5μ Width
5	5	14.5	5.0	1.4
	223	3.6	1.1	−1.8
15	5	16.7	6.1	2.9
	223	4.8	1.4	−1.8

[a] Data from Waggoner and Zelitch, 1965. (Copyright 1965 by the American Association for the Advancement of Science).

[b] These examples pertain to a leaf in sunlight that would gain 0.6 cal cm⁻² min⁻¹ by radiation on its two surfaces in air at a temperature of 30° and 50% humidity. The calculations are shown for a 5 cm sec⁻¹ (0.1 mile/hr) and a 223 cm sec⁻¹ breeze (5 miles/hr). The stomata in the example are 10 μ deep, have elliptical openings 25 μ long, are of variable width, and have 10,000 stomata/cm².

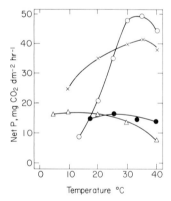

FIG. 8.5. Effect of temperature on net photosynthesis in single leaves of various species at 300 ppm of CO_2 in air at high illuminance. ($\bigcirc$), Maize (from D. N. Moss, 1963, and Hofstra and Hesketh, 1969a); ($\times$), Bermuda grass (from Murata and Iyama, 1963); ($\bullet$), tobacco (from Decker, 1959a, and J. Hesketh, 1967); ($\triangle$), wheat (from Murata and Iyama, 1963, and Joliffe and Tregunna, 1968).

with these calculations (Waggoner and Shaw, 1952; Slatyer and Bierhuizen, 1964; Drake *et al.*, 1970).

Some representative examples of the effect of temperature on net photosynthesis are illustrated in Fig. 8.5. Most species such as wheat, tobacco, sugar beet, soybean, and orchard grass do not have a very pronounced temperature optimum at high light intensities in air between 10 and 35°

(Murata and Iyama, 1963; Hofstra and Hesketh, 1969a; Table 5.1), the maximum being only about 20% greater than the minimum in this temperature range. These species all have high rates of photorespiration, and the increasing photorespiration as the leaf warms probably masks the increase in gross photosynthesis that might otherwise be observed (Table 6.5). However, net photosynthesis in species with a low photorespiration, such as maize, approximately doubles for every 10° rise between 15 and 35°, indicating that gross CO_2 uptake is very temperature-dependent when photorespiration is low. In maize and Bermuda grass the temperature optimum is about 35°, and there is a rapid decline of CO_2 absorption at higher temperatures. At higher temperatures, decreases in net photosynthesis may be related to stomatal closing associated with losses in leaf hydration and turgor. Thus the photosynthetically efficient species become relatively more efficient at warmer temperatures, while at temperatures of about 15° the rate of CO_2 absorption may be lower in maize than in the ordinarily less-efficient species.

E. INHIBITION BY O_2 (WARBURG EFFECT)

In 1920 O. Warburg observed that the rate of photosynthesis in *Chlorella* fell off with increasing O_2 concentration and was only 65% of the maximal rate in 21% O_2 and 55% of the maximum in nearly 100% O_2. The inhibition by O_2, the so-called Warburg effect, has several typical characteristics: (*1*) there is an inverse relation between the inhibition by O_2 and the CO_2 concentration, so that the effect is diminished at high CO_2 levels; (*2*) the inhibition is rapidly reversed by reducing the partial pressure of O_2 or restored by increasing it; (*3*) the effect is associated with the synthesis of glycolate and high rates of photorespiration; and (*4*) therefore not all species show the Warburg effect. There is some evidence that part of this phenomenon may also be caused by an inhibition by O_2 of enzymes of the photosynthetic carbon reduction cycle, photosynthetic electron transport, and photophosphorylation (J. S. Turner and Brittain, 1962).

Some examples of the increase in net CO_2 assimilation in different species in about 2% O_2 compared with 21% O_2 are given in Table 8.5. The photosynthesis in species with a high rate of photorespiration increases about 40% in low partial pressures of O_2 at temperatures not greater than 30°. At 35 to 40°, the increases in 2% O_2 vary from 70 to 100% in leaves of wheat and tobacco as might be expected with increasing photorespiration at higher temperatures, but it was apparently not raised proportionately in sugar

TABLE 8.5   Effect of O_2 Concentration on Net Photosynthesis of Single Leaves of Various Species Under High Illuminance at 300 ppm CO_2.

Species	Temp. (°C)	Net photosynthesis (mg CO_2 dm^{-2} hr^{-1})		% Increase	Reference
		21% O_2	2% O_2		
Ribgrass, *Plantago lanceolata*	22	28	40	43	Björkman (1966)
Barley	21	21	29	40	Apel and Lehmann (1969)
Wheat	30	31	47	52	J. Hesketh (1967)
Wheat	40	12	24	100	J. Hesketh (1967)
Wheat	25	17	23	35	Joliffe and Tregunna (1968)
Wheat	35	14	24	71	Joliffe and Tregunna (1968)
Sunflower	30	37	53	43	J. Hesketh (1967)
Tobacco	30	17	24	41	J. Hesketh (1967)
Tobacco	40	16	30	88	J. Hesketh (1967)
Sugar beet	25	25	35	40	Hofstra and Hesketh (1969a)
Sugar beet	35	28	38	36	Hofstra and Hesketh (1969a)
Maize	30	49	52	6	J. Hesketh (1967)
Maize	26	46	46	0	Bull (1969)
Maize	35	50	50	0	Hofstra and Hesketh (1969a)
Sugarcane	26	52	52	0	Bull (1969)
Amaranthus viridis	21	44	46	4	Bull (1969)
Atriplex nummularia	35	32	32	0	Hofstra and Hesketh (1969a)

beet. Nevertheless, in species with high rates of photorespiration, lowering the ambient O_2 concentration changes the products of photosynthesis (Chapter 4.K), inhibits glycolate synthesis and oxidation as well as photorespiration. Such stimulations in net photosynthesis at low O_2 increase the rates to approach those of the photosynthetically more efficient species. Table 8.5 also shows that maize, sugarcane, and species of *Amaranthus* and *Atriplex,* species with low rates of photorespiration (Tables 5.4, 5.6), also are essentially unaffected in net photosynthesis by lowering the O_2 levels from 21 to 2%.

In a study of the Warburg effect on grasses, twenty-five species of tropical origin (all members of the tribes *Panicoideae* and *Chloridoidea*) showed essentially no enhancement of photosynthesis in a low O_2 environment. In thirty-three other species in the temperate subfamily *Festucoideae* and the isolated tribe *Oryzeae,* enhancement from 23 to 60% with an average of 44% was observed (Downes and Hesketh, 1968). Leaves of *Atriplex patula* have a high photorespiration and *Atriplex rosea* a low photorespiration (Table 5.6). Lowering the O_2 level from 21% to less than 2% increased net photosynthesis 53% in *A. patula* and only 4% in *A. rosea* (Gauhl and Björkman, 1969). Since transpiration rates were unaffected by the change in O_2, internal factors and not stomatal diffusion must have been changed by the O_2 concentration, and very likely reduction of photorespiration was the primary effect.

The action spectrum for net photosynthesis in radish leaves was compared in 2% and 21% O_2 at a constant irradiance of 4×10^4 ergs sec^{-1} cm^{-2} at wavelengths from 402 to 694 nm. The stimulation of uptake of CO_2 by the low O_2 was about 100% and was constant at all wavelengths (Bulley *et al.*, 1969). The percent enhancement of CO_2 uptake by low O_2 was proportionately greater in soybean leaves at a CO_2 concentration of 75 ppm of CO_2 than in 275 ppm, where there was a 54% increase (Forrester *et al.*, 1966a). In maize leaves, there was no inhibition of CO_2 assimilation with increasing O_2 levels up to 40% O_2, but photosynthesis was inhibited 50% in 80% O_2 and 70% in about 100% O_2 (Forrester *et al.*, 1966b). Once 100% O_2 was reached, on returning to an atmosphere of 21% O_2 there was still a 30 to 50% inhibition of photosynthesis, so that very high concentrations eliminate the reversible nature of the Warburg effect, and must cause additional and irreversible inhibitions not directly associated with the usually reversible effects observed on glycolate synthesis and photorespiration.

The history of the Warburg effect has been briefly summarized (Gibbs, 1969b), and it was shown to occur in isolated spinach chloroplasts. The magnitude of the inhibition by O_2 was inversely related to the bicarbonate concentration, and high levels of CO_2 in this system could be replaced by substrates such as ribose-5-phosphate to eliminate the inhibitory effect of O_2. Some O_2 sensitivity could be shown for several isolated enzymes that participate in the photosynthetic carbon reduction cycle and in the electron transport system of the chloroplast, but these inhibitions did not seem to be of sufficient magnitude to account for the Warburg effect, and it was concluded that glycolate metabolism is a key to the understanding of this phenomenon. In another study, a supernatant fraction devoid of plastids was obtained from *Euglena* which showed the main features of the Warburg effect in darkness (Ellyard and San Pietro, 1969), thus the inhibition of CO_2 uptake by O_2 may occur partly in the photosynthetic carbon reduction cycle. If this occurs *in vivo,* the rapid reversibility of this phenomenon would still have to be explained. At the present time, however, the properties of the inhibitory effect of O_2 on CO_2 assimilation, its apparent dependence on temperature, and the species specificity all suggest that gly-colate synthesis and oxidation and the process of photorespiration largely account for the known facts about this phenomenon.

F. LEAF HYDRATION

Since CO_2 must diffuse through open stomata which depend on guard cell turgor for their opening (Chapter 7.C), a sufficient decrease in the relative leaf water content will close stomata and result in a lowered net photo-synthesis.

The relative leaf water content describes the hydration status of a leaf ("relative turgidity"), and is determined by quickly weighing samples, such as leaf disks, and weighing them again after they have become fully turgid when floated on water in the dark. Moisture stress in leaves results when the water required by the plant, as determined by the atmosphere and the diffusive resistances to transpiration, exceeds the supply furnished by soil water. Leaves of many species wilt when their water content is about 80 to 85% of their fully turgid condition, and stomata usually begin to close as the water content decreases even before wilting is observed. Stomata are generally fully closed when the wilting point is reached.

The hydration status of a leaf is also often expressed as the "water po-tential." The water potential of the leaf is defined as a measure of the ac-tivity of the water compared with pure free water which has an activity of 1.0 (Slatyer, 1967). It is expressed in units of atmospheres, or bars, of ten-sion. The units are negative in leaves that have increased solute or a de-creased water activity. Thus the wilting point of leaves usually occurs at a water potential of -15 to -20 atmospheres. Stomata of tobacco plants grown in the laboratory begin to close at -4 to -5 atmospheres and close completely at -16 to -17 atmospheres, while the stomata in tobacco in the field do not begin to close until a leaf water potential of -11 to -12 atmospheres is reached and then they close rapidly at only a slightly greater dehydration (Begg and Turner, 1969).

Several methods are widely used to determine the water potential besides the common measurement of the freezing point depression of cell sap and determination of the concentration of a sucrose solution that is in equi-librium and will thus maintain the normal turgor of the cytoplasm of the leaf cells. Precise measurements can be made with a thermocouple psy-chrometer, a device that consists of a closed container with the tissue and a thermocouple containing a droplet of water inside, all at a constant temperature. The cooling of the thermocouple by water vapor transfer from droplet to tissue is measured, and the rate of water vapor transfer is proportional to the difference in potential between the thermocouple and the plant tissue. Since the stomatal diffusive resistance of leaf tissue affects this rate, solutions of various potential have been placed on the thermocouple so that the potential at which there is no water movement can be determined and thus the leaf resistance error is eliminated in this manner (Boyer and Knipling, 1965).

A second method of determining the water potential of plant tissue involves the use of a pressure chamber, and this method is particularly valuable because it provides accurate determinations that can be made easily in the field (Scholander *et al.,* 1964; De Roo, 1969). The equilibrium pressure required to precisely balance the water potential in a leaf is established by increasing the atmospheric pressure on the outside of the leaf maintained in the chamber by means of a cylinder of compressed nitrogen until a droplet of xylem sap is just forced out of the cut end of the petiole outside the chamber. The pressure required is just equal to the water potential of the leaf.

The effect of dehydration on the extent of stomatal closure varies with the species, leaf surface (upper or lower), and the prior environment. The portion of the decreased CO_2 assimilation caused by stomatal closure under conditions of water stress also varies. Maize growing in soil was allowed to deplete the soil moisture to various contents from 24 to 10% several days before experiments were carried out, and CO_2 uptake and transpiration were then compared (Shimshi, 1963). The leaves had not reached the wilting point even at the lowest soil moisture (stomatal width about 1 μ), and photosynthesis was inhibited 80% and transpiration only 60% compared with plants maintained at the highest soil moisture (stomatal widths 3.6 μ). Therefore, as the soil dries the internal resistances to CO_2 assimilation increased more rapidly than the stomatal diffusive resistance.

In other experiments with cotton leaves, a reduction in net photosynthesis occurred on decreasing the relative leaf water content, and this decrease coincided with increasing stomatal closure until about 75% of the relative leaf water content was reached. Greater dehydration then caused decreases in net photosynthesis that were attributable to increases in the internal diffusive resistances (Troughton, 1969). No changes in net photosynthesis or stomatal aperture were observed down to 90% relative turgidity in soybeans grown in varying soil moisture concentrations at high light intensities. A sharp decline in CO_2 assimilation then occurred and resulted in a 50% inhibition at 85% relative turgidity, which coincided with closure of stomata (Shaw and Laing, 1966).

Stomata and CO_2 uptake were more sensitive to decreasing leaf water potential in maize than in soybean leaves when plants were dehydrated by maintaining a lower soil moisture before the experimental period (Boyer, 1970b). Differences in CO_2 assimilation solely due to stomatal behavior were found in maize leaves down to leaf water potentials of -10 atmospheres and in soybean down to -16 atmospheres, while at lower potentials other internal factors inhibited CO_2 uptake in addition to stomatal closure. There was some inhibition of net photosynthesis in maize at -6 to -8 atmospheres, while in soybean there was none until -13 atmospheres was reached, and in sunflower no inhibition was observed until -8 atmospheres.

The first recognizable symptom of water stress in leaves is a decrease in cell enlargement and a rapid reduction in rate of leaf area expansion, hence turgor pressure must be essential for the plastic extension of cell walls (Shaw and Laing, 1966; Slatyer, 1969). Only a small water deficit, leaf water potential of -4 atmospheres, was sufficient to prevent leaf expansion completely for several days in sunflower and greatly inhibited leaf enlargement in soybean and maize, although 50% inhibition of CO_2 assimilation required deficits of -14 atmospheres in sunflower, -17 atmospheres in soybean, and -14 atmospheres in maize leaves (Boyer, 1970a).

Part of the increase in the internal diffusive resistance to CO_2 uptake that accompanies stomatal closure may result from an increase in leaf temperature (Table 8.4), since dark respiration and especially photorespiration increase as the leaf warms. Most recent investigations show that dark respiration at a fixed temperature is relatively unaffected by water deficits until at least a moderate stress exists (Slatyer, 1969). The direct effect of water stress on photorespiration has not yet been examined, and experiments showing increases in the CO_2 compensation point during water stress may merely indicate that the internal physical resistances to CO_2 fixation are increased by dehydration. Internal resistances to the evaporation of water may occur in mesophyll cells during water deficits; hence it seems possible that similar physical diffusive resistances may be induced by dehydration that would increase the size of the physical resistances to the passage of CO_2 in the leaf interior.

G. BIOCHEMICAL CONTROL AND TRANSLOCATION

The principle of the regulation of the rates of biochemical sequences by the action of an end product of the sequence on an enzyme functioning in an earlier step has been established in a number of examples particularly in bacterial systems. This demonstrates that enzymes may be controlled by their response to external signals. Besides this type of feedback control, many small molecules not necessarily directly involved in a metabolic sequence modify enzymatic activities at low concentrations. An outstanding example is the inhibition and stimulation of pathways by the adenylate pool, which has led to the concept of "energy charge," a term derived by analogy with the charging of a battery (Atkinson, 1966, 1968). This concept explains how small molecules (called "modifiers"), such as adenine nucleotides, interact with enzymes to result in the modulation of their activities, presumably by altering the conformation of the protein molecule and thus changing the affinity of the enzyme for its substrate.

The equilibrium of the adenosine pool is maintained by the enzyme adenylate kinase, which catalyzes the reaction $AMP + ATP \leftrightarrows 2\ ADP$. The energy charge, which is expressed on a scale from 0 to 1, is represented by the sum of the high-energy bonds ($\sim P$), or twice the ATP concentration plus the ADP concentration, divided by twice the total $AMP + ADP + ATP$ concentration, all on a molar basis. Thus if only AMP were present, the energy charge would be 0, and if ATP were exclusively present, the energy charge would be 1. At equilibrium the energy charge would be 0.5. If analysis of a tissue showed relative molar ratios, for example, of 10% AMP, 30% ADP, and 60% ATP, it would have an energy charge of $[(0.3 \times 1) + (0.6 \times 2)]/2 = 0.75$. Although at equilibrium the adenylate kinase reaction would have an ATP/AMP ratio of 1.0 and an energy charge of 0.5, the example above illustrates that changing the ATP/AMP ratio to 6.0 only increases the energy charge to 0.75. Hence the adenylate kinase reaction serves as a buffer in preventing large oscillations in the energy charge.

Most of the concepts about energy charge controlling biochemical sequences has been tested with isolated enzymes. In metabolic processes that generate ATP (as in oxidative or photosynthetic phosphorylation) a high energy charge will be produced, and this condition causes inhibition of enzymes of the tricarboxylic acid cycle (such as citrate synthetase and isocitrate dehydrogenase) especially when the energy charge exceeds about 0.75, and the enzymatic rates increase when the AMP concentration increases. On the other hand, enzymatic reactions that utilize ATP (such as phosphofructokinase in glycolysis) are markedly inhibited when the energy charge is less than about 0.75, or when the AMP concentration rises. Thus the energy charge tends to stabilize rates of biochemical sequences. The inhibition of CO_2 fixation by AMP in spinach leaf is probably caused by an inhibition of this type on the ribulose-5-phosphate kinase reaction (Preiss and Kosuge, 1970). These effects of energy charge would regulate activities *in vivo* only if they were present in the cytoplasmic organelles or the compartment where various reactions may occur, and such regulation probably exists because enzymes possess regulatory sites that affect their kinetic properties as well as catalytic sites that bind with substrates.

Adenylate kinase activity was found in wheat leaves, and Atkinson's hypothesis that this enzyme tends to buffer the energy charge was supported by experiments in which the leaves were placed in different environments for periods of 0.75 to 3 hours after taking up ^{32}P-orthophosphate, and the concentrations of adenine nucleotides were then determined (Bomsel and Pradet, 1968). The energy charge in the entire leaf tissue can be calculated from their results to be as follows: light, N_2, 0.34 to 0.47; light, air, 0.60 to 0.64; dark, air, 0.63 to 0.85; dark, N_2, 0.13 to 0.22. In spite of large shifts in the energy load in which the ATP/AMP ratio varied from extremes of 0.05 (dark, N_2) to 15 (dark, air), the ADP concentration was fairly stable so that the energy charge changed relatively little in the entire leaf. In another study, adenylate kinase activity was detected in the chloroplasts of beet and spinach leaves, and during light–dark transitions the ATP/AMP ratio in the chloroplasts was high in the light, low in the dark, and little change appeared in the ADP concentration (Santarius and Heber, 1965). This suggests that the energy charge tends to stabilize the rates of metabolic processes in the chloroplasts under both conditions.

The enhancement by light of enzymatic reactions associated with photosynthesis has been shown in several instances such as for ribulose diphosphate carboxylase (Chapter 4.E) and the phosphopyruvate synthetase reaction (Chapter 4.F). The activation of certain key enzymes in the light and their inactivation in darkness may control whether the pentose phosphate cycle is oxidative (respiratory) or reductive (photosynthetic) in its direction (Krause and Bassham, 1969). It is not known whether there is some direct interaction on enzymes by light, as on sulfhydryl groups, to activate them, or whether a metabolite formed in the light causes such changes in enzyme activity, or whether relief of inhibition is brought about by a mechanism like a changed adenylate concentration. Nevertheless, it seems that subtle changes may alter metabolic pathways in living cells and that inhibitions need not be caused by mass action effects. Mass action is probably of least importance in controlling enzymatic processes such as photosynthesis *in vivo*, because the end products largely accumulate in the vacuole (Chapter 4.M), are translocated, or are present as insoluble substances (starch) which are largely metabolically inert over short periods of time.

As early as 1868 T. B. Boussingault believed that the accumulation of assimilates in an illuminated leaf could be responsible for a reduction in the net photosynthesis of that leaf, but considerable work has produced conflicting results under different conditions and in various tissues (Neales and Incoll, 1968). Perhaps this is partly because most studies have only examined changes in the final end product concentration (sucrose). Also the concept of photosynthetic control in terms of "sources" and "sinks" has mostly been invoked, and it implies that mass action regulates CO_2 assimilation. This is least likely in view of the discussion above. Several examples of an apparent inhibition of photosynthesis by end product accumulation have, however, been described. Perhaps ATP required for carboxylation reacts instead with excess glucose in the reaction catalyzed by hexokinase (Scheme 5.1).

Removal of the ears of maize plants caused a 25% inhibition of CO_2 uptake within one week and a 75% inhibition 11 days after the removal compared with normal plants (D. N. Moss, 1962b), and at this time the concentration of sucrose was twice as great in the stalks of plants without ears. The detachment of tomato fruits likewise caused an increase in the sucrose concentration of the leaves and a severe inhibition of photosynthesis. Removal of apple fruits from intact shoots in early summer resulted in a 33% inhibition of net photosynthesis in the leaves that lasted for several months, but the inhibition was not observed soon after a period of darkness (Hansen, 1970). During the period of grain filling in wheat, removal of the ear diminished net photosynthesis about 50% within 3 to 15 hours in the flag leaf; this leaf usually translocates about 45% of its photosynthate to the ear and only 12% to the roots and young shoots (King *et al.,* 1967).

Decreases in net photosynthesis can be induced by changes in the biochemistry of carboxylation or photosynthetic electron transport, diffusive resistances, respiration, or translocation, but these possibilities have not yet been clarified in examples like the ones above. Such experiments may sometimes be misleading for other reasons as illustrated by data on the severe inhibition of CO_2 absorption in sugarcane leaves that was apparently correlated with increases in their sucrose concentration (Hartt, 1963). These observations may have resulted mainly from stomatal closure under the environmental conditions used, hence the results with sugarcane were only incidentally related to the sucrose content of the leaves (J. C. Waldron *et al.,* 1967). Similarly, an apparent stimulation of net photosynthesis observed by some workers on the remaining leaves when some leaves were removed from a plant has been ascribed to hormonal effects (Wareing *et al.,* 1968). It is equally likely that the enhancement was caused by greater stomatal opening resulting from a higher leaf water content when the total leaf area exposed to irradiation was decreased.

The effect of shading parts of plants has given variable results in a number of experiments on determination of the net assimilation (Thorne, 1965), and removing the florets or shading the ears of barley did not greatly change the net CO_2 uptake of the upper flag leaf (Nössberger and Thorne, 1965). Sugar beet has a large storage root, and when spinach beet (also a *Beta vulgaris*) was grafted onto sugar beet roots there was a greater net assimilation by the tops than when it was on its natural smaller root (Thorne and Evans, 1964).

Rates of translocation may control net photosynthesis and respiration, and translocation is essential for such processes as the growth of roots, the development of fruits, the filling of grain, and the growth of tubers. Sucrose is the primary form of organic transport in nearly all plants (Stoy, 1963; H. K. Porter, 1966). The sugar is collected in the minor veins of the leaf and is then transmitted mainly to the larger veins in the phloem for export (Trip, 1969). The movement of sucrose into the phloem requires a supply of metabolic energy (Milthorpe and Moorby, 1969). Generally, the proportion of carbon assimilated by any one leaf which is exported for utilization elsewhere increases with age of the leaf and then decreases with the onset of senescence. Upward and downward movement in the phloem usually proceeds in different sieve tubes although bidirectional flow also occurs (Trip and Gorham, 1968). The amounts moving up and down depend on the position of the leaf. Material in leaves nearer the apex moves mostly upward while leaves lower down the stem mainly supply the roots (H. K. Porter, 1966). However, roots of sugar beet do not control the rate of translocation as evidenced by removal of portions of the root, nor do the roots control the quantity exported when measured over an extended period (Winter and Mortimer, 1967). Only the uppermost one or two leaves contributes photosynthate directly in large amount to the filling of grain in species like wheat and barley (Stoy, 1963), and some aspects of this most important process are discussed in Chapter 9.A.

Species with high rates of net photosynthesis, such as maize, also have rapid rates of translocation compared with the less-efficient species. Translocation rates of maize leaves were 2.5 times those in sugar beet, and when $^{14}CO_2$ was supplied for 2 minutes to maize and the tissue was frozen 1.5 minutes later, autoradiography showed most of the ^{14}C had already been translocated to the bundle-sheath cells (D. N. Moss and Rasmussen, 1969). In contrast, ^{14}C in sugar beet leaves was evenly distributed throughout the mesophyll cells. Similar results were obtained in a comparison of the ^{14}C distribution in cross-sections of maize and barley (Pristupa, 1964), but no information is given about how quickly the tissue was killed to avoid an artifact resulting from the rapid translocation of labeled material from the mesophyll to the bundle-sheath cells of maize.

Between 80 and 90% of the assimilated ^{14}C was translocated from the fed area of maize leaves within 24 hours, and 50% in the first 30 minutes

(Hofstra and Nelson, 1969), and these results are similar to those found by other workers for sugarcane and were about twice as rapid as has been found in soybean and tobacco. Rates of translocation down the petiole of sugar beet leaves were as fast as 150 cm hr^{-1} (Mortimer, 1965). This was nearly equivalent to 18 mg CO_2 fixed dm^{-2} hr^{-1}, a value similar to the net photosynthetic rate, although usually about 50% of the assimilated CO_2 was exported during the day and an additional 40% was exported at night (Mortimer and Terry, 1969). Rates of translocation down the petiole of 60 cm hr^{-1} and the equivalent of 5 mg CO_2 fixed dm^{-2} hr^{-1} have also been measured for sugar beet (Geiger and Saunders, 1969). The more rapid translocation shown by the more efficient species in net CO_2 absorption may play a role in the somewhat greater CO_2 uptake in these species observed even under nearly saturating conditions of light and CO_2 (Table 8.3).

H. OTHER CONTROLLING FACTORS

1. NUTRIENTS

Photosynthesis is responsible for 90 to 95% of the dry weight of a plant, and only 5 to 10% is ash plus nitrogen. The mineral status, and especially the potassium concentration, may nevertheless have a large effect on determining rates of CO_2 assimilation (Chapter 7.C). In rice and several other species, nitrogen fertilization is clearly one of the most important factors regulating photosynthesis and productivity. A close positive correlation was found between the CO_2 assimilation of rice leaves and their total nitrogen content on a dry weight basis, which was regulated by the quantity of nitrogen supplied (Murata, 1969). Leaves containing 2% nitrogen fixed CO_2 at a rate of 8.5 mg dm^{-2} hr^{-1} while leaves with 5% nitrogen had a photosynthetic rate of 15 mg CO_2 dm^{-2} hr^{-1}. Since a large portion of the leaf protein is found in the chloroplasts (Chapter 2.C), the increased nitrogen is undoubtedly reflected in improved structure and enzymatic activity in these organelles.

2. PATHOGENS

Many plant pathogens inhibit photosynthesis by damaging chloroplasts, inhibiting chlorophyll synthesis, or by destroying specific chloroplast proteins. Sugar beet chloroplasts from plants infected with virus yellows, and chloroplasts from tobacco plants infected with tobacco mosaic virus had decreased rates of the Hill reaction and photophosphorylation on a chlorophyll basis (Goodman *et al.,* 1967). There is also evidence that tobacco mosaic virus is formed at the expense of Fraction I protein (Chapter 4.E) in the chloroplast (Wildman *et al.,* 1949). The yellow dwarf virus of barley leaves caused a decrease of photosynthesis of at least 50% on a chlorophyll basis 10 to 15 days after inoculation (Jensen, 1968). Bacterial infections are known frequently to result in chlorophyll destruction or inhibition of synthesis. This was found with the leaf-spotting xanthomonads, and for several *Pseudomonas* species that produce toxins that cause chlorotic halos on susceptible leaves. The extracellular toxin most studied is produced by the tobacco wildfire organism, *P. tabaci,* which acts like a methionine antagonist and brings about the breakdown of chloroplast structure and function (Durbin, 1971).

Bean rust (*Uromyces phaseoli*) and wheat rust (*Puccinia graminis*) cause leaves to become chlorotic, but chlorophyll is retained in "green islands" at the edge of the infection sites. These green areas resemble tissue that remains in the juvenile state in senescing leaves spotted with cytokinins, and these are photosynthetically active (Bushnell, 1967). A gradual decline in photosynthetic activity of rust-infected bean and wheat leaves was observed beginning with the first symptoms of infection (Livne, 1964), but rust-free trifoliate leaves on rust-infected bean plants showed a CO_2 assimilation increase of 58% compared to healthy plants. Perhaps this was related to a greater leaf water content in the remaining leaves of infected plants and thus larger stomatal opening. Biochemical changes in proteins of rice leaves infected by the rice blast fungus, *Piricularia oryzae,* have been examined, and evidence was obtained that Fraction I protein of the chloroplast is modified by the fungus (Akazawa and Ramakrishnan, 1967), although it is uncertain whether this is a primary or a secondary effect of the infection.

3. GENETIC REGULATION

Man has encountered limitations imposed by the genetic control of photosynthesis in his efforts to stabilize and improve plant productivity by selection and breeding. The extent of these limitations and knowledge of the nature of the controls and how they work is now beginning to emerge. The hybrid vigor, or heterosis, resulting when certain inbred lines are intercrossed has been known for many years and has been successfully exploited by maize breeders. Net photosynthesis showed properties of heterosis in several examples of single crosses of inbred maize (G. H. Heichel and Musgrave, 1969), and one example is illustrated in Table 5.8. Heterosis was also found in CO_2 assimilation and increased dry weight accumulation by leaf disks of single-cross and double-cross hybrids of maize, and the enhanced photosynthesis of the hybrids was well correlated with increases in grain yield in these combinations (Fousova and Avratovscukova, 1967). On the other hand, there was more dominance for low photosynthetic rates in the progeny of two varieties of red kidney bean (Izhar and Wallace, 1967).

Among hundreds of rice selections under trial, two of the high-yielding varieties had as one of the parents a variety whose leaves showed the highest rate of net photosynthesis among fifty varieties tested (Chandler, 1969a). In a comparison of two wheat varieties whose grain yields differed by about 30%, this difference was accounted for by an approximately 30% greater assimilation and translocation in the flag leaf and second leaf of the more efficient variety (Lupton, 1969).

Net photosynthesis by single leaves of twelve soybean varieties was compared with their seed yields outdoors, and the average yield was generally higher in varieties with higher rates of CO_2 absorption, but there were exceptions in which high seed yield was associated with lower net photosynthesis rates (Curtis *et al.*, 1969). Although there is generally an excellent correlation between net photosynthesis and productivity (Chapter 9.A), this is not a necessary condition if, for example, some other factor counteracted the beneficial effect. Factors negating photosynthesis could be high rates of dark respiration, poor root growth, mutual shading of leaves, low total leaf area, and lower assimilation rates during periods of grain filling or seed production. Thus a superior photosynthesis must be present in a genetic background that permits its potential to be utilized.

Experiments attempting to examine the genetic basis and breeding characteristics of low photorespiration have been started (Table 5.8), but there has still not been much work on the genetic basis for obtaining greater rates of net photosynthesis in other promising areas (Chapter 7.E). Mutations that affect chloroplast structure, chlorophyll and carotenoid synthesis, the photosynthetic electron transport chain, and enzymes of the photosynthetic carbon reduction cycle are known (Levine, 1969). These mutations have been helpful in evaluating the role of these aspects of photosynthesis when they are eliminated, but they have not been directed to increasing the rate of CO_2 uptake. Plant breeders using empirical methods have made considerable progress in obtaining efficient plant structures and enhancing the ability of plants to utilize added soil nutrients, and undoubtedly they have selected for high rates of CO_2 absorption during the critical stages of development concerned with agricultural yield. It must now be determined whether productivity can be increased faster by logical rather than by empirical methods. Genetic improvements will probably eventually concentrate on the activities of single enzymatic reactions and in methods of assaying these in large populations of plants. Diminishing photorespiration and dark respiration, increasing stomatal numbers and the rate and extent of their opening, and diminishing internal diffusive resistances—all by genetic means—or simply incorporating lines with higher than usual rates of net photosynthesis per unit leaf area into otherwise ideal plant types should provide such increases providing the problems encountered by single leaves in a stand are also considered.

9 Relation of Photosynthesis, Total Respiration, and other Factors to Control of Productivity in Stands

The net productivity of plants and stands is determined by an integration of all of the factors affecting the photochemistry, biochemistry, and physical diffusion of CO_2 for chloroplasts, leaf cells, and single leaves discussed in the previous chapters, as well as the activities of nonphotosynthetic tissues. Economic productivity, or yield, also usually involves the ultimate distribution of the products derived almost exclusively from CO_2 assimilation into the materials and tissues that comprise yield; hence yield mainly involves additional biochemistry and translocation that frequently takes place only during a portion of the total growth period. Plant productivity is affected (1) by the soil, including soil moisture, nutrient availability, and CO_2 released during soil respiration; (2) by the properties of the leaves comprising the stand, such as their stomatal numbers and behavior, response of the mesophyll cells to irradiance, reflectance and transmittance properties, effect of temperature on dark respiration and photorespiration, and their physical resistances and carboxylation characteristics; (3) by the architecture of the stand, including the total leaf area covering a unit area

TABLE 9.1   *Estimated Average Yields of Several Crops*[a]

Crop	Average market yield (lbs fresh weight/acre)	Estimated dry weight[b] (lbs/acre)	Estimated growing season (weeks)	Average yield (lbs dry weight acre^{-1} week^{-1})	Average crop growth rate (gm dry weight m^{-2} week^{-1})
Maize silage	23,600	7,080	17	443	50
Sorghum silage	21,600	6,480	17	405	45
Sugarcane (cane)	54,000	16,200	36	450	50
Spinach	5,800	580	5	102	11
Tobacco (leaf)	1,945	3,140[c]	14	224	25
Hay	4,000	3,600	20	180	16

[a]Data from Agricultural Statistics, 1969. U. S. Dept. of Agriculture.

[b]The average yield values in the first column were obtained from *Agricultural Statistics* It was assumed that the maize silage, sorghum silage, and cane contained 30% dry weight; spinach and cabbage, 10% dry weight; and hay, 90% dry weight. To convert pounds dry weight per acre to grams per square meter (1 acre = 4047 square meters, 1 pound = 454 gm), multiply by the factor 454/4047 = 0.112.

[c]For tobacco, a yield of stalk was added equal to 90% of the yield of leaf, to give a total of 3696 lbs fresh weight per acre, or 3140 lbs dry weight.

of soil, leaf distribution along the stem, and the angle of leaf elevation from the horizontal; (*4*) by ambient climatic factors, such as the air temperature, wind speed, CO_2 concentration, relative humidity, angle of the sun, and whether irradiation is diffuse or direct. (*5*) Finally, productivity is ultimately determined by the duration of photosynthesis, changes of photosynthesis with leaf size, the efficiency of CO_2 assimilation, the rapidity with which the leaf area enlarges to absorb the available irradiation, plant height (in stands of mixed species), photosynthesis by organs other than leaves, and the efficiency of transport of photosynthate to tissues of economic importance. In that last category are such diverse products as roots, stems, grain, tubers, fibers, fruits, and specific organic compounds.

The manner in which many of these factors operate in a stand will already be apparent from the description of simpler systems in previous chapters. Therefore only those aspects most affected by the interaction of plants in communities will be emphasized in this chapter, and finally the use of mathematical models to simulate the activity of plants in a stand will be used to demonstrate how many of these factors may be integrated.

A. CORRELATION OF NET PHOTOSYNTHESIS WITH PRODUCTIVITY

1. THE PRODUCTIVITY OF VARIOUS STANDS

The efficiency of net CO_2 absorption by single leaves differs greatly among species (Table 8.1) and among varieties within a species (Table 8.2). Some examples showing that higher productivity is sometimes closely related to higher rates of CO_2 assimilation by single leaves in a stand have been given in Chapter 8.H. An estimate of the *average* productivity of several leafy crops in the United States is presented in Table 9.1. The examples were selected to include only crops where the entire aerial portions are harvested so that other aspects of yield that are dependent on translocation such as the filling of grain (as discussed below) are not included. These estimates are thus almost entirely dependent on the total quantity of CO_2 assimilated less that lost by respiration. Since they are *average* values, they are also diminished from maximal values of productivity by losses incurred because of inadequate cultural practices, limitations of nutrients and water, plant pests and diseases, and the vagaries of irradiation and climate. The average crop growth rate is about twice as great for the photosynthetically efficient species such as maize, sorghum, and sugarcane compared with spinach, tobacco, and grass or legumes. These estimated average growth rates determined over the entire growing season are about 14% of the maxi-

TABLE 9.2　Maximal Growth Rates by Various Species in the Field

Species	Maximal growth rates (gm dry wt m^{-2} ground area day^{-1})[a]	Location	Average total solar radiation, (cal cm^{-2} day^{-1})	Reference
Bulrush millet (Pennisetum typhoides)	54	Australia	510	Begg (1965)
Sorghum	51	U.S.A.	600[b]	R. S. Loomis and Williams (1963)
Maize	51	U.S.A.	600[b]	R. S. Loomis and Williams (196
Maize	52,33[c]	Japan		Togari et al. (1969b)
Maize	42	Israel	600[b]	Westlake (1963); Monteith (1965)
Sugarcane	42	Hawaii	400	Monteith (1965)
Rice	29	Phillipines	270	Chandler (1969b)
Rice	36,29[c]	Japan		Togari et al. (1969b)
Sugar beet	28,23[c]	Japan		Togari et al. (1969b)
Sugar beet	31	U.S.A.	600[b]	R. S. Loomis and Williams (1963)
Sugar beet	32	England	265	Monteith (1965)
Townsville lucerne (Stylosanthes humilis)	28	Australia		Begg and Jarvis (1968)
Soybean	18,17[c]	Japan		Togari et al. (1969b)
Various crops	20	Netherlands		de Wit (1968)
Various established forests	5–17	England, West Indies, Africa		Westlake (1963)
Algae	30[d]	Japan		Tamiya (1957)

[a] Most investigators include only the growth rate of the aerial portion of the plants.
[b] Estimated.
[c] Results of maximal rates obtained at sixteen different stations in 1967 and 1968 respectively.
[d] Apparently when supplemented with additional CO_2.

mal growth rates obtained for similar species under optimal conditions (Table 9.2).

The productivity of species also differs even when measured under optimal conditions of growth, and once again the photosynthetically efficient plants (maize, sugarcane, sorghum) with low rates of photorespiration have the highest productivity rates, about 50 gm dry weight m^{-2} ground area day^{-1} in a number of world-wide locations (Table 9.2). In terms of the more familiar units of photosynthesis, this is equivalent to a net CO_2 uptake of about 735 mg CO_2 dm^{-2} ground area day^{-1}. The net rate of photosynthesis of single leaves of the efficient bulrush millet stand or its rate of photorespiration has not been described. Most of the other crop species listed had maximal crop growth rates of 20 to 30 gm m^{-2} day^{-1} in various localities, and include species with high rates of photorespiration and lower rates of photosynthesis in single leaves (Table 8.1). The *mean* productivity, presumably determined in well-established stands, was generally 35 to 45% of the *maximum* values for several species (R. S. Loomis and Williams, 1963). In a study of the maximum productivity in trials carried out with four species at seventeen stations in Japan in two successive years, the most productive (gm dry weight m^{-2} ground area day^{-1}) was maize, followed by rice, sugar beet, and soybean. The size of the largest standing crop (Chapter 9.A.2) at the end of the entire summer growing season was also greatest for maize (26,500 kg ha^{-1}), intermediate for rice and sugar beet, and smallest for soybean (9000 kg ha^{-1}) (Togari *et al.*, 1969a,b). The total productivity of trees appears to be about the same as most crop species when expressed on a daily basis, but the longer growing season because of the early and rapid expansion of leaves by deciduous species may make trees more productive in dry weight than some crops on a yearly basis. Coniferous trees are still better, and a vigorously growing stand of red pine (*Pinus resinosa*) trees growing in a temperate continental climate had a total annual productivity of 22,000 kg dry weight ha^{-1} (the roots contributed about 2000 kg ha^{-1}) during a growing season of at least 200 days (G. R. Stephens, Conn. Agr. Expt. Station).

It is not surprising that total crop growth yields (including roots) of 83,000 kg dry weight ha^{-1} $year^{-1}$ have been obtained for sugarcane in Hawaii (Westlake, 1963). In this example we have a species of high rate of photosynthesis (low rate of photorespiration) growing under high irradiance in a long growing period. In an environment such as an arid desert the annual producitivty may be only 1000, for ocean phytoplankton 2000, deciduous forests 12,000, and agricultural crops about 30,000 kg ha^{-1} $year^{-1}$ (Westlake, 1963). Similar yields of agricultural productivity have been estimated by de Wit (1968) for the Netherlands for a season of about 150 days. This is equivalent to about 12,500 kg of organic matter as food, and is sufficient to provide the minimal caloric requirements (2500 kcal day^{-1}) for 50 people for 1 year (de Wit, 1967).

2. ANALYSIS OF GROWTH RATE

Attempts to evaluate the interdependent processes in the environment and within the plant that determine yield has led to the development of several different concepts of growth analysis. F. F. Blackman in 1919 considered the increase in dry weight of plants a process of continuous compound interest in which the increment produced was added to the capital. This evolved as R "the relative growth rate" (D. J. Watson, 1952). Thus $R = (1/W)(dW/dt)$ expressed in units of gm dry weight gm^{-1} day^{-1}, where W is the dry weight of the plant at any time t. The sequence of highest relative growth rate in a comparison of stands of rice, soybean, maize, and sunflower varied considerably in two successive years, and maize never produced the highest R although it had the greatest yield (Togari et al., 1969b).

The growth rate of plants has also been expressed as the rate of increase in dry weight per unit leaf area, or the "net assimilation rate" on a leaf area basis, NAR, a concept developed by R. Gregory in 1917. Hence $NAR = (1/L)(dW/dt)$ where L is the total leaf area of the plant. The net assimilation rate is usually expressed in units of gm dry weight m^{-2} leaf area day^{-1}. The net assimilation rate thus measures the excess of dry matter gained over loss by respiration, therefore not all changes in the net assimilation rate will be determined by variations in the rate of photosynthesis.

In measuring the net assimilation rate, it is usually not practical to make continuous records of the dry weight and leaf area, thus samples are generally taken at about weekly intervals, and simplifications have been made in the equations to enable the relative growth rate and the net assimilation rate to be determined more easily. The following mean NAR values (in gm dry weight m^{-2} leaf area day^{-1}) were obtained in crop stands in England: wheat (5.0), barley (4.4), and oats (4.0), and the maximal NAR values were: sugar beet (12.6), potatoes (8.4), and wheat (7.0) (D. J. Watson, 1952). The maximal NAR values observed in Japan were: rice (30.2), maize (23.7), sunflower (20.5), and soybean (18.0) (Togari et al., 1969b). The latter collaborative studies showed that the net assimilation rate was closely correlated with the total solar radiation. For sunflower the maximal net assimilation rate observed at different localities varied from 10.2 to 17.6, but in clear weather at midsummer when the plants were widely spaced, rates of 28 gm m^{-2} leaf area day^{-1} were obtained (Warren Wilson, 1966).

The *NAR* is the "average leaf in the canopy" shown in Fig. 9.1; hence this measure alone is an inadequate indicator of productivity because it fails to account properly for differences in respiratory losses, total leaf area available for photosynthesis, or the response to irradiation and light interception when various plants are grown in a stand. Moreover, productivity must be compared on a ground area basis in order to be meaningful. Thus differences in the net assimilation rate or relative growth rate between and within species are not as great, or necessarily in the expected relation to each other, as are the known differences in photosynthesis and productivity.

The importance of the variation in leaf area was introduced into growth analysis by D. J. Watson in 1947, who used the term "leaf area index" (*LAI*) to express the area of leaf surface (one surface only) in a stand covering a unit area of land surface. In practice the *LAI* is frequently determined by obtaining the dry weight of the leaves over a measured area of soil, and the leaf area is estimated by determining the dry weight per unit area in samples (such as leaf disks). The *LAI* increases with time as plants grow, mainly because of increases in the leaf area per plant rather than because of increases in shoot numbers or size. Such factors as availability of water and nutrients affect the *LAI*, and the leaf area usually decreases with increasing solar radiation and increases with warmer temperatures providing there are no contrary temperature effects on the dormancy of leaf buds.

D. J. Watson (1958) altered the leaf area index of kale and sugar beet stands in the field by removing different fractions of the plant population (the whole plant), and determined the net assimilation rate for subsequent 10 to 14 day periods. He found that the net assimilation rate (on a leaf area basis) for kale *decreased* linearly with an increase in the *LAI* from 1 to 5, while the net assimilation rate of sugar beet decreased until the *LAI* rose above approximately 3.

Because of the dependence of the net assimilation rate on the leaf area index, the rate of dry matter production per unit area of land was examined, and this parameter was called the crop growth rate *C* (D. J. Watson, 1958). $C = (NAR)(LAI)$, and it is usually expressed as gm dry weight m^{-2} ground area day^{-1}. This measure of productivity has already been utilized in Table 9.1 and Table 9.2 since it is the most meaningful for comparing species, varieties, or the effect of environment. The crop growth rate of the kale stand rose from 8.5 at a leaf area index of 1 to a maximum of 17 at a *LAI* of 3 to 4, and decreased to 11 gm m^{-2} ground area day^{-1} at a *LAI* of 5 (D. J. Watson, 1958). For sugar beet, the maximum crop growth rate was not obtained until the leaf area index was 6 to 9. The crop growth rate and the leaf area index are thus closely connected.

The optimum leaf area index depends on the net photosynthesis in the canopy which is determined by the stand architecture including the average leaf angle (Chapter 9.B), light interception, mutual shading that may occur so as to affect dark respiration and photorespiration, and internal and external diffusive resistances. The optimum leaf area index will vary because of differences in these properties between different plant species and in varieties (Chapter 9.B). The total productivity depends on the time required to initiate optimal conditions for the net assimilation rate and on the leaf area index.

3. EFFICIENCY OF LIGHT ENERGY CONVERSION TO CARBOHYDRATE

Under the best conditions of photosynthesis a maximum quantum yield of 8 corresponds to a photochemical efficiency of about 12% (Chapter 3.A). The average total solar radiation in the United States for the 100-day growing period in the summer beginning June 1 differs from 390 to 680 cal cm^{-2} day^{-1} depending on location (R. S. Loomis and Williams, 1963). Using a modest value of 500 cal cm^{-2} day^{-1} for total solar radiation for a bright day, 222 cal cm^{-2} of photosynthetically useful radiation between 400 and 700 nm (Chapter 8.B) would reach a canopy of leaves. If this radiation were entirely absorbed (as it would be by a dense stand), then about 12% of this radiation would be available to provide chemical energy (26.6 cal cm^{-2} day^{-1} or 266 kcal m^{-2} ground area day^{-1}). Since 3.7 kcal are needed per gram of carbohydrate synthesized [112 kg per 30 gm of (CH_2O) in Equation (3.1)], the photochemical energy available might produce $266/3.7 = 72$ gm of carbohydrate m^{-2} ground area day^{-1}. This is about 40% greater than the maximum growth rates shown in Table 9.2 even though no corrections for losses by respiration have been included. Calculations of maximum productivity have also been made on a quantum rather than an energy basis. Assuming the respiration loss to be 33% of gross photosynthesis, R. S. Loomis and Williams (1963) estimated maximum growth rates to be 71 gm dry weight m^{-2} ground area day^{-1} for a species like maize.

The values of crop growth rate of about 50 gm dry weight m^{-2} ground area day^{-1} obtained experimentally (Table 9.2) were generally achieved with a total solar irradiation more than the 500 cal cm^{-2} day^{-1} used in the above examples. The observations on bulrush millet, however, were made with a mean of 510 cal cm^{-2} day^{-1} and a photochemical efficiency of about 10% of the incident visible radiation (Begg, 1965), while the high growth rates with sorghum were obtained with an irradiation of 690 cal cm^{-2} day^{-1}. The lower growth rate observed for most crop species probably results from greater losses by dark respiration and photorespiration than by great differences in photochemical efficiency. The maximum daily efficiency of light energy conversion in a stand of maize approached 5% without considering respiratory losses, and the mean for the entire day was close to 3% (C. S. Yocum *et al.*, 1964). Similar results have been obtained for other field crops (Wassink, 1959; Altman and Dittmer, 1968).

Algae grown in the laboratory produced photochemical efficiencies of 12 to 15% at high irradiances and 20 to 24% at low irradiances, while outdoors a yield of 20 gm dry weight m^{-2} day^{-1} was obtained (presumably with CO_2-enriched air) at full sunlight with a photochemical efficiency of 8% (Wassink, 1959). Algae (*Chlorella*) do not have maximal growth rates superior to higher plants on an area basis (Table 9.2; Tamiya, 1957). Because of difficulties in avoiding contamination by other microorganisms during their growth and the engineering problems associated with their culture, raising algae does not offer advantages over higher plants as a source of food.

The efficiency of light energy conversion (number of calories captured by the leaf as organic matter per number of calories of visible radiation received) is greatest at low irradiance (Chapter 8.B), hence photochemical efficiency is not directly related to productivity. Clearly the total yield increases greatly with increasing irradiation (and decreasing photochemical efficiency).

4. CONTRIBUTION OF ORGANS BESIDES LEAVES AND BY SPECIFIC LEAVES IN THE CANOPY

The lamina is the only significant organ of photosynthesis in land plants, but under certain conditions other organs can contribute major proportions to the total photosynthesis. Stems do not usually fix much CO_2. For example, the stems of a stand of Townsville lucerne contributed only 3 to 5% of the total in the upper half of the canopy (Begg and Jarvis, 1968). The CO_2 uptake by the green skin of apple fruits during their development was also negligible compared with that of the leaves (Hansen, 1970). The contribution by the bark of trees to total photosynthesis is variable, but in a desert tree species (*Cercidium floridum*) that is without leaves for much of the year and contains a large number of small-diameter stems having chlorophyllous corticle tissue, net photosynthesis by the stems contributed more than 40% of the total of the trees with leaves (Adams and Strain, 1969).

The contribution to photosynthesis by various organs in grasses producing small grains (wheat, rice, barley, oats) has received considerable study. In wheat and barley, the leaf sheaths, peduncles (flower stems), and ears are important producers of photosynthetic products especially during the grain filling period (Stoy, 1965). The ears in barley are more important than in wheat, and account for about 50% of the net photosynthesis. In wheat, an appreciable fraction of the carbohydrate of the grain is derived from photosynthate assembled in the stem immediately before and during anthesis (full bloom), and the stored products are then translocated to the grain.

Various estimates of the relative contribution of photosynthesis by the ear to grain yield for wheat and barley have appeared in the literature. The contribution by varieties possessing awns (beard-like structures in the flowers) is generally greater than in awnless varieties. In wheat, 17% of the dry weight of the grain was derived from ear photosynthesis and 83% was translocated from the uppermost or flag leaf, while in barley 60% was derived from the ear and 40% was translocated from the flag leaf (Thorne, 1965). However, net photosynthesis in the flag leaf and the ears of wheat and barley was approximately equal so that large losses by respiration in the ear or during translocation must have occurred. Another investigation also concluded that photosynthesis by the ear in wheat was as great as that of the flag leaf, and that 49% of the carbon assimilated by the flag leaf was translocated to the grain (Carr and Wardlaw, 1965). Kriedemann (1966) found that the contribution of the wheat ear to grain yield varied from 10 to 44%, depending on the method of analysis and the environment, with a mean of about 30%, hence about 70% of the dry weight was derived from leaf photosynthesis.

An interesting analysis has been made of the relation of photosynthesis to grain yield in two varieties of wheat that varied in their productivity of grain by about 30% (Lupton, 1969). Photosynthesis occurring before anthesis contributed about 12% of the grain carbohydrate, and the largest supply of carbohydrate to the grain therefore occurred during 21 days after anthesis or longer. The increased yield in the higher yielding variety was accounted for by the 35% greater net CO_2 assimilation by the flag leaf (and the sheath) as well as by the second leaf (plus its sheath). In both varieties ear photosynthesis contributed 10 to 14%, flag leaf plus sheath 60 to 63%, and the second leaf plus sheath 26 to 27% of the CO_2 converted into grain yield.

In contrast to the small grain cereals, the photosynthetic contribution by the inflorescence is negligible in crops producing large grain. In sorghum, the panicle contributed about 1% of the leaf photosynthesis on a dry weight basis during the early bloom stages, and by the time the grain developed to the "soft dough" stage CO_2 release occurred in the light and there was no detectable assimilation (Eastin and Sullivan, 1969). The situation in maize is also different from that found in the small grains where the upper leaves and the sheath as well as the ear make substantial contributions. An analysis of various portions of a maize canopy disclosed that the different portions of the total leaf area after flowering contributed to the total dry matter production (the grain would comprise about 50% of the total dry weight) as follows: top five leaves 26% of the leaf area and 40% of productivity; middle four leaves, 42% of the leaf area and 35 to 50% of productivity; and bottom six leaves, 32% of the leaf area and 5 to 25% of the productivity (Allison and Watson, 1966). The leaf sheaths contained about 20% of the total leaf area and presumably contributed some photosynthate, but photosynthesis by the ear in maize was negligible since its surface comprised only 2% of that of the leaves. Thus photosynthesis by the upper two-thirds of the maize canopy contributes substantially to the productivity of grain formation in contrast to observations on the small-grained cereals.

B. VARIATIONS IN AVAILABLE IRRADIATION

1. TOTAL IRRADIATION AND SPECTRAL DISTRIBUTION

The total irradiation available for photosynthesis is undoubtedly the most important factor in determining the final productivity, and it is regulated by climate as well as the plant. In eight crops of turnips planted at various times from mid-May until the beginning of July, so that each crop had a different environment, the final yield was highly positively correlated only with the visible radiation received (Stanhill, 1958). The number of days in the summer receiving a total solar radiation of 500 cal cm^{-2} day^{-1} varies from 29 to 94 at different locations in the United States (R. S. Loomis and Williams, 1963), hence this limitation would surely affect the potential yields in different localities. Monteith (1965) analyzed the light distribution in field crops and concluded that differences in the leaf arrangement are relatively unimportant in determining the maximal photosynthesis, providing the crop intercepts the incident radiation completely, and that a small irradiation over a long day provides as much CO_2 absorption as a larger radiation accumulated in a shorter day.

The spectral distribution of solar radiation received by and passing through a single maize leaf (Chapter 8.B) is similar in a dense stand consisting of a succession of leaves (64,000 plants ha^{-1}, $LAI = 4.3$, crop height 300 cm). About 10% of the incident radiation of similar wavelengths was transmitted down to 40 cm above the ground in the visible spectrum while the transmission in the near infrared (750 to 900 nm) range was about 30 to

40% (C. S. Yocum *et al.*, 1964; Cowan, 1968). On a clear day, when the visible radiation received by the crop was about four times that of an overcast period, the spectral distribution did not differ greatly. A comparison of clear and overcast skies on the crop rate of photosynthesis by plants such as maize has been analyzed in relation to the angle of the sun (de Wit and Brouwer, 1969), and when the sun is directly overhead a cloudy sky provides relatively less radiation than does an overcast sky when the sun is at a lower angle from the horizontal. Differences in the spectral composition of radiation from the sun plus sky with the sun at 7 to 10 degrees above the horizon has been compared with 17 to 18 degrees above the horizon, and although from 590 nm upward the light was slightly redder at the higher sun angles, these differences only approached 1% (G. C. Evans, 1969). A geometric analysis of light penetration in stands disclosed that with increasing depths in the canopy the differences between diffuse light and direct sunlight (when the sun is not obscured by clouds) becomes less important because of light scattering and effects of shading by the foliage (Warren Wilson, 1967).

2. RELATION OF LEAF AREA INDEX TO CO_2 ASSIMILATION

The effect of increasing light intensity on net photosynthesis by single leaves has been illustrated in Figs. 8.1 and 8.2, and the manner in which this differs from the effect of the irradiation on CO_2 assimilation by stands is shown in Fig. 9.1. At low light intensities the single leaf shows a greater rate of photosynthesis than does a canopy covering 3.4 times the soil area, because the shaded leaves in a stand fix CO_2 more slowly than brightly illuminated leaves and may show net respiration in deep shade. Although the single leaf shows light saturation at about one-fourth of full sunlight, with increasing radiation the stand assimilates considerably more CO_2 than would a single leaf or an "average" leaf in the stand. Because of shading effects, an average leaf in the canopy would have low rates of CO_2 assimilation at all light intensities, and since in this example the *LAI* was not too great, 3.4, the canopy assimilates at a rate about 3.4 that of an average leaf. It is also clear that the canopy demands more light than a single horizontal leaf in order to exceed the light compensation point. Similarly, CO_2 uptake in single leaves of alfalfa was saturated at 0.12 cal cm^{-2} min^{-1} of visible irradiation, but an alfalfa crop was not saturated until 0.45 cal cm^{-2} min^{-1} of incident radiation reached the top of the canopy (Gaastra, 1962).

As with single leaves (Fig. 8.1), the total irradiation received by a crop that is sufficient to cause saturation varies with the species. Single leaves of maize do not show light saturation, and the stimulation of photosynthesis is also nearly linear with light intensity for maize growing in a stand (Fig.

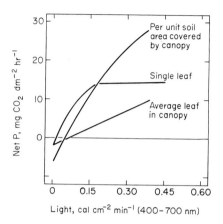

FIG. 9.1. Comparison of effect of irradiance at the top of the canopy (400 to 700 nm) on net photosynthesis of a single exposed leaf, an average leaf in the canopy, and a stand of mustard plants (adapted from de Wit, 1965). The leaf area in the canopy was 3.4 times the ground area (leaf area index = 3.4).

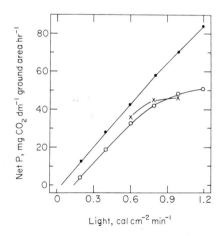

FIG. 9.2. Effect of total solar irradiance at the top of the canopy on net photosynthesis in stands of maize, wheat, and cotton. (●) Maize, $LAI = 4.3$ (data from Baker and Musgrave, 1964); (×) wheat, $LAI = 3.9$ (data from Puckridge, 1969; (○) cotton, $LAI = 5.7$ (data from Baker, 1965).

9.2). The maximal rate of CO_2 assimilation by the stand is also greater in maize than for species whose leaves have high rates of photorespiration (Table 9.2), although the photosynthetically less efficient species fix considerably more CO_2 in brighter light. Stands of wheat and cotton show light

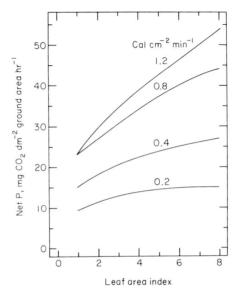

FIG. 9.3. Effect of total solar irradiance at the top of the canopy on net photosynthesis of field-grown soybeans with differing leaf area index (from Jeffers and Shibles, 1969). Rates of photosynthesis for the crop would decrease at higher values of the leaf area index than are shown.

saturation at higher light intensities than do single leaves, and their maximal rates of CO_2 absorption are lower than the photosynthetically more efficient species. These results further emphasize the importance of total irradiation on the productivity of plants in a community growing outdoors (D. N. Moss *et al.*, 1961). The initial slope of CO_2 assimilation with light intensity by crops does not vary very much between species at low light intensities (de Wit, 1967) (as is true with single leaves, Chapter 8.B), and the differences in productivity between species are thus greatest at high light intensities as is shown in Fig. 9.2.

The optimal leaf area index differs among species, and although it changes during the growth of a crop it can also differ greatly in nature from values of 2 to 12 in stands of trees in tropical and mountainous rain forests (Stephens and Waggoner, 1970). The optimum leaf area index may not be the same for varieties within a species. Thus a less productive variety of rice had an optimum *LAI* of 5 to 6 while in a higher yielding variety the optimum *LAI* was 7 to 8 (Chandler, 1969b).

The optimum leaf area index will also depend on the intensity of the incident irradiation. The rates of net photosynthesis were determined in communities of wheat, alfalfa, and clover at an illumination of 1100, 2200 and 3300 ft-c (King and Evans, 1967). Net photosynthesis in all species rose rapidly with increasing *LAI* to 3 to 4, while in wheat and alfalfa assimilation increased slowly up to the highest *LAI* of 10. The results of another study on the effect of irradiance on CO_2 uptake at different leaf area indexes is illustrated for field-grown soybeans in Fig. 9.3. Highest rates of photosynthesis occurred with full sunlight with a leaf area index greater than 8, and as the light intensity decreased net assimilation was diminished and the optimum *LAI* decreased. Clearly with a higher value of the leaf area index than is shown in Fig. 9.3, at lower light intensities net assimilation would begin to decrease as dark respiration becomes increasingly important because of mutual shading of the lower part of the canopy.

3. EFFECT OF PLANT HEIGHT, TRANSIENT LIGHT, AND FOLIAGE DISTRIBUTION

Stems do not usually have very high rates of photosynthesis (Chapter 9.A), and since they respire photosynthate they tend to diminish productivity. Although stems undoubtedly elongate in nature to enable plants to compete for light, such competitive elongation confers no advantage in uniform crops. Hence stems should be as small a part of the canopy as possible to support an optimum leaf area. It has been estimated that a plant 10 cm shorter than those surrounding it would have about 20% less dry matter at low planting densities and about 50% less dry matter at high population; if the height of a single plant were 30 cm below its neighbors, the deficiency in dry matter production would be 40% and 80% at low and high population densities respectively (Duncan, 1969).

The relative contribution of transient irradiation, or sunflecks, to total irradiation in agricultural crops and in forests has recently been investigated. The irradiation by sunflecks deep in the canopies of soybean, corn, and brome-alfalfa, was measured at wavelengths between 400 and 700 nm, and in all stands including maize (which has a more open canopy) the contribution of sunflecks to total irradiation in the upper portions of the canopy was negligible (Norman and Tanner, 1969). At ground level in a hardwood stand in a forest with a closed canopy as well as in a heavily thinned red pine stand, most of the radiant energy was contained in sunflecks of long duration. Therefore short-duration sunflecks do not contribute greatly to total photosynthetic assimilation (Reifsnyder and Furnival, 1970).

The vertical distribution of foliage differs in various species, but the leaf area distribution with changing height is usually symmetrical and follows the normal distribution. Stephens (1969) observed that the vertical distribution of foliage of red pine trees was similar despite differences in the size and age of the trees, stand density, and site quality. For this species the mean of the maximum canopy depth occurred about 50% of the way down from the top of the crown and had a relative standard deviation of 20% of the crown length. Stephens compared the distribution of foliage in various species and found a similar distribution in other coniferous species, while in deciduous trees (*Acacia, Betula, Castanopsis*) the maximum foliage area occurred closer to the top of the tree (37 to 40% from the top of the crown). The relative mean of the maximum canopy depth, as percent from the top of crops was 62% for maize and 36% for red clover (*Trifolium pratense*). Thus the foliage distribution of grasses more nearly resembles conifers and that of dicotyledons is like the foliage distribution of hardwoods.

A comparison has been made of the vertical distribution of the leaves of sorghum and tobacco in the field in relation to the net photosynthesis of individual leaves as measured by exposure to short periods with $^{14}CO_2$ (N. C. Turner and Incoll, 1971). With sorghum (80 cm height, $LAI = 2.3$) the point of maximum leaf density occurred 63% from the top of the plants and the greatest $^{14}CO_2$ uptake was close to the region of maximum leaf density. On the other hand in tobacco (140 cm height, $LAI = 8.5$), the maximum leaf density occurred half-way down the canopy, but maximal photosynthesis occurred at a point considerably closer to the top of the canopy. These results suggest that the differences in leaf density (and perhaps also in leaf angles) between these two types of stands alters the pattern of light penetration and hence the depth at which photosynthesis occurs. The method of measuring crop photosynthesis with $^{14}CO_2$ may be a useful alternative to the aerodynamic or momentum balance method (Chapter 9.D).

4. LEAF ANGLES AND THE CROP EXTINCTION COEFFICIENT (α_L)

The leaf area index determines the efficiency of the utilization of solar energy by plants in communities, but the mean leaf angle from the horizontal in a stand also regulates the degree of penetration of incident light. This was fully appreciated by M. Monsi and T. Saeki in 1953 when they used an ingenious method of describing the light penetration by a crop by analogy with Beer's law for the absorption of light by solutions:

$$I/I_0 = e^{-(\alpha_L)(LAI)} \tag{9.1}$$

where I, I_0, LAI, and α_L are the light intensities measured inside the canopy, at the surface of the top of the plant community, the leaf area index, and the crop extinction coefficient (Saeki, 1963). Therefore,

$$\log_e (I/I_0) = -(\alpha_L)(LAI) \tag{9.2}$$

Equation (9.2) shows that if less of the incident light is absorbed per unit height by the leaves in a stand (or more penetrates the canopy) then α_L

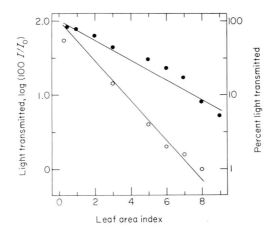

FIG. 9.4. Determination of the crop extinction coefficient (α_L) in plant communities composed largely of grass (●——●) or clover (○——○) by measurement of the light penetration at different depths from the top of the canopy (data from Stern and Donald, 1962). The slope of the line (× 2.303) gives an α_L of 0.30 for the grass sward and 0.62 for the clover sward.

will have a low value. If a great deal of the incident irradiation is absorbed by leaves at the top of a canopy (and less penetrates to lower depths) α_L will be larger. The extinction coefficient α_L will appear to be smallest about noon when the light intensity is highest, and a single horizontal leaf ($LAI = 1$) would have an α_L value of 1 if reflection and transmission of light were negligible.

In most cases grass communities have values of α_L of 0.3 to 0.5 while in dicotyledons the values are generally between 0.7 and 1.0 (Saeki, 1963; R. S. Loomis *et al.*, 1967). From Equation (9.1), it follows that if a canopy with a leaf area index of 1 has an α_L of 1.0, then 37% of the incident radiation would pass to the leaves below it; with a leaf area index of 1.0 and an α_L of 0.5, 61% of the incident radiation received by the leaf would pass through to lower layers.

The crop extinction coefficient can be determined by measuring the incident irradiation at the top of the canopy and at different heights within the canopy with a photocell that is sensitive to visible radiation. The leaf area index is determined at these same heights. The α_L is then obtained from the slope of $\log_e (I/I_0)$ plotted against the leaf area index. An example of such a determination was made from a grass community and one consisting primarily of clover (Fig. 9.4), and the estimated α_L was 0.30 for the grass and 0.62 for the clover. In a similar manner, ryegrass ($\alpha_L = 0.41$) was shown to have more vertically disposed leaves that intercepted less light per unit of leaf area index than clover ($\alpha_L = 0.83$) having more horizontal leaves (Brougham, 1958). Monteith (1969) has summarized the extinction coefficients for sixteen species and lists values of α_L of 0.8 to 1.0 for sunflower, clover, cotton, and beans; 0.5 to 0.7 for maize, rice, and sorghum; and 0.25 to 0.40 for ryegrass.

From geometrical considerations it was concluded that at large values of the leaf area index more erect leaves give greater productivity (R. S. Loomis and Williams, 1969). Experimentally, a variety of maize with horizontal leaves and a *LAI* of 4.1 normally intercepted 99% of the incident light at noon compared with 90% when only the leaves above the ear were maintained upright by mechanical means (Pendleton *et al.*, 1968). The grain yields were only increased from 10,700 to 12,200 kg ha^{-1} by positioning the upper portion of the canopy in a more upright angle. In a comparison of the crop growth rate of barley varieties with erect and horizontal leaves in which both had a high *LAI* of about 9.5, the variety with erect leaves had a 19% greater productivity during a season in which there was normal rainfall and above-normal temperatures (J. W. Tanner, 1969). Thus there is some evidence that a crop architecture consisting of more erect leaves will cause increases in productivity at high planting densities, but the increases do not seem to be very large.

C. RELATION OF STAGE OF PLANT DEVELOPMENT TO PRODUCTIVITY

1. PLANT POPULATION DENSITY AND LEAF AREA INDEX DURING GROWTH

It is necessary that the leaf area of a crop should expand and reach its optimal leaf area index as rapidly as possible in order to obtain maximum productivity. The effect of differences in the population density of maize on light interception and its relation to productivity has been studied intensively. The dry matter that fills the grain in maize is assimilated by the leaves after flowering, and the leaf area remains fairly constant for much of the time between flowering and maturity. In a thorough analysis by Allison (1969) plants were compared at population densities ranging from 23,000 to 73,800 per hectare, and for the 6 to 10 weeks after sowing, the crop growth rate increased about 40% with an increase in the leaf area index from 2 to 5.5. At the lowest planting density a maximal leaf area index of 2.5 was achieved 8 weeks after planting, and a maximal leaf area index of 6.5 was reached 10 weeks after planting at the highest population. The maximum amount of dry matter production after flowering was obtained with a leaf area index of 5 to 6 at the time of flowering (50,000 to 60,000 plants ha^{-1}) (Figs. 9.9 and 9.10), and when the area of leaves covering the ground was smaller much of the light was not intercepted at noon. The grain yield and total dry weight at harvest increased 50% and 30% respectively when the plant populations increased from 23,000 to 48,000 plants ha^{-1}, but did not increase further with populations up to 74,000 ha^{-1}. Light profiles throughout the growing season have also been examined in stands of maize with varying population densities by R. S. Loomis and colleagues (1968).

Thus the optimum leaf area index is determined by rates of CO_2 assimilation and respiration as well as the relation of the leaf architecture and the manner in which the light interception affects photosynthetic and respiratory capacities of the leaves within the canopy.

2. SEASONAL VARIATIONS

Net photosynthesis may also differ for individual leaves within a stand at different stages of development, and this factor may also influence productivity. The net assimilation by tobacco leaf disks taken from leaves harvested at different stages of the growing season has been examined for five tobacco varieties during three successive growing seasons, and the mean values (as mg dry weight dm^{-2} leaf area hr^{-1}) for samples taken at 10-day intervals were 10.2, 10.6, 9.0, 13.0, and 10.6 (Avratovscukova, 1968). Differences in assimilation of as much as 45% were thus obtained in this species at different times. Increases in the net photosynthesis of single soybean leaves were observed in most varieties examined at approximately the stage of seed filling, and leaves in one variety that was sampled throughout the season increased from 32 to 45 mg CO_2 dm^{-2} hr^{-1} (Dornhoff and Shibles, 1970). Progressive decreases in rate beginning 10 to 20 days after unfolding were found in leaves of field-grown alfalfa so that after 30 days the leaves practically ceased to fix CO_2; leaves from a growth chamber, however, were very active 40 to 60 days after unfolding (Pearce *et al.*, 1968; Pearce and Lee, 1969). No explanations based on changes in climate, diffusive resistance, or biochemistry have been put forward to account for these patterns of change of photosynthesis during development.

Many observations indicate that dark respiration decreases continuously in leaves as they age, so that by the end of the leaf cycle dark respiration may only be 10% of that in a young leaf. In rice leaves, for example, when the growth of a new leaf is complete its dark respiration drops to 40% of the previous rate (Ishizuka, 1969). It is unknown whether the rate of dark respiration decreases proportionately with decreasing photosynthesis during the approach of leaf senescence, nor is it known if there is an effect of age on the rate of photorespiration.

The importance of adequate leaf hydration on productivity has already been indicated from the large direct and indirect effects of water stress on leaf expansion and CO_2 assimilation by single leaves (Chapter 8.F). Besides these general effects, there are also more specific effects of water stress that curtail yield when the inadequate water supply occurs at critical periods of the growing season, especially in cereal plants during flowering or grain filling. Thus water stress applied only during the limited periods indicated, diminished grain yield in drought susceptible maize as follows: tassel emerging period, 6%; pollination period, 67%; ear initiation period, 48% (Shaw and Laing, 1966). Water stress therefore gave the greatest reduction in yield when it occurred during the 5-day period of pollination. Even a slight water stress can diminish the rate of the appearance of floral primordia, and the final weight per grain is generally influenced by the state of hydration before and after flowering, but the post-flowering stage is the more critical (Slatyer, 1969).

D. DIFFUSION RESISTANCES AND CO_2 CONCENTRATIONS AND GRADIENTS IN STANDS

1. STOMATAL DIFFUSIVE RESISTANCE

Irradiation and leaf water content are the most important factors regulating stomatal apertures in single leaves (Chapter 7.C), and this is also true in stands when nutrient supply is not limiting. Leaf temperatures, which affect stomatal opening, do not usually vary greatly from the ambient air temperature in illuminated leaves within the canopy (Table 8.4), and the CO_2 concentration in stands is also within 50 ppm of the atmospheric concentration (as shown below) and would thus not influence stomatal opening. Warmer air temperatures would tend to most benefit those species that have lower rates of total respiration in the light (Fig. 8.5). Hence under optimal conditions, solar irradiation within the canopy largely controls stomatal aperture of individual leaves and the stomatal diffusive resistance R_s.

In a dense barley field in the early morning when dew was on the leaves, R_s for the crop was very small (about 0.2 to 0.4 sec cm^{-1}), but it increased to 1.5 sec cm^{-1} as the leaves lost moisture during the day, and finally rose to 3.0 sec cm^{-1} late in the afternoon when irradiance decreased (Monteith *et al.*, 1965). When the leaf area index exceeded 7, there was insufficient illumination reaching the lower barley leaves to allow stomata to open, thus R_s was twice as great in the third leaf as in the uppermost flag leaf. The stomata in the upper leaves of field-grown maize began to close at an irradiance below 60% of full sunlight, and R_s increased sharply when the irradiance decreased to about 10% of the maximal solar irradiation (N. C. Turner, 1969), while R_s doubled at 25% of full sunlight (Waggoner, 1969b). At levels of illumination in the lower canopy which caused the greatest stomatal closure, photosynthetic carboxylation rates would already be low. Stomatal closure in the lower canopy is therefore probably more important in conserving water (because the wavelengths in the infrared penetrate more effectively than the photosynthetically active wavelengths) than in slowing CO_2 uptake.

Since the R_s of a crop depends on the stomatal apertures of both leaf surfaces, it is of interest that stomata on upper and lower surfaces of leaves in sorghum and tobacco stands may differ in their response to illumination (N. C. Turner, 1970). The lower stomata usually receive dimmer light (since only about 20% of the incident light would pass through a horizontal leaf), and when the irradiance was diminished lower stomata closed at 10% of full sunlight in sorghum and 2% of full sunlight in tobacco. Stomata on the upper surface of both species, however, closed at irradiations of 20% of maximal sunlight in both species. This adaptation would tend to prevent the stomata on the lower surface from closing at low light intensities frequently encountered in a canopy.

2. EFFECT OF WIND SPEED

The boundary layer resistance R_a is primarily regulated by the wind speed and it does not greatly limit plant growth when plants have an adequate supply of water (Chapter 7.B). The R_a is generally only a small fraction of the total diffusive resistance to photosynthesis in single leaves. The turbulence of the bulk air within a canopy of leaves will decrease exponentially as does light intensity [Equation (9.1)], and the best evidence therefore indicates that poor ventilation does not ordinarily restrict CO_2 assimilation because the light intensity is simultaneously low at positions in the canopy where R_a might become large. In maize plants (height, 285 cm; 64,000 plants ha^{-1}; $LAI = 4.5$) when the wind speed 424 cm above the ground was 241 cm sec^{-1}, at 216, 140, and 51 cm above ground the wind speed was 99, 32, and 14 cm sec^{-1}. When the wind speed at 424 cm was only 125 cm sec^{-1}, the corresponding values were 51, 17, and 7 cm sec^{-1} (E. R. Lemon *et al.*, 1969). Thus the air in the canopy is not still even when the wind speeds above the crop are slow.

A naturally ventilated maize leaf near the top of the canopy had an R_a not greater than 0.3 sec cm^{-1}, or less than 10% of the estimated total diffusive resistance (Impens *et al.*, 1967). The crop height was 250 cm, and 150 cm from the ground R_a was 0.48 sec cm^{-1}, and at a height of 65 cm, it increased only to 0.54 sec cm^{-1}. Although the other diffusive resistances increased greatly on moving down the canopy, R_a increased only slightly, showing that the stomatal and internal diffusive resistance are much more important in limiting CO_2 uptake than the atmospheric boundary resistances under conditions commonly found in a stand.

3. CO_2 GRADIENTS IN CANOPIES IN RELATION TO OTHER DIFFUSIVE RESISTANCES

Since air within the canopy is moving when wind speeds above the canopy are slow, the diffusive resistance of the bulk air between the layers of leaves in a stand is likewise not very great. Thus the CO_2 transfer in the bulk air surrounding the leaves is probably very rapid in comparison with the rate at which CO_2 can be removed from the ambient atmosphere. This is consistent with the previous suggestion that decreasing stomatal and internal physical diffusive resistances are areas where large increases in net CO_2 assimilation might be achieved (Chapter 7.D and 7.E).

Much of the older literature suggesting that large variations in CO_2 concentration can exist outdoors between the air and near the leaves within the canopy has been misleading. Knowing the CO_2 concentration near photosynthesizing leaves is rather important, because of the effect on net photosynthesis in single leaves brought about by large changes in the CO_2 concentration in the air (Fig. 8.3; Table 8.3). The CO_2 available to a crop is received almost entirely from the atmosphere, partly from the soil, and in small part perhaps from dark respiration at lower levels in the canopy.

The CO_2 concentration in the atmosphere and at different heights in stands was determined at high irradiance at the canopy surface by Inoue (1965). In a wheat canopy (height, 90 cm) the CO_2 concentration at the uppermost layer was 285 ppm, and the lowest concentration was observed 50 cm from the ground (280 ppm), while the concentration immediately above the soil was 294 ppm. The gradients were slightly steeper in a rice nursery bed (height, 20 cm), where the lowest concentration was 257 ppm near the top of the canopy, and just above the soil it was 267 ppm. Hence the largest decrease within the stand from the atmospheric concentration was no greater than 50 ppm of CO_2.

Gradients of CO_2 concentrations similar to those in the rice seedlings were observed in a dense field of maize during the brightest part of the day

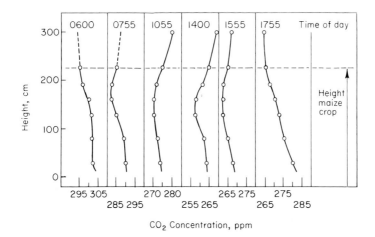

FIG. 9.5. Profiles of CO_2 concentrations at various heights in a stand of maize taken at different times during the day (adapted from E. Lemon, 1967). The field had 26,000 plants per acre (64,200 ha^{-1}) and a leaf area index of 4.48.

(Fig. 9.5), although the CO_2 concentrations were usually less than 50 ppm below the normal atmospheric concentration. The CO_2 levels near the soil were also close to the atmospheric concentration. Crops with very different geometric structures have similar (and small) CO_2 gradients, indicating that the diffusive resistance to CO_2 transport in the air between the plants and the leaf layers must be less than the internal resistances of the leaf (Gaastra, 1969). Hence CO_2 fertilization cannot likely be profitably applied in the field (as discussed below), since added CO_2 would largely escape to the atmosphere. Such small changes in CO_2 concentration as are encountered in a canopy of leaves could not greatly affect the stomatal diffusive resistance in the light (Chapter 7.D) or the rate of net photosynthesis, although the CO_2 concentration will always be limiting outdoors.

An outgrowth of studies on the profiles of CO_2 concentration in canopies was the development of methods of determining the net photosynthesis of crops by taking advantage of certain meteorological parameters. The net photosynthesis of crops is usually estimated by placing a transparent chamber over a portion of a stand and measuring the CO_2 exchange by methods used conventionally for single leaves. The measurement of $^{14}CO_2$ uptake in the field has also been used (Chapter 9.B). From measurements of CO_2 concentrations in different strata above and within a canopy, as in Fig. 9.5, a calculation of net photosynthesis has been made using an "aerodynamic" or "momentum balance" method (E. R. Lemon and Wright, 1969). The wind speed of the bulk air within and outside the canopy and the CO_2 profiles permit an analysis of the rate of CO_2 absorption by means of calculated diffusivity coefficients at different strata in the aerodynamic method. In the second technique, the diffusivity coefficients are calculated by consideration of the vertical profiles of the density of the foliage surface in the canopy. By either technique, the result depends on measurements of these small changes in the CO_2 concentration gradients and somewhat cumbersome calculations. Hence the $^{14}CO_2$ method may become a more versatile one for evaluating net photosynthesis in the field.

The contribution of CO_2 for photosynthesis by the soil derived from the respiration of soil microorganisms and roots is negligible compared with that of the atmosphere. Lundegårdh (1927) determined that the respiration of various bare soils ranged from 1.3 to 4.0 mg CO_2 dm^{-2} soil surface hr^{-1}, and from soil containing oat roots he found that the roots increased the soil respiration about 50%, from 2.7 to 4.0 mg dm^{-2} hr^{-1}. Similar values have been observed more recently. In a maize field soil respiration was 3.5 mg dm^{-2} hr^{-1}, therefore it accounted for only 5% of the CO_2 assimilation on a sunny day (D. N. Moss *et al.*, 1961). A summer maximum of 3.0 mg dm^{-2} hr^{-1} was found for bare soil respiration with a Q_{10} of 3, and the contribution of root respiration was similar to Lundegårdh's (Monteith *et al.*, 1964). Thus in most weather conditions the atmospheric mixing of CO_2 is so vigorous that the CO_2 concentration in the canopy and the atmosphere above it are very similar, and CO_2 assimilation is largely independent of the flux of CO_2 derived from the soil.

4. CO_2 FERTILIZATION

Unlike the situation outdoors, increasing the CO_2 concentration in the atmosphere in an enclosed space such as a greenhouse can have a prodound effect on productivity because the higher CO_2 concentrations can be maintained under these circumstances. A clear demonstration of this principle was described by E. Demoussy in 1904 when he enriched a greenhouse with 1500 ppm of CO_2 and obtained large increases in dry weight of several flowering species. The subject of CO_2 enrichment of greenhouses has been thoroughly discussed by Wittwer and Robb (1964), and they have described mean increases in dry weight of lettuce of 120% during a 42-day period in the low-incident irradiation occurring during the winter months with 800 to 2000 ppm of CO_2 during daylight hours compared with the 125 to 500 ppm in the normal greenhouse. Large increases in the mean dry weight of tomato plants (115%) and fruits (43%) were obtained by these workers by CO_2 enrichment. P. M. Bishop and Whittingham (1968) observed increases in CO_2 uptake of 100% or more by tomato plants in 1000 ppm

CO_2 compared with normal air even at low irradiances, 5 to 20% of full
sunlight (Fig. 8.2), but at low light intensities plants needed a period of
adaptation before they could benefit from the higher CO_2 concentrations.
At low light intensities tomato plants had 41% more dry weight when
grown in 1000 ppm compared with 350 ppm of CO_2 (Hurd, 1968). Perhaps
the greater rates of growth at low light intensities than might be expected
occur in these species because photorespiration is inhibited by the higher
levels of CO_2. Other reports of productivity enhanced by CO_2 fertilization
in greenhouses or growth chambers include an increase of 45% in the dry
weight yield of soybean straw and a 37% increase in the seed at 1350 ppm
(R. L. Cooper and Brun, 1967), and a doubling of the stem lengths of
seedlings of birch (*Betula*) and crabapple (*Malus toringoides*) at 2000 ppm
of CO_2 (Krizek *et al.*, 1969).

E. REGULATION OF DARK RESPIRATION
AND PHOTORESPIRATION

1. MAGNITUDE OF LOSSES BY DARK RESPIRATION

Since the rate of dark respiration in photosynthetic tissues is only 5 to
10% of the CO_2 assimilation in bright light, it is often assumed that dark
respiration is an unimportant factor in plant productivity, and that losses
may occur of about 20 to 25% of crop photosynthesis on a daily basis.
The periods of maximal irradiation exist a brief part of the day, and the
hours of darkness far exceed those of rapid photosynthesis. Moreover,
dark respiration probably continues unabated in the light (Chapter 5.B).
The lower leaves in a stand are shaded to varying degrees depending on
the leaf area index and the crop extinction coefficient (Chapter 9.B.4) and
carry out little photosynthesis. There are also tissues such as stems, roots,
and fruits that only respire and carry out little or no CO_2 absorption.

A budget for the CO_2 uptake and release by various tissues has been
constructed for maize plants growing for 30 days in a chamber, including
the contributions of leaves, stems, and roots (G. Heichel, 1971b). It is re-
assuring that the instantaneous measurements of CO_2 exchange agreed
with the dry weight gain within 1%.

Some examples of the minimal losses of total photosynthetic CO_2 uptake,
caused only by dark respiration, are listed for different species in Table 9.3.
The percent loss by respiration increases as the leaf density becomes greater,
but decreases of 40 to 70% of the fixed carbon are more common than
lower values, which mainly represent examples where only the upper part

TABLE 9.3   *Minimal Quantity of Gross Photosynthesis Lost by Dark Respiration in Various Species Growing in Stands*[a]

Species	Ratio of total dark respiration to "gross" photosynthesis (%)	Reference
Alfalfa	33–49	Thomas and Hill (1949)
Alfalfa	70	King and Evans (1967)
Maize	16–29[b]	E. Lemon (1969)
Maize	22,33[c]	G. Heichel (1971b)
Wheat	48	King and Evans (1967)
Wheat	27[b]	E. Lemon (1969)
Rice	52,71[c]	Tanaka *et al.* (1966)
Beans	26,160[c]	Monteith (1962)
Various terrestrial communities	40–70	Westlake (1963)
Tropical rain forest	66	E. Lemon *et al.* (1970)
African rain forest	75	Stephens and Waggoner (1970)

[a] "Gross" photosynthetic assimilation was assumed to equal total net photosynthesis plus dark respiration in the light, but any losses resulting from photorespiration were neglected in all of these examples.

[b] Upper leaves only.

[c] The two values represent measurements taken at different times during the growing season. The last value for beans was taken after the plants already had pods, and the loss was greater than the photosynthetic assimilation. The maize experiment was carried out in a growth chamber.

of the canopy was measured, or where plants were raised in a growth chamber rather than outdoors. These losses are minimal ones for many species, because the refixation of CO_2 generated by photorespiration within tissues (Table 6.5) has not even been considered in estimating the "gross" photosynthesis in Table 9.3. Certainly dark respiration is essential for many growth processes, but the question of how much is essential and what proportion may be a "waste" respiration in various tissues is not known (Chapter 6.E). Since such a large part of the assimilated carbon is released by this process, it is obviously important to establish how much is coupled to useful synthetic processes and what proportion might be eliminated and hence result in gains of productivity.

2. CONTROL OF PHOTORESPIRATION

Photorespiration increases with greater irradiation in a manner similar to photosynthesis, and increases greatly at warmer temperatures (Q_{10} about 3) in species that have high rates (Chapters 5 and 6). Thus photo-

TABLE 9.4   *Effect of O_2 Concentration on Dry Matter Production at Different CO_2 Levels in Mimulus and Maize[a]*

Species	% CO_2 Concentration	Dry weight increase in 10 days (mg/plant)		% Increase in 4% O_2; growth in 21% $O_2 = 100$
		21% O_2	4% O_2	
Mimulus cardinalis	0.011	< 10[b]	150[b]	> 1000
	0.032	565[c]	1076[c]	190
	0.064	804[c]	1144[c]	142
Zea mays	0.011	218[d]	196[d]	94
	0.032	1269[e]	1473[e]	116

[a] Data from Björkman *et al.*, 1969.
[b] Difference between means very highly significant. $P < 0.01$.
[c] Difference between means highly significant. $P < 0.05$.
[d] Difference between means not significant. $P > 0.5$.
[e] Difference between means slightly significant. $P < 0.4$.

respiration may account for 50% or more of the gross photosynthetic assimilation (Table 6.5) at high light intensities in single leaves. Since photorespiration occurs simultaneously with photosynthesis, its overall effect on a daily basis (light plus darkness) in a stand may not be any greater than that of dark respiration (Table 9.3), but it would nevertheless constitute a substantial part of the CO_2 budget.

Evidence of the importance of photorespiration in diminishing productivity in stands has already been presented in Table 9.2, which shows that species with high rates of photorespiration have crop growth rates about half of those which do not. The growth of three tobacco varieties that differ in their rates of photorespiration has also been examined (Zelitch and Day, 1968), and the incremental increase in leaf area over a 13-day period was greatest in the variety with a low rate of photorespiration. Hence this line of investigation continues to remain a promising one for increasing crop productivity.

Photorespiration in single leaves iš greatly dependent on the ambient O_2 concentration, and the rate of CO_2 release is negligible at about 2% O_2 (Chapter 5, Chapter 6.E). This characteristic of photorespiration probably accounts for the inhibition of net photosynthesis by O_2 in species showing the Warburg effect (Chapter 8.E, Table 8.5, Fig. 8.4). Björkman and his associates (1969) grew *Mimulus* (high photorespiration) and maize (low photorespiration) plants with their roots in normal air and the aerial portions in an atmosphere containing various concentrations of CO_2 in

4% or 21% O_2 (Table 9.4). After 10 days of continuous illumination, measurement of the increases in dry weight showed that the growth of *Mimulus* was inhibited by the higher O_2 concentration while that of maize was not. At 4% O_2 the dry weight accumulation by *Mimulus* plants at 320 ppm of CO_2 approximately doubled and approached the dry weight gain of maize. This experiment demonstrates that large increases in crop growth rates should be anticipated in stands if photorespiration were inhibited by biochemical or genetic means, and the effect of decreasing photorespiration would probably be greatest (as it is in maize) at warmer temperatures and high total irradiance.

F. SIMULATING THE ACTIVITIES OF STANDS WITH COMPUTERS

The availability of high-speed digital computers has permitted the construction of models that simulate how various parts of the canopy function in a system that ultimately determines the productivity of a stand. The simulation of CO_2 assimilation by single leaves and its usefulness in determining where the barriers to CO_2 uptake probably occur has already been described briefly in Chapter 7.E. Stands of plants have also been subjected to analysis by models of several different types. The distribution of light within a canopy of leaves was analyzed with a geometrical model as a function of the solar leaf angles and the leaf area distribution (de Wit, 1965). This simulator disclosed that with a maximum *LAI* of five for a grass or a small grain crop, the crop growth rate on a perfectly clear day was a reasonable 28, 40 or 44 gm dry weight m^{-2} ground area day^{-1} (Table 9.2) with the sun at a fixed height of 30, 60, or 90 degrees from the horizontal (de Wit, 1967).

Duncan (1967) has also created "light" models of stands to analyze CO_2 assimilation using such parameters as leaf area, leaf angle, leaf position, transmission of light, the effect of irradiance on net photosynthesis, and the brightness and position of the sun. For a species such as maize, at a *LAI* of 8, and the sun a constant 90 degrees from the horizontal, the simulated crop growth rate was 60 gm dry weight m^{-2} ground area day^{-1}, and at an angle of 45 degrees it was 45 gm dry weight m^{-2} ground area day^{-1} (R. S. Loomis *et al.*, 1967). However, the light model suggested that the crop growth rate would continue to increase up to a *LAI* of 16, and this is an unlikely result even for a stand with erect leaves.

None of the models mentioned above considers any physiological parameters associated with CO_2 assimilation, or the climatological changes within canopies. Inoue (1965) simulated the ventilation in a canopy and thereby estimated the CO_2 concentration profiles within stands and calculated photosynthetic and respiration rates. The previously mentioned model for single leaves (Waggoner, 1969a; Chapter 7.E) which separates the "meso-phyll resistance" into its component physical and biochemical parts (Fig. 5.1) has been expanded to simulate the activities of plants in stands (Waggoner, 1969b).

Waggoner's model considers light, ventilation, plant physiology and bio-chemistry and how these all interact. Leaf temperatures are estimated from the temperature and humidity of the air at the top of the canopy and near the soil, the ventilation and radiation in the canopy, and its architecture and stomatal resistance. Then the effect of temperature on the carboxyla-tion reaction and total respiration is included. Profiles of net radiation and its effect on stomatal diffusive resistance, the effect of leaf angles, photo-synthetic carboxylation, and respiration are also considered. The physical resistances in the atmosphere, at the leaf boundary layer, and within the leaf are also accounted for. The canopy is conceived as a number of strata of leaves whose rates of CO_2 assimilation are controlled by potentials and resistances in a manner shown in Fig. 5.1. The canopy is then simulated as a ladder of conductors between strata of single leaves, each under the control of the various environmental or physiological parameters prescribed. The functioning of the entire stand under different stated circumstances can then be evaluated.

A comparison of the crop growth rates of an efficient species (maize, with zero photorespiration in the model) and an inefficient one (tobacco, with a rapid photorespiration) has been described for a leaf area index of 4 at $20°$ and a total solar irradiation of 690 cal cm^{-2} day^{-1}. The crop growth was a reasonable 46 gm dry weight m^{-2} ground area day^{-1} for the efficient species (Table 9.2) and 35 gm dry weight m^{-2} ground area day^{-1} for the less efficient one (Waggoner, 1969b). The simulator also showed that decreasing the wind speed at the top of the canopy from 225 to 22 cm sec^{-1} decreased photosynthesis of the canopy by less than 2% (Chapter 7.B).

In the standard case, single leaves of both types of species at $20°$ have a dark respiration of 2 mg dm^{-2} hr^{-1} (Q_{10} of 2), a K_m for CO_2 and incident total irradiation of 300 ppm and 0.5 cal cm^{-2} min^{-1}, and a maximal gross photosynthesis (saturation light and CO_2 and no resistances) of 180 mg CO_2 dm^{-2} hr^{-1} (Q_{10} of 2). In addition to these characteristics the less efficient species has a photorespiration in bright light of 10 mg dm^{-2} hr^{-1} at $20°$ (Q_{10} of 3) and a K_m of 1.0 cal cm^{-2} min^{-1} for the increase of photo-respiration with radiation. The stomatal resistance of both species has a

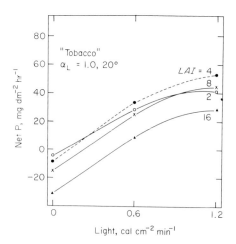

FIG. 9.6. Simulated net photosynthesis of a stand of an inefficient species with a crop extinction coefficient α_L of 1.0 (provided by P. E. Waggoner). The CO_2 assimilation is given per ground area. The total solar irradiation and the leaf area index have been varied.

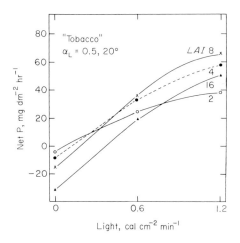

FIG. 9.7. Simulated net photosynthesis of a stand of an inefficient species with a crop extinction coefficient α_L of 0.5. As Fig. 9.6 except α_L was 0.5 rather than 1.0.

minimal value R_s of 2.0 sec cm^{-1} which doubled between full sunlight and 0.3 cal cm^{-2} min^{-1} of total irradiation. At the top of the canopy the wind speed was 225 cm sec^{-1}, and the wind and diffusivity decreased exponentially within the canopy.

Further results obtained from this simulator (Waggoner, 1969b) that have not yet been published are presented to demonstrate its usefulness, and to help summarize the experimental findings and principles relating to photosynthesis and productivity that are described in this and earlier chapters.

The simulated CO_2 uptake of tobacco when the leaves are arranged horizontally as is implied by the crop extinction coefficient (Chapter 9.B.4) is shown in Fig. 9.6, and the results when they are more erect are illustrated in Fig. 9.7. The CO_2 exchange is shown in darkness, one-half of full sunlight, and full sunlight. At maximal irradiation, the greatest photosynthesis occurs with a *LAI* of 4 for horizontal leaves and 8 for more erect leaves. In addition, when the leaves are more erect, increasing the *LAI* from 2 to 8 makes the increment of assimilation per increment of radiation greater. At half-maximal solar irradiation, however, there is little difference between a *LAI* of 4 and 8 whether the leaves are horizontal or erect. At a *LAI* of 16, the CO_2 uptake is slight at any irradiation.

The crop growth rates on a daily basis while receiving 690 cal cm^{-2} day^{-1} incident irradiation at the top of the canopy in both types of crops are simulated in Fig. 9.8 (with the leaves in a horizontal position) and in Fig. 9.9 (with the leaves more erect). The temperature at the top of the canopy is assumed to be constant day and night. The accumulation of dry matter with increasing leaf area increases to a maximum of about 45 in maize and 35 in tobacco at a *LAI* of 4 when the leaves are horizontal; when the leaves are more erect, the optimal *LAI* is increased to about 6 and the productivity in maize or tobacco changes very little in spite of the greater photosynthesis suggested above. Presumably this is because both crops have considerable rates of dark respiration. Similar optimal values of the *LAI* on the crop growth observed in the field (Chapter 9.B.2) further sustain the realism of this model.

If the temperature at the top of the canopy is increased to a constant 30° (Fig. 9.10) compared with 20° (Fig. 9.9) with the leaves in a more erect position, the optimal *LAI* remains the same but the productivity revealed in the simulator increases in maize from 45 to 65 and in tobacco only from 35 to 41. This suggests, as we have already seen for single leaves (Fig. 8.5), that warmer temperatures greatly increase the productivity of the efficient

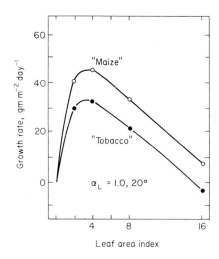

FIG. 9.8. Simulated crop growth rate of a photosynthetically efficient and inefficient species with a crop extinction coefficient of 1.0 at different values of the leaf area index. The growth rate is given as gm dry weight m^{-2} ground area day^{-1}. A "day" is defined as one part full sun, two parts half sun, and two parts of darkness, where total solar irradiation is 1.2 cal cm^{-2} min^{-1}. The daily temperature cycle was ignored and temperature was a steady 20° at the top of the canopy.

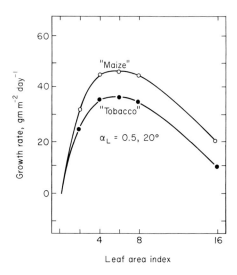

FIG. 9.9. Simulated crop growth rate of a photosynthetically efficient and inefficient species with a crop extinction coefficient of 0.5 at different values of the leaf area index. As Fig. 9.8 except α_L was 0.5 rather than 1.0.

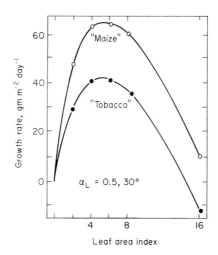

FIG. 9.10. Simulated crop growth rate of a photosynthetically efficient and inefficient species with a crop extinction coefficient of 0.5 at different values of the leaf area index. As Fig. 9.9 except that temperature was a steady 30° at the top of the canopy rather than 20°.

species (because photorespiration rates are still low). Hence the crop growth rate may be increased by 50% in maize when the temperature is raised from 20° to 30° largely because of the biochemically diminished rate of photorespiration.

Thus Waggoner's model improves greatly on earlier ones for determining the effect of various parameters on net photosynthesis and productivity, and it demonstrates how our knowledge of physiology and meteorology can be integrated realistically. This is a useful beginning, and future models will no doubt attempt to analyze such problems as the growth of the crop and of specific organs in relation to the total productivity. However, considerably more biological data will be required before computations of such complex interactions can be undertaken.

Bibliography

Adams, M. S., and Strain, B. R. (1969). Seasonal photosynthetic rates in stems of *Cercidium floridum* Benth. *Photosynthetica* **3**, 55–62.

Akazawa, J., and Ramakrishnan, L. (1967). Change in chloroplast proteins of the rice plant infected by the blast fungus, *Piricularia oryzae*. *In* "The Dynamic Role of Molecular Constituents in Plant-Parasite Interaction" (C. J. Mirocha and I. Uritani, eds.), pp. 329–341. Am. Phytopathol. Soc., St. Paul, Minnesota.

Akazawa, T., and Sugiyama, T. (1969). Subunit structure of carboxydismutase. *Abstr. Int. Bot. Congr.; 11th, 1969,* p. 1.

Allen, M. B., Arnon, D. I., Capindale, J. B., Whatley, F. R., and Durham, L. J. (1955). Photosynthetically isolated chloroplasts. III. Evidence for complete photosynthesis. *J. Amer. Chem. Soc.* **77**, 4149–4155.

Allison, J. C. S. (1969). Effect of plant population on the production and distribution of dry matter in maize. *Ann. Appl. Biol.* **63**, 135–144.

Allison, J. C. S., and Watson, D. J. (1966). The production and distribution of dry matter in maize after flowering. *Ann. Bot. (London)* [N. S.] **30**, 365–381.

Altman, P. L., and Dittmer, D. S., eds. (1968). "Metabolism." Fed. Am. Soc. Exptl. Biol., Bethesda, Maryland.

Alvim, P. de T. (1965). A new type of porometer for measuring stomatal opening and its use in irrigation studies. *UNESCO Arid Zone Res.* **25**, 325–329.

Andersen, W. R., Wildner, G. F., and Criddle, R. S. (1970). Ribulose diphosphate carboxylase. III. Altered forms of ribulose diphosphate carboxylase from mutant tomato plants. *Arch. Biochem. Biophys.* **137**, 84–90.

Anderson, L., and Fuller, R. C. (1967). The rapid appearance of glycolate during photosynthesis in *Rhodospirillum rubrum*. *Biochim. Biophys. Acta* **131**, 198–201.

Anderson, J. M., and Pyliotis, N. A. (1969). Studies with manganese-deficient spinach chloroplasts. *Biochim. Biophys. Acta* **189**, 280–293.

Apel, P., and Lehmann, C. O. (1969). Variabilität und Sortenspezifität der Photosyntheserate bei Sommergerste. *Photosynthetica* **3**, 255–262.

Arnon, D. I. (1958). Chloroplasts and photosynthesis. *Brookhaven Symp. Biol.* **11**, 181–235.

Arnon, D. I., Tsujimoto, H. Y., and McSwain, B. D. (1967). Ferredoxin and photosynthetic phosphorylation. *Nature (London)* **214**, 562–566.

Asada, K., and Kasai, Z. (1962). Inhibition of the photosynthetic CO_2 fixation of green plants by α-hydroxysulfonates, and its effects on the assimilation products. *Plant Cell Physiol.* **3**, 125–136.

Asada, K., Saito, K., Kitoh, S., and Kasai, Z. (1965a). Photosynthesis of glycine and serine in green plants. *Plant Cell Physiol.* **6**, 47–59.

Asada, K., Kitoh, S., Deura, R., and Kasai, Z. (1965b). Effect of α-hydroxysulfonates on photochemical reactions of spinach chloroplasts and participation of glyoxylate in photophosphorylation. *Plant Cell Physiol.* **6**, 615–629.

Atkinson, D. E. (1966). Regulation of enzyme activity. *Annu. Rev. Biochem.* **35**, 85–124.

Atkinson, D. E. (1968). The energy charge of the adenylate pool as a regulatory parameter. Interaction with feedback modifiers. *Biochemistry* **7**, 4030–4034.

Avratovscukova, N. (1968). Differences in photosynthetic rate of leaf disks in five tobacco varieties. *Photosynthetica* **2**, 149–160.

Avron, M. (1960). Photophosphorylation by Swiss-chard chloroplasts. *Biochim. Biophys. Acta* **40**, 257–272.

Avron, M., and Chance, B. (1967). Relation of phosphorylation to electron transport in isolated chloroplasts. *In* "Energy Conversion by the Photosynthetic Apparatus," pp. 149–160. Brookhaven Natl. Lab., Upton, New York.

Avron, M., and Neumann, J. (1968). Photophosphorylation in chloroplasts. *Annu. Rev. Plant Physiol.* **19**, 137–166.

Avron, M., and Shavitt, N. (1965). Inhibitors and uncouplers of photophosphorylation. *Biochim. Biophys. Acta* **109**, 317–331.

Axelrod, B., and Beevers, H. (1956). Mechanisms of carbohydrate breakdown in plants. *Annu. Rev. Plant Physiol.* **7**, 267–298.

Baker, D. N. (1965). Effects of certain environmental factors on net assimilation in cotton. *Crop Sci.* **5**, 53–56.

Baker, D. N., and Musgrave, R. B. (1964). Photosynthesis under field conditions. V. Further plant chamber studies of the effects of light on corn (*Zea mays*). *Crop Sci.* **4**, 127–131.

Baldry, C. W., Coombs, J., and Gross, D. (1968). Isolation and separation of chloroplasts from sugar cane. *Z. Pflanzenphysiol.* **60**, 78–81.

Baldry, C. W., Bucke, C., and Coombs, J. (1969). Light/phosphoenolpyruvate dependent carbon dioxide fixation by isolated sugar cane chloroplasts. *Biochem. Biophys. Res. Commun.* **37**, 828–832.

Balegh, S. E., and Biddulph, O. (1970). The photosynthetic action spectrum of the bean plant. *Plant Physiol.* **46**, 1–5.

Bamji, M. S., and Jagendorf, A. T. (1966). Amino acid incorporation by wheat chloroplasts. *Plant Physiol.* **41**, 764–770.

Bandurski, R. S. (1955). Further studies on the enzymatic synthesis of oxalacetate from phosphorylenolpyruvate and carbon dioxide. *J. Biol. Chem.* **217**, 137–150.

Bandurski, R. S., and Greiner, C. M. (1953). The enzymatic synthesis of oxalacetate from phosphoryl-enolpyruvate and carbon dioxide. *J. Biol. Chem.* **204**, 781–786.

Bange, G. G. J. (1953). On the quantitative explanation of stomatal transpiration. *Acta Bot. Neer.* **2**, 255–297.

Bassham, J. A. (1964). Kinetic studies of the photosynthetic carbon reduction cycle. *Annu. Rev. Plant Physiol.* **15**, 101–120.

Bassham, J. A. (1968). The Calvin photosynthetic cycle in algae and leaves. *In* "Photosynthesis in Sugar Cane," (J. Coombs, ed.) pp. 1–14. Imperial College, London.

Bassham, J. A., and Calvin, M. (1957). "The Path of Carbon in Photosynthesis." Prentice-Hall, Englewood Cliffs, New Jersey.

Bassham, J. A., and Jensen, R. G. (1967). Photosynthesis of carbon compounds. *In* "Harvesting the Sun" (A. San Pietro, F. A. Greer, and T. J. Army, eds.), pp. 79–110. Academic Press, New York.

Bassham, J. A., and Kirk, M. (1960). Dynamics of the photosynthesis of carbon compounds. I. Carboxylation reactions. *Biochim. Biophys. Acta* **43**, 447–464.

Bassham, J. A., and Kirk, M. (1962). The effect of oxygen on the reduction of CO_2 to glycolic acid and other products during photosynthesis by *Chlorella*. *Biochem. Biophys. Res. Commun.* **9**, 376–380.

Bassham, J. A., Kirk, M., and Jensen, R. G. (1968). Photosynthesis by isolated chloroplasts. I. Diffusion of labeled photosynthetic intermediates between isolated chloroplasts and suspending medium. *Biochim. Biophys. Acta* **153**, 211–218.

Becker, J.-D., Döhler, G., and Egle, K. (1968). Die Wirkung monochromatischen Lichts auf die extrazelluläre Glykolsäure-Ausscheidung bei der Photosynthese von *Chlorella*. *Z. Pflanzenphysiol.* **58**, 212–221.

Beevers, H. (1961). "Respiratory Metabolism in Plants." Harper & Row, New York.

Begg, J. E. (1965). High photosynthetic efficiency in a low-latitude environment. *Nature (London)* **205**, 1025–1026.

Begg, J. E., and Jarvis, P. G. (1968). Photosynthesis in Townsville lucerne (*Stylosanthes humilis*). *Agr. Meteorol.* **5**, 91–109.

Begg, J. E., and Turner, N. C. (1971). Profiles of stomatal resistance and leaf water potential in three field crops. *Annu. Rep. CSIRO Div. Land Res.,* 1969–1970, p. 71.

Bendall, D. S. (1968). Oxidation-reduction potentials of cytochromes in chloroplasts from higher plants. *Biochem. J.* **109**, No. 3, 46P–47P.

Bendall, D. S., and Hill, R. (1968). Haem-proteins in photosynthesis. *Annu. Rev. Plant Physiol.* **19**, 167–186.

Bender, M. M. (1968). Mass spectrometric studies of carbon 13 variations in corn and other grasses. *Radiocarbon* **10**, 468–472.

Ben Hayyim, G., and Avron, M. (1970). Cytochrome b of isolated chloroplasts. *Eur. J. Biochem.* **14**, 205–213.

Benson, A. A. (1961). Lipid function in the photosynthetic structure. *In* "Light and Life" (W. D. McElroy and B. Glass, eds.), pp. 392–396. Johns Hopkins Press, Baltimore, Maryland.

Benson, A. A. (1963). Chlorophyll's lipid environment. *In* "Photosynthetic Mechanisms of Green Plants," pp. 571–574. Natl. Acad. Sci.—Natl. Res. Council, Washington, D. C.

Benson, A. A., and Calvin, M. (1950). The path of carbon in photosynthesis. VII. Respiration and photosynthesis. *J. Exp. Bot.* **1**, 63–68.

Berry, J. A., Downton, J. S., and Tregunna, E. B. (1970). The photosynthetic carbon metabolism of *Zea mays* and *Gomphrena globosa*: The location of the CO_2 fixation and the carboxyl transfer reactions. *Can. J. Bot.* **48**, 777–786.

Bidwell, R. G. S., Levin, W. B., and Shephard, D. C. (1969). Photosynthesis, photorespiration, and respiration of chloroplasts from *Acetabularia mediterrania*. *Plant Physiol.* **44**, 946–954.

Biggins, J., and Park, R. B. (1964). Nucleic acid content of chloroplasts of spinach isolated by a non-aqueous technique. *Nature (London)* **203**, 425–426.

Bishop, N. I. (1958). The influence of the herbicide, DCMU, on the oxygen-evolving system of photosynthesis. *Biochim. Biophys. Acta* **27**, 205–206.

Bishop, P. M., and Whittingham, C. P. (1968). The photosynthesis of tomato plants in a carbon dioxide enriched atmosphere. *Photosynthetica* **2**, 31–38.

Björkman, O. (1966). The effect of oxygen concentration on photosynthesis in higher plants. *Physiol. Plant.* **19**, 618–633.

Björkman, O. (1968). Carboxydismutase activity of shade-adapted and sun-adapted species of higher plants. *Physiol. Plant.* **21**, 1–10.

Björkman, O., and Gauhl, E. (1969). Carboxydismutase activity in plants with and without β-carboxylation photosynthesis. *Planta* **88**, 197–203.

Björkman, O., Gauhl, E., Hiesey, W. M., Nicholson, F., and Nobs, M. A. (1969). Growth of *Mimulus, Marchantia* and *Zea* under different oxygen and carbon dioxide levels. *Annu. Rep., Carnegie Inst. Wash., 1967–1968* pp. 477–479.

Black, C. C., Jr. (1967). Evidence supporting the theory of two sites of photophosphorylation in green plants. *Biochem. Biophys. Res. Commun.* **28**, 985–990.

Black, C. C., Chen, T. M., and Brown, R. H. (1969). Biochemical basis for plant competition. *Weed Sci.* **17**, 338–344.

Boardman, N. K. (1968). The photochemical systems of photosynthesis. *Advan. Enzymol.* **30**, 1–79.

Boardman, N. K., and Anderson, J. M. (1966). Fractionation of the photochemical systems of photosynthesis. II. Cytochrome and carotenoid contents of particles isolated from spinach chloroplasts. *Biochim. Biophys. Acta* **143**, 187–203.

Boardman, N. K., Francki, R. I. B., and Wildman, S. G. (1966). Protein synthesis by cell-free extracts of tobacco leaves. III. Comparison of the physical properties and protein-synthesizing capacities of 70 S chloroplast and 80 S cytoplasmic ribosomes. *J. Mol. Biol.* **17**, 470–487.

Bogorad, L. (1965). Chlorophyll biosynthesis. *In* "Chemistry and Biochemistry of Plant Pigments" (T. W. Goodwin, ed.), pp. 29–74. Academic Press, New York.

Bogorad, L. (1967). Chloroplast structure and development. *In* "Harvesting the Sun" (A. San Pietro, F. A. Greer, and T. J. Army, eds.), pp. 191–210. Academic Press, New York.

Böhme, H., and Trebst, A. (1969). On the properties of ascorbate photooxidation in isolated chloroplasts. Evidence for two ATP sites in noncyclic photophosphorylation. *Biochim. Biophys. Acta* **180**, 137–148.

Bomsel, J.-L., and Pradet, A. (1968). Study of adenosine 5'-mono-, di-, and triphosphates in plant tissues. IV. Regulation of the level of nucleotides, *in vivo*, by adenylate kinase: Theoretical and experimental study. *Biochim. Biophys. Acta* **162**, 230–242.

Bonner, J. (1965). Ribosomes. *In* "Plant Biochemistry" (J. Bonner and J. E. Varner, eds.), 2nd ed., pp. 21–37. Academic Press, New York.

Bonner, W., and Hill, R. (1963). Light induced optical changes in green leaves. *In* "Photosynthetic Mechanisms of Green Plants," pp. 82–90. Natl. Acad. Sci.—Natl. Res. Council, Washington, D. C.

Bottrill, D. E., and Possingham, J. V. (1969). Isolation procedures affecting the retention of water-soluble nitrogen by spinach chloroplasts in aqueous media. *Biochim. Biophys. Acta* **189**, 74–79.

Bové, J., and Raacke, I. D. (1959). Amino acid-activating enzymes in isolated chloroplasts from spinach leaves. *Arch. Biochem.* **85**, 521–531.

Bové, J. M., Bové, C., Whatley, F. R., and Arnon, D. I. (1963). Chloride requirement for oxygen evolution in photosynthesis. *Z. Naturforsch. B* **18**, 683–688.

Boyer, J. S. (1970a). Leaf enlargement and metabolic rates in corn, soybean, and sunflower at various leaf water potentials. *Plant Physiol.* **46**, 233–235.

Boyer, J. S. (1970b). Differing sensitivity of photosynthesis to low leaf water potentials in corn and soybean. *Plant Physiol.* **46**, 236–239.

Boyer, J. S., and Knipling, E. B. (1965). Isopiestic technique for measuring leaf water potentials with a thermocouple psychrometer. *Proc. Nat. Acad. Sci. U. S.* **54**, 1044–1051.

Bradbeer, J. W., and Anderson, C. M. A. (1967). Glycollate formation in chloroplast preparations. *In* "The Biochemistry of Chloroplasts" (T. W. Goodwin, ed.), Vol. 2, pp. 175–179. Academic Press, New York.

Bradbeer, J. W., and Racker, E. (1961). Glycolate formation from fructose-6-phosphate by cell-free preparations. *Fed. Proc., Fed. Amer. Soc. Exp. Biol.* **20**, 88d.

Bradbeer, J. W., Ranson, S. L., and Stiller, M. (1958). Malate synthesis in crassulacean leaves. I. The distribution of C^{14} in malate of leaves exposed to $C^{14}O_2$ in the dark. *Plant Physiol.* **33**, 66–70.

Bravdo, B.-A. (1968). Decrease in net photosynthesis caused by respiration. *Plant Physiol.* **43**, 479–483.

Brawerman, G., and Eisenstadt, J. M. (1967). The nucleic acids associated with the chloroplasts of *Euglena gracilis* and their role in protein synthesis. *In* "Organizational Biosynthesis" (H. J. Vogel, J. O. Lampen, and V. Bryson, eds.), pp. 419–434. Academic Press, New York.

Breidenbach, R. W., and Beevers, H. (1967). Association of the glyoxylate cycle enzymes in a novel subcellular particle from castor bean endosperm. *Biochem. Biophys. Res. Commun.* **27**, 462–469.

Breidenbach, R. W., Kahn, A., and Beevers, H. (1968). Characterization of glyoxysomes from castor bean endosperm. *Plant Physiol.* **43**, 705–713.

Brix, H. (1968). Influence of light intensity at different temperatures on rate of respiration of Douglas-fir seedlings. *Plant Physiol.* **43**, 389–393.

Brody, S. S., and Brody, M. (1963). Aggregated chlorophyll *in vivo. In* "Photosynthetic Mechanisms of Green Plants," pp. 455–478. Natl. Acad. Sci. — Natl. Res. Council, Washington, D. C.

Brooks, J. L., and Stumpf, P. K. (1965). A soluble fatty acid synthesizing system from lettuce chloroplasts. *Biochim. Biophys. Acta* **98**, 213–216.

Brooks, J. L., and Stumpf, P. K. (1966). Fat metabolism in higher plants. XXXIX. Properties of a soluble fatty acid synthesizing system from lettuce chloroplasts. *Arch. Biochem. Biophys.* **116**, 108–116.

Brougham, R. W. (1958). Interception of light by the foliage of pure and mixed stands of pasture plants. *Aust. J. Agr. Res.* **9**, 39–52.

Broun, G., Selegny, E., Tran Minh, C., and Thomas, D. (1970). Facilitated transport of CO_2 across a membrane bearing carbonic anhydrase. *FEBS Lett.* **7**, 223–226.

Brown, A. H. (1953). The effects of light on respiration using isotopically enriched oxygen. *Amer. J. Bot.* **40**, 719–729.

Brown, D. L., and Tregunna, E. B. (1967). Inhibition of respiration during photosynthesis by some algae. *Can. J. Bot.* **45**, 1135–1143.

Brown, S. A. (1966). Lignins. *Annu. Rev. Plant Physiol.* **17**, 223–244.

Brown, W. V. (1958). Leaf anatomy in grass systematics. *Bot. Gaz.* **119**, 170–178.

Brown, W. V., and Johnson, S. C. (1962). The fine structure of the grass guard cell. *Amer. J. Bot.* **49**, 110–115.

Bruinsma, J. (1961). A comment on the spectrophotometric determination of chlorophyll. *Biochim. Biophys. Acta* **52**, 576–578.

Buchanan, B. B., and Arnon, D. I. (1965). Ferredoxin-dependent synthesis of labeled pyruvate from acetyl coenzyme A and carbon dioxide. *Biochem. Biophys. Res. Commun.* **20**, 163–168.

Buchanan, B. B., and Evans, M. C. W. (1965). The synthesis of α-ketoglutarate from succinate and carbon dioxide by a subcellular preparation of a photosynthetic bacterium. *Proc. Nat. Acad. Sci. U. S.* **54**, 1212–1218.

Bull, T. A. (1969). Photosynthetic efficiencies and photorespiration in Calvin cycle and C_4-dicarboxylic acid plants. *Crop Sci.* **9**, 726–729.

Bulley, N. R., Nelson, C. D., and Tregunna, E. B. (1969). Photosynthesis: Action spectra for leaves in normal and low oxygen. *Plant Physiol.* **44**, 678–684.

Bültemann, V., Rüppel, H., and Witt, H. T. (1964). Intermediary reactions in the water-splitting part of photosynthesis. *Nature* (*London*) **204**, 646–648.

Bunt, J. S. (1969). The CO_2 compensation point, Hill activity and photorespiration. *Biochem. Biophys. Res. Commun.* **35**, 748–753.

Burk, L. G., and Menser, H. H. (1964). A dominant aurea mutation in tobacco. *Tobacco Sci.* **8**, 101–104.

Burkard, G., Eclancher, B., and Weil, J. H. (1969). Presence of *N*-formylmethionyl-transfer RNA in bean chloroplasts. *FEBS Lett.* **4**, 285–287.

Bushnell, W. R. (1967). Symptom development in mildewed and rusted tissues. *In* "The Dynamc Role of Molecular Constituents in Plant-Parasite Interaction" (C. J. Mirocha and I. Uritani, eds.), pp. 21–39. Am. Phytopathol. Soc., St. Paul, Minnesota.

Butt, V. S., and Peel, M. (1963). Participation of glycollate oxidase in glucose uptake by illuminated *Chlorella* suspensions. *Biochem. J.* **88**, 31P.

Calvin, M., and Bassham, J. A. (1962). "The Photosynthesis of Carbon Compounds." Benjamin, New York.

Calvin, M., and Benson, A. A. (1948). The path of carbon in photosynthesis. *Science* **107**, 476–480.

Calvin, M., and Benson, A. A. (1949). The path of carbon in photosynthesis. IV. The identity and sequence of the intermediates in sucrose synthesis. *Science* **109**, 140–142.

Calvin, M., and Massini, P. (1952). The path of carbon in photosynthesis. XX. The steady state. *Experientia* **8**, 445–457.

Calvin, M., and Pon, N. G. (1959). Carboxylations and decarboxylations. *In* "Symposium on Enzyme Reaction Mechanisms," pp. 51–74. Oak Ridge Natl. Lab., Oak Ridge, Tennessee.

Cannell, R. Q., Brun, W. A., and Moss, D. N. (1969). A search for high net photosynthetic rate among soybean genotypes. *Crop Sci.* **9**, 840–841.

Carpenter, W. D., and Beevers, H. (1959). The distribution and properties of isocitratase in plants. *Plant Physiol.* **34**, 403–409.

Carr, D. J., and Wardlaw, I. F. (1965). The supply of photosynthetic assimilates to the grains from the flag leaf and ear of wheat. *Aust. J. Biol. Sci.* **18**, 711–719.

Carter, H. E., Johnson, P., and Weber, E. J. (1965). Glycolipids. *Annu. Rev. Biochem.* **34**, 109–142.

Chan, H. W. S., and Bassham, J. A. (1967). Metabolism of ^{14}C-labeled glycolic acid by isolated spinach chloroplasts. *Biochim. Biophys. Acta* **141**, 426–429.

Chandler, R. F., Jr., (1969a). New horizons for an ancient crop. *11th Int. Bot. Congr. Symp. World Food Supply.*

Chandler, R. F., Jr. (1969b). Plant morphology and stand geometry in relation to nitrogen. *In* "Physiological Aspects of Crop Yield," (J. D. Eastin, *et al.,* eds.) pp. 265–285. A.S.A. and C.S.S.A., Madison, Wisconsin.

Chang, S. B., and Lundin, K. (1965). Specificity of galactolipids in photochemical reactions coupled with cytochrome c reduction. *Biochem. Biophys. Res. Commun.* **21,** 424–431.

Chartier, P., Chartier, M., and Catsky, J. (1970). Resistances for carbon dioxide diffusion and for carboxylation as factors in bean leaf photosynthesis. *Photosynthetica* **4,** 48–57.

Cheniae, G. M. (1965). Phosphatidic acid and glyceride synthesis by particles from spinach leaves. *Plant Physiol.* **40,** 235–243.

Cheniae, G. M., and Martin, I. F. (1967). Photoreactivation of manganese catalyst in photosynthetic oxygen evolution. *Biochem. Biophys. Res. Commun.* **28,** 89–95.

Chun, E. H. L., Vaughan, M. H., and Rich, A. (1963). The isolation and characterization of DNA associated with chloroplast preparations. *J. Mol. Biol.* **7,** 130–141.

Clagett, C. O., Tolbert, N. E., and Burris, R. H. (1949). Oxidation of α-hydroxy acids by enzymes from plants. *J. Biol. Chem.* **178,** 977–987.

Clayton, R. A. (1959). Pentose cycle activity in cell-free extracts of tobacco leaves and seedlings. *Arch. Biochem. Biophys.* **79,** 111–123.

Clayton, R. K. (1965). "Molecular Physics in Photosynthesis." Ginn (Blaisdell), Boston, Massachusetts.

Clendenning, K. A. (1957). Biochemistry of chloroplasts in relation to the Hill reaction. *Annu. Rev. Plant Physiol.* **8,** 137–152.

Codd, E. A., Lord, J. M., and Merrett, M. J. (1969). The glycollate oxidizing enzyme of algae. *FEBS Lett.* **5,** 341–342.

Constantopoulos, G., and Bloch, K. (1967). Effect of light intensity on the lipid composition of *Euglena gracilis. J. Biol. Chem.* **242,** 3538–3542.

Coombs, J., and Whittingham, C. P. (1966a). The mechanism of inhibition of photosynthesis by high partial pressures of oxygen in *Chlorella. Proc. Roy. Soc., Ser. B* **164,** 511–520.

Coombs, J., and Whittingham, C. P. (1966b). The effect of high partial pressures of oxygen on photosynthesis in *Chlorella.* I. *Phytochemistry* **5,** 643–651.

Cooper, R. L., and Brun, W. A. (1967). Response of soybeans to a carbon dioxide-enriched atmosphere. *Crop Sci.* **7,** 455–457.

Cooper, T. G., and Beevers, H. (1969). β-Oxidation in glyoxysomes from castor bean endosperm. *J. Biol. Chem.* **244,** 3514–3520.

Cooper, T. G., Filmer, D., Wishnick, M., and Lane, M. D. (1969). The active species of "CO_2" utilized by ribulose diphosphate carboxylase. *J. Biol. Chem.* **244,** 1081–1083.

Corbett, J. R., and Wright, B. J. (1970). Mechanism of inhibition of glycollate oxidase by aldehyde bisulphite addition compounds and by related structures. *Biochem. J.* **118,** No. 3, 51P.

Cossins, E. A., and Sinha, S. K. (1966). The interconversion of glycine and serine by plant tissue extracts. *Biochem. J.* **101,** 542–549.

Cowan, I. R. (1968). The interception and adsorption of radiation in plant-stands. *J. Appl. Ecol.* **5,** 367–379.

Crane, F. L., Henninger, M. D., Wood, P. M., and Barr, R. (1966). Quinones in chloroplasts. *In* "The Biochemistry of Chloroplasts" (T. W. Goodwin, ed.), Vol. 1, pp. 133–151. Academic Press, New York.

Criddle, R. S. (1966). Protein and lipoprotein organization in the chloroplast. *In* "The Biochemistry of Chloroplasts" (T. W. Goodwin, ed.), Vol. 1, pp. 203–231. Academic Press, New York.

Criddle, R. S. (1969). Structural proteins of chloroplasts and mitochondria. *Annu. Rev. Plant Physiol.* **20**, 239–252.

Curtis, P. E., Ogren, W. L., and Hageman, R. H. (1969). Varietal effects in soybean photosynthesis and photorespiration. *Crop Sci.* **9**, 323–327.

Dalton, J., and Dougherty, R. C. (1969). Formation of the macrocyclic ring in tetrapyrrole biosynthesis. *Nature (London)* **223**, 1151–1153.

Dalziel, K., and Londesborough, J. C. (1968). The mechanisms of reductive carboxylation reactions. Carbon dioxide or bicarbonate as substrate of nicotinamide-adenine dinucleotide phosphate-linked isocitrate dehydrogenase and 'malic' enzyme. *Biochem. J.* **110**, 223–230.

Davenport, D. C. (1967). Effects of chemical antitranspirants on transpiration and growth of grass. *J. Exp. Bot.* **18**, 332–347.

Deamer, D. W. (1969). ATP synthesis: The current controversy. *J. Chem. Educ.* **46**, 198–206.

Decker, J. P. (1947). The effect of air supply on apparent photosynthesis. *Plant Physiol.* **22**, 561–571.

Decker, J. P. (1955). A rapid, postillumination deceleration of respiration in green leaves. *Plant Physiol.* **30**, 82–84.

Decker, J. P. (1957). Further evidence of increased carbon dioxide production accompanying photosynthesis. *J. Sol. Energy Sci. Eng.* **1**, 30–33.

Decker, J. P. (1958). The effects of light on respiration using isotopically enriched oxygen: An objection and alternate interpretation. *Plant Sci. Bull.* **4**, 3.

Decker, J. P. (1959a). Comparative responses of carbon dioxide outburst and uptake in tobacco. *Plant Physiol.* **34**, 100–102.

Decker, J. P. (1959b). Some effects of temperature and carbon dioxide concentration on photosynthesis of *Mimulus*. *Plant Physiol.* **34**, 103–106.

Decker, J. P. (1970). "Early History of Photorespiration," Bioeng. Bull. No. 10. Eng. Res. Center, Arizona State University, Tempe, Arizona.

Decker, J. P., and Tió, M. A. (1959). Photosynthetic surges in coffee seedlings. *J. Agr. Univ. Puerto Rico* **43**, 50–55.

Decker, J. P., and Wien, J. D. (1958). Carbon dioxide surges in green leaves. *J. Sol. Energy Sci. Eng.* **2**, 39–41.

de Duve, C., and Baudhuin, P. (1966). Peroxisomes (microbodies and related particles). *Physiol. Rev.* **46**, 323–357.

Del Campo, F. F., Ramirez, J. M., and Arnon, D. I. (1968). Stoichiometry of photosynthetic phosphorylation. *J. Biol. Chem.* **243**, 2805–2809.

De Roo, H. C. (1969). Leaf water potentials of sorghum and corn, estimated with the pressure bomb. *Agron. J.* **61**, 969–970.

Devor, K. A., and Mudd, J. B. (1968). Acetate binding of spinach chloroplasts as a facet of fatty acid synthesis. *Plant Physiol.* **43**, 853–858.

de Wit, C. T. (1965). Photosynthesis of leaf canopies. *Agr. Res. Rep., Wageningen* **663**.

de Wit, C. T. (1967). Photosynthesis: Its relationship to overpopulation. *In* "Harvesting the Sun" (A. San Pietro, F. A. Greer, and T. J. Army, eds.), pp. 315–320. Academic Press, New York.

de Wit, C. T. (1968). Plant production. *Meded. Landbouwhogesch., Wageningen, Misc. Ser.* pp. 25–50.

de Wit, C. T., and Brouwer, R. (1969). The simulation of photosynthetic systems. *IBP/PP Tech. Meet. Trebon* Suppl., pp. 3–12.

Dilley, R. A., and Vernon, L. P. (1965). Ion and water transport processes related to the light-dependent shrinkage of spinach chloroplasts. *Arch. Biochem. Biophys.* **111**, 365–375.

Doman, N. G. (1959). The interrelation of photosynthesis and respiration in plants. *Biokhimiya* **24**, 19–24.

Dornhoff, G. M., and Shibles, R. M. (1970). Varietal differences in net photosynthesis of soybean leaves. *Crop Sci.* **10**, 42–45.

Downes, R. W., and Hesketh, J. D. (1968). Enhanced photosynthesis at low oxygen concentrations: Different response of temperate and tropical grasses. *Planta* **78**, 79–84.

Downton, W. J. S., and Tregunna, E. B. (1968). Photorespiration and glycolate metabolism: A re-examination and correlation of some previous studies. *Plant Physiol.* **43**, 923–929.

Drake, B. G., Raschke, K., and Salisbury, F. B. (1970). Temperatures and transpiration resistances of *Xanthium* leaves as affected by air temperature, humidity, and windspeed. *Plant Physiol.* **46**, 324–330.

Dreger, R. H., Brun, W. A., and Cooper, R. L. (1969). The effect of genotype on the photosynthetic rate of soybean (*Glycine max* (L.) Merr.). *Crop Sci.* **9**, 429–431.

Ducet, G., and Rosenberg, A. J. (1962). Leaf respiration. *Annu. Rev. Plant Physiol.* **13**, 171–200.

Duncan, W. G. (1967). Model building in photosynthesis. *In* "Harvesting the Sun" (A. San Pietro, F. A. Greer, and T. J. Army, eds.), pp. 309–320. Academic Press, New York.

Duncan, W. G. (1969). Cultural manipulation for higher yields. *In* "Physiological Aspects of Crop Yield," (J. D. Eastin, *et al.*, eds.) pp. 327–339. A.S.A. and C.S.S.A., Madison, Wisconsin.

Duncan, W. G., and Hesketh, J. D. (1968). Net photosynthetic rates, relative leaf growth rates, and leaf numbers of 22 races of maize grown at eight temperatures. *Crop Sci.* **8**, 670–674.

Durbin, R. D. (1971). Chlorosis-inducing pseudomonad toxins: Their mechanism of action and structure. *In* "Morphological and Biochemical Events in Plant-Parasite Interaction" (S. Akai and S. Ouchi, eds.), pp. 369–386. Phytopath. Soc. Jap., Tokyo.

Duysens, L. N. M., and Amesz, J. (1962). Function and identification of two photochemical systems in photosynthesis. *Biochim. Biophys. Acta* **64**, 243–260.

Eastin, J. D., and Sullivan, C. Y. (1969). Carbon dioxide exchange in compact and semi-open sorghum inflorescences. *Crop Sci.* **9**, 165–166.

Edelman, J., Ginsburg, V., and Hassid, W. Z. (1955). Conversion of monosaccharides to sucrose and cellulose in wheat seedlings. *J. Biol. Chem.* **213**, 843–854.

Edwards, G. E., Lee, S. S., Chen, T. M., and Black, C. C. (1970). Carboxylation reactions and photosynthesis of carbon compounds in isolated mesophyll and bundle sheath cells of *Digitaria sanguinalis. Biochem. Biophys. Res. Commun.* **39**, 389–395.

Egle, K., and Fock, H. (1967). Light respiration-correlations between CO_2 fixation, O_2 pressure and glycollate concentration. *In* "The Biochemistry of Chloroplasts" (T. W. Goodwin, ed.), Vol. 2, pp. 79–87. Academic Press, New York.

Egle, K., and Schenk, W. (1953). Der Einfluss der Temperatur auf die Lage des CO_2-Kompensationspunktes. *Planta* **43**, 83–97.

Eglinton, E., and Hamilton, R. J. (1967). Leaf epicuticular waxes. *Science* **156**, 1322–1335.

Ekern, P. C. (1965). Evapotranspiration of pineapple in Hawaii. *Plant Physiol.* **40**, 736–739.

Ellyard, P. W., and Gibbs, M. (1969). Inhibition of photosynthesis by oxygen in isolated spinach chloroplasts. *Plant Physiol.* **44**, 1115–1121.

Ellyard, P. W., and San Pietro, A. (1969). The Warburg effect in a chloroplast-free preparation from *Euglena gracilis. Plant Physiol.* **44**, 1679–1683.

El-Sharkawy, M., and Hesketh, J. (1965). Photosynthesis among species in relation to characteristics of leaf anatomy and CO_2 diffusion resistances. *Crop Sci.* **5**, 517–521.

El-Sharkawy, M. A., Hesketh, J. D., and Muramoto, H. (1965). Leaf photosynthetic rates and other growth characteristics among 26 species of *Gossypium. Crop Sci.* **5**, 173–175.

310   Bibliography

El-Sharkawy, M. A., Loomis, R. S., and Williams, W. A. (1967). Apparent reassimilation of respiratory carbon dioxide by plant species. *Physiol. Plant.* **20**, 171–186.

Emerson, R. (1958). The quantum yield of photosynthesis. *Annu. Rev. Plant Physiol.* **9**, 1–24.

Emerson, R., and Arnold, W. (1932). A separation of the reactions of photosynthesis by means of intermittent light. *J. Gen. Physiol.* **15**, 391–420.

Emerson, R., and Lewis, C. M. (1943). The dependence of the quantum yield of *Chlorella* photosynthesis on wavelength of light. *Amer. J. Bot.* **30**, 165–178.

Emerson, R., Chalmers, R., and Cederstrand, C. (1957). Some factors influencing the long-wave limit of photosynthesis. *Proc. Nat. Acad. Sci. U. S.* **43**, 133–143.

Epel, B., and Butler, W. L. (1969). Cytochrome a_3: Destruction by light. *Science* **166**, 621–622.

Epstein, E. (1969). Ion absorption by plant cells: The dual transport function of the plasmalemma. *Abstr. Int. Bot. Congr., 11th, 1969,* p. 55.

Evans, G. C. (1969). The spectral composition of light in the field. I. Its measurement and ecological importance. *J. Ecol.* **57**, 109–125.

Evans, M. C. W., Buchanan, B. B., and Arnon, D. I. (1966). A new ferredoxin-dependent carbon reduction cycle in a photosynthetic bacterium. *Proc. Nat. Acad. Sci. U. S.* **55**, 928–934.

Everson, R. G., and Gibbs, M. (1967). Photosynthetic assimilation of carbon dioxide and acetate by isolated chloroplasts. *Plant Physiol.* **42**, 1153–1154.

Everson, R. G., and Slack, C. R. (1968). Distribution of carbonic anhydrase in relation to the C_4 pathway of photosynthesis. *Phytochemistry* **7**, 581–584.

Ficken, G. E., Johns, R. B., and Linstead, R. R. (1956). Chlorophyll and related compounds. IV. The position of the extra hydrogens in chlorophyll. The oxidation of pyrophaeophorbide-*a*. *J. Chem. Soc.* pp. 2273–2280.

Filner, B., and Klein, A. O. (1968). Changes in enzymatic activities in etiolated bean seedling leaves after a brief illumination. *Plant Physiol.* **43**, 1587–1596.

Fincham, J. R. S., and Day, P. R. (1971). "Fungal Genetics." Blackwell, Oxford.

Finkle, B. J., and Arnon, D. I. (1954). Metabolism of isolated cellular particles from photosynthetic tissues. II. Oxidative decarboxylation of oxalic acid. *Physiol. Plant.* **7**, 614–624.

Fischer, R. A. (1968). Stomatal opening: Role of K uptake by guard cells. *Science* **160**, 784–785.

Flowers, H. M., Batra, K. K., Kemp, J., and Hassid, W. Z. (1969). Biosynthesis of cellulose *in vitro* from guanosine diphosphate D-glucose with enzymic preparations from *Phaseolus aureus* and *Lupinus albus*. *J. Biol. Chem.* **244**, 4969–4974.

Fock, H., and Egle, K. (1967). Uber die Beziehungen zwischen dem Glykolsäure-Gehalt und dem Photosynthese-Gaswechsel von Bohnenblättern. *Z. Pflanzenphysiol.* **57**, 389–397.

Fock, H., and Krotkov, G. (1969). Relation between photorespiration and glycolate oxidase activity in sunflower and red kidney bean leaves. *Can. J. Bot.* **47**, 237–240.

Fock, H., Egle, K., Schaub, H., and Hilgenberg, W. (1969a). Der Einfluss des Sauerstoff-Partialdrucks auf die Radioaktivität der nach $^{14}CO_2$-Zufütterung entstehenden Photosynthese-Pudukte. *Z. Pflanzenphysiol.* **61**, 261–263.

Fock, H., Schaub, H., Hilgenberg, W., and Egle, K. (1969b). Über den Einfluss niedriger und hoher O_2-Partialdrucke auf den Sauerstoff- und Kohlendioxidumsatz von *Amaranthus* und *Phaseolus* während der Lichtphase. *Planta* **86**, 77–83.

Forrester, M. L., Krotkov, G., and Nelson, C. D. (1966a). Effect of oxygen on photosynthesis, photorespiration and respiration in detached leaves. I. Soybean. *Plant Physiol.* **41**, 422–427.

Forrester, M. L., Krotkov, G., and Nelson, C. D. (1966b). Effect of oxygen on photosynthesis, photorespiration and respiration in detached leaves. II. Corn and other monocotyledons. *Plant Physiol.* **41**, 428–431.

Forti, G. (1968). The stoichiometry of NADP dependent photosynthetic phosphorylation. *Biochem. Biophys. Res. Commun.* **32**, 1020–1024.

Forti, G., Bertolè, M. L., and Zanetti, G. (1965). Purification and properties of cytochrome *f* from parsley leaves. *Biochim. Biophys. Acta* **109**, 33–40.

Fousova, S., and Avratovscukova, N. (1967). Hybrid vigor and photosynthetic rate of leaf disks in *Zea mays* L. *Photosynthetica* **1**, 3–12.

Franck, J. (1951). A critical survey of the physical background of photosynthesis. *Annu. Rev. Plant Physiol.* **2**, 53–86.

Frederick, S. E., and Newcomb, E. H. (1969a). Microbody-like organelles in leaf cells. *Science* **163**, 1353–1355.

Frederick, S. E., and Newcomb, E. H. (1969b). Cytochemical localization of catalase in leaf microbodies (peroxisomes). *J. Cell Biol.* **43**, 343–353.

Frederick, S. E., Newcomb, E. H., Vigil, E. L., and Wergin, W. P. (1968). Fine-structural characterization of plant microbodies. *Planta* **81**, 229–252.

French. C. S. (1961). Light, pigments, and photosynthesis. *In* "Light and Life" (W. D. McElroy and B. Glass, eds.), pp. 447–471. Johns Hopkins Press, Baltimore, Maryland.

French, C. S., and Young, V. M. K. (1956). The absorption, action, and fluorescence spectra of photosynthetic pigments in living cells and in solutions. *Radiat. Biol.* **3**, 343–391.

Frigerio, N. A., and Harbury, H. A. (1958). Preparation and some properties of crystalline glycolic acid oxidase. *J. Biol. Chem.* **231**, 135–157.

Frydman, R. B., and Frydman, B. (1970). Purification and properties of porphobilinogen deaminase from wheat germ. *Arch. Biochem. Biophys.* **136**, 193–202.

Fujino, M. (1967). Role of adenosinetriphosphate and adenosinetriphosphatase in stomatal movement. *Sci. Bull., Fac. Educ., Nagasaki Univ.* **18**, 1–47.

Fuller, R. C., and Nugent, N. A. (1969). Pteridines and the function of the photosynthetic reaction center. *Proc. Nat. Acad. Sci. U. S.* **63**, 1311–1318.

Gaastra, P. (1959). Photosynthesis of crop plants as influenced by light, carbon dioxide, temperature, and stomatal resistance. *Meded. Landbouwhogesch. Wageningen* **59**, 1–68.

Gaastra, P. (1962). Photosynthesis of leaves and field crops. *Neth. J. Agr. Sci.* **10**, 311–324.

Gaastra, P. (1963). Climatic control of photosynthesis and respiration. *In* "Environmental Control of Plant Growth" (L. T. Evans, ed.), pp. 113–140. Academic Press, New York.

Gaastra, P. (1969). *In* "Physiological Aspects of Crop Yield," (J. D. Eastin, *et al.,* eds.) pp. 140–142. A.S.A. and C.S.S.A., Madison, Wisconsin.

Gaffron, H. (1960). Energy storage: Photosynthesis *In* "Plant Physiology," (F. C. Steward, ed.), Vol. 1B, pp. 3–277. Academic Press, New York.

Gaffron, H., and Fager, E. W. (1951). The kinetics and chemistry of photosynthesis. *Annu. Rev. Plant Physiol.* **2**, 87–114.

Gamborg, O. L., Wetter, L. R., and Neish, A. C. (1962). The oxidation of some aromatic α-hydroxy acids by glycollate: O_2 oxidoreductase. *Phytochemistry* **1**, 159–168.

Gates, D. M. (1965). Radiant energy, its receipt and disposal. *In* "Agricultural Meteorology," (P. E. Waggoner, ed.) pp. 1–26. Am. Meteorol. Soc., Boston, Massachusetts.

Gauhl, E., and Björkman, O. (1969). Simultaneous measurements on the effect of oxygen concentration on water vapor and carbon dioxide exchange in leaves. *Planta* **88**, 187–191.

Geiger, D. R., and Saunders, M. A. (1969). Path parameters of a translocating sugar beet leaf. *Abstr. Int. Bot. Congr., 11th, 1969*, p. 69.

Gerhardt, B. P., and Beevers, H. (1969). Occurrence of RNA in glyoxysomes from castor bean endosperm. *Plant Physiol.* **44**, 1475–1477.

Geronimo, J., and Beevers, H. (1964). Effect of aging and temperature on respiratory metabolism of green leaves. *Plant Physiol.* **39**, 786–793.

Gibbs, M. (1954). The respiration of the pea plant. Oxidation of hexose phosphate and pentose phosphate by cell-free extracts of pea leaves. *Plant Physiol.* **29,** 34–39.

Gibbs, M. (1969a). Photorespiration, Warburg effect and glycolate. *Ann. N. Y. Acad. Sci.* **168,** 356–368.

Gibbs, M. (1969b). Control of photosynthesis by oxygen. *Conn., Agr. Exp. Stat., New Haven, Bull.* **708,** 63–79.

Gibbs, M., and Beevers, H. (1955). Glucose dissimilation in the higher plant. Effect of age of tissue. *Plant Physiol.* **30,** 343–347.

Gibbs, M., and Calo, N. (1959). Factors affecting light induced fixation of carbon dioxide by isolated spinach chloroplasts. *Plant Physiol.* **34,** 318–323.

Gibbs, M., Latzko, E., Everson, R. G., and Cockburn, W. (1967). Carbon mobilization by the green plant. *In* "Harvesting the Sun" (A. San Pietro, F. A. Greer, and T. J. Army, eds.), pp. 111–130. Academic Press, New York.

Gibbs, M., Latzko, E., O'Neal, D., and Hew, C.-S. (1970). Photosynthetic carbon fixation by isolated maize chloroplasts. *Biochem. Biophys. Res. Commun.* **40,** 1356–1361.

Gibor, A. (1965). Chloroplast heredity and nucleic acids. *Amer. Natur.* **99,** 229–239.

Gibor, A., and Granick, S. (1962). Ultraviolet sensitive factors in the cytoplasm that affect the differentiation of *Euglena* plastids. *J. Cell Biol.* **15,** 559–603.

Glass, B. (1961). Summary. *In* "Light and Life" (W. D. McElroy and B. Glass, eds.), pp. 817–911. Johns Hopkins Press, Baltimore, Maryland.

Goldberg, M. L., and Racker, E. (1962). Formation and isolation of a glycolaldehyde-phosphoketolase intermediate. *J. Biol. Chem.* **237,** PC3841–PC3842.

Goldsworthy, A. (1966). Experiments on the origin of CO_2 released by tobacco leaf segments in the light. *Phytochemistry* **5,** 1013–1019.

Goldsworthy, A. (1968). Comparison of the kinetics of photosynthetic carbon dioxide fixation in maize, sugarcane and tobacco, and its relation to photorespiration. *Nature (London)* **217,** 849.

Goldsworthy, A., and Day, P. R. (1970). Further evidence for reduced role of photorespiration in low compensation point species. *Nature (London)* **228,** 687–688.

Good, N., Izawa, S., and Hind, G. (1966). Uncoupling and energy transfer inhibition in photophosphorylation. *Curr. Top. Bioenerg.* **1,** 75–112.

Good, N. E. (1963). Carbon dioxide and the Hill reaction. *Plant Physiol.* **38,** 298–304.

Good, N. E., and Brown, A. H. (1961). The contribution of endogenous oxygen to the respiration of photosynthesizing *Chlorella* cells. *Biochim. Biophys. Acta* **50,** 544–554.

Good, N. E., Winget, E. D., Wilhelmina, W., Connolly, T. N., Izawa, S., and Singh, R. M. M. (1966). Hydrogen ion buffers for biological research. *Biochemistry* **5,** 467–477.

Goodman, R. N., Kiraly, Z., and Zaitlin, M. (1967). "The Biochemistry and Physiology of Infectious Plant Disease." Van Nostrand, Princeton, New Jersey.

Goodwin, T. W. (1965). The biosynthesis of carotenoids. *In* "Biosynthetic Pathways in Higher Plants" (J. B. Pridham and T. Swain, eds.), pp. 37–55. Academic Press, New York.

Gorman, D. S., and Levine, R. P. (1966). Photosynthetic electron transport chain of *Chlamydomonas reinhardi*. VI. Electron transport in mutant strains lacking either cytochrome 553 or plastocyanin. *Plant Physiol.* **41,** 1648–1656.

Goulding, K. H., and Merrett, M. J. (1967). The role of glycollic acid in the photoassimilation of acetate by *Chlorella pyrenoidosa*. *J. Exp. Bot.* **18,** 620–630.

Graham, D., and Walker, D. A. (1962). Some effects of light on the interconversion of metabolites in green leaves. *Biochem. J.* **82,** 554–560.

Graham, D., and Whittingham, C. P. (1968). The path of carbon during photosynthesis in *Chlorella pyrenoidosa* at high and low carbon dioxide concentrations. *Z. Pflanzenphysiol.* **58,** 418–427.

Granick, S. (1965). Cytoplasmic units of inheritance. *Amer. Natur.* **99**, 193–199.

Green, D. E., and MacLennan, D. H. (1969). Structure and function of the mitochondrial cristae membrane. *BioScience* **19**, 213–222.

Griffith, T., and Byerrum, R. U. (1959). Biosynthesis of glycolate and related compounds from ribose-1-C^{14} in tobacco leaves. *J. Biol. Chem.* **234**, 762–764.

Grime, J. P. (1965). Shade tolerance in flowering plants. *Nature (London)* **208**, 161–163.

Grodzinski, B., and Colman, B. (1970). Glycolic acid oxidase activity in cell-free preparations of blue-green algae. *Plant Physiol.* **45**, 735–737.

Hadziyev, D., and Zalik, S. (1970). Amino acid incorporation by ribosomes and polyribosomes from wheat chloroplasts. *Biochem. J.* **116**, 111–124.

Hall, D. O., and Palmer, J. M. (1969). Mitochondrial research today. *Nature (London)* **221**, 717–723.

Hansen, P. (1970). ^{14}C-Studies on apple trees. VI. The influence of the fruit on the photosynthesis of the leaves, and the relative photosynthetic yields of fruits and leaves. *Physiol. Plant.* **23**, 805–810.

Hanson, K. R., and Havir, E. A. (1970). L-Phenylalanine ammonia lyase. IV. Evidence that the prosthetic group contains a dehydroalanyl residue and mechanism of action. *Arch. Biochem. Biophys.* **141**, 1–17.

Harris, R. V., and James, A. T. (1965). Linoleic and α-linolenic acid biosynthesis in plant leaves and a green alga. *Biochim. Biophys. Acta* **106**, 456–464.

Harris, R. V., James, A. T., and Harris, P. (1967). Synthesis of unsaturated fatty acids by green algae and plant leaves. *In* "The Biochemistry of Chloroplasts" (T. W. Goodwin, ed.), Vol. 2, pp. 241–253. Academic Press, New York.

Hartt, C. E. (1963). Translocation as a factor in photosynthesis. *Naturwissenschaften* **50**, 666.

Hassid, W. Z. (1967). Transformation of sugars in plants. *Annu. Rev. Plant Physiol.* **18**, 253–280.

Hatch, M. D., and Slack, C. R. (1966). Photosynthesis by sugar-cane leaves. A new carboxylation reaction and the pathway of sugar formation. *Biochem. J.* **101**, 103–111.

Hatch, M. D., and Slack, C. R. (1968). A new enzyme for the interconversion of pyruvate and phosphopyruvate and its role in the C_4 dicarboxylic acid pathway of photosynthesis. *Biochem. J.* **106**, 141–146.

Hatch, M. D., and Slack, C. R. (1969a). Studies on the mechanism of activation and inactivation of pyruvate, phosphate dikinase. A possible regulatory role for the enzyme in the C_4 dicarboxylic acid pathway of photosynthesis. *Biochem. J.* **112**, 549–558.

Hatch, M. D., and Slack, C. R. (1969b). Mode of operation of the C_4-dicarboxylic acid pathway of photosynthesis and the regulation of the process. *Abstr. Int. Bot. Congr., 11th. 1969*, p. 86.

Hatch, M. D., and Slack, C. R. (1969c). NADP-specific malate dehydrogenase and glycerate kinase in leaves and evidence for their location in chloroplasts. *Biochem. Biophys. Res. Commun.* **34**, 589–593.

Hatch, M. D., and Slack, C. R. (1970). Photosynthetic CO_2-fixation pathways. *Annu. Rev. Plant Physiol.* **21**, 141–162.

Hatch, M. D., Slack, C. R., and Bull, T. A. (1969). Light-induced changes in the content of some enzymes of the C_4-dicarboxylic acid pathway of photosynthesis and its effect on other characteristics of photosynthesis. *Phytochemistry* **8**, 697–706.

Haverkate, F., and van Deenen, L. L. M. (1965). Isolation and chemical characterization of phosphatidyl glycerol from spinach leaves. *Biochim. Biophys. Acta* **106**, 78–92.

Havir, E. A., and Gibbs, M. (1963). Studies on the reductive pentose phosphate cycle in intact and reconstituted chloroplast systems. *J. Biol. Chem.* **238**, 3183–3187.

Hawker, J. S. (1967). The activity of uridine diphosphate glucose-D-fructose 6-phosphate 2-glucosyltransferase in leaves. *Biochem. J.* **105,** 943–946.

Heath, O. V. S., and Orchard, B. (1957). Temperature effects on the minimum intercellular space CO_2 concentration "Γ." *Nature (London)* **180,** 180–182.

Heath, O. V. S., and Russell, J. (1954). An investigation of the light responses of wheat stomata with the attempted elimination of control by the mesophyll. Part 2. Interactions with external carbon dioxide, and general discussion. *J. Exp. Bot.* **5,** 269–292.

Heath, O. V. S., Mansfield, T. A., and Meidner, H. (1965). Light-induced stomatal opening and the postulated role of glycollic acid. *Nature (London)* **207,** 960–962.

Heber, A., Pon, N. G., and Heber, M. (1963). Localization of carboxydismutase and triose-phosphate dehydrogenases in chloroplasts. *Plant Physiol.* **38,** 355–360.

Heber, U. (1969). Conformational changes of chloroplasts induced by illumination of leaves *in vivo. Biochim. Biophys. Acta* **180,** 302–319.

Heber, U. W., and Santarius, K. A. (1965). Compartmentation and reduction of pyridine nucleotides in relation to photosynthesis. *Biochim. Biophys. Acta* **109,** 390–408.

Heichel, G. (1970). Prior illumination and the respiration of maize leaves in the dark. *Plant Physiol.* **46,** 359–362.

Heichel, G. H. (1971a). Response of respiration of tobacco leaves in light and darkness and the CO_2 compensation concentration to prior illumination and oxygen. *Plant Physiol.* (in press).

Heichel, G. (1971b). Confirming measurements of respiration and photosynthesis with dry matter accumulation. *Photosynthetica* **5,** 93–95.

Heichel, G. H., and Musgrave, R. B. (1969). Relation of CO_2 compensation concentration to apparent photosynthesis in maize. *Plant Physiol.* **44,** 1724–1728.

Helmsing, P. J. (1969). Purification and properties of galactolipase. *Biochim. Biophys. Acta* **178,** 519–533.

Herbin, E. A., and Robins, P. A. (1969). Patterns of variation and development in leaf wax alkanes. *Phytochemistry* **8,** 1985–1998.

Hesketh, J. (1967). Enhancement of photosynthetic CO_2 assimilation in the absence of oxygen, as dependent upon species and temperature. *Planta* **76,** 371–374.

Hesketh, J. D. (1963). Limitations to photosynthesis responsible for differences among species. *Crop Sci.* **3,** 493–496.

Hesketh, J. D., and Moss, D. N. (1963). Variation in the response of photosynthesis to light. *Crop Sci.* **3,** 107–110.

Hess, J. L., and Tolbert, N. E. (1966). Glycolate, glycine, serine, and glycerate formation during photosynthesis by tobacco leaves. *J. Biol. Chem.* **241,** 5705–5711.

Hess, J. L., and Tolbert, N. E. (1967). Glycolate pathway in algae. *Plant Physiol.* **42,** 371–379.

Hew, C.-S., and Gibbs, M. (1969). A study of chloroplasts of corn, sorghum and sugarcane. *Plant Physiol.* **44S,** 10.

Hew, C.-S., and Krotkov, G. (1968). Effect of oxygen on the rates of CO_2 evolution in light and in darkness by photosynthesizing and nonphotosynthesizing leaves. *Plant Physiol.* **43,** 464–466.

Hew, C.-S., Krotkov, G., and Canvin, D. T. (1969a). Determination of the rate of CO_2 evolution by green leaves in light. *Plant Physiol.* **44,** 662–670.

Hew, C.-S., Krotkov, G., and Canvin, D. T. (1969b). Effects of temperature on photosynthesis and CO_2 evolution in light and darkness by green leaves. *Plant Physiol.* **44,** 671–677.

Heytler, P. G. (1963). Uncoupling of oxidative phosphorylation by carbonyl cyanide phenyl-hydrazones. I. Some characteristics of m-Cl-CCP action on mitochondria and chloroplasts. *Biochemistry* **2,** 357–361.

Highkin, H. R., Boardman, N. K., and Goodchild, D. J. (1969). Photosynthetic studies on a pea-mutant deficient in chlorophyll. *Plant Physiol.* **44,** 1310–1320.

Hill, R. (1954). The cytochrome *b* component of chloroplasts. *Nature (London)* **174,** 501–503.

Hill, R. (1965). The biochemists' green mansions: The photosynthetic electron-transport chains in plants. *Essays Biochem.* **1,** 121–151.

Hill, R., and Bendall, F. (1960). Function of the two cytochrome components in chloroplasts: A working hypothesis. *Nature (London)* **186,** 136–137.

Hill, R., and Bonner, W. D., Jr. (1961). The nature and possible function of chloroplast cytochromes. *In* "Light and Life" (W. D. McElroy and B. Glass, eds.), pp. 424–435. Johns Hopkins Press, Baltimore, Maryland.

Hiller, R. G., and Walker, D. A. (1961). Formation of labelled amino acids by exchange transamination. *Biochem. J.* **78,** 56–60.

Hind, G. (1968). Light-induced changes in cytochrome b-559 in spinach chloroplasts. *Photochem. Photobiol.* **7,** 369–375.

Hind, G., and Nakatani, H. Y. (1970). Determination of the concentration and the redox potential of chloroplast cytochrome 559. *Biochim. Biophys. Acta* **216,** 223–225.

Hind, G., Nakatani, H. Y., and Izawa, S. (1969). The role of Cl^- in photosynthesis. I. The Cl^- requirement of electron transport. *Biochim. Biophys. Acta* **172,** 277–289.

Hoch, G., and Knox, R. S. (1968). Primary processes in photosynthesis. *In* "Photophysiology" (A. C. Giese, ed.), Vol. 3, pp. 225–251. Academic Press, New York.

Hoch, G., Owens, O. v. H., and Kok, B. (1963). Photosynthesis and respiration. *Arch. Biochem. Biophys.* **101,** 171–180.

Hofstra, G., and Hesketh, J. D. (1969a). Effect of temperature on the gas exchange of leaves in the light and dark. *Planta* **85,** 228–237.

Hofstra, G., and Hesketh, J. D. (1969b). The effect of temperature on stomatal aperture in different species. *Can. J. Bot.* **47,** 1307–1310.

Hofstra, G., and Nelson, C. D. (1969). The translocation of photosynthetically assimilated ^{14}C in corn. *Can. J. Bot.* **47,** 1435–1442.

Holmgren, P. (1968). Leaf factors affecting light-saturated photosynthesis in ecotypes of *Solidago virgaurea* from exposed and shaded habitats. *Physiol. Plant.* **21,** 676–698.

Holmgren, P., and Jarvis, P. G. (1967). Carbon dioxide efflux from leaves in light and darkness. *Physiol. Plant.* **20,** 1045–1051.

Holmgren, P., Jarvis, P. G., and Jarvis, M. S. (1965). Resistances to carbon dioxide and water vapour transfer in leaves of different plant species. *Physiol. Plant.* **18,** 557–573.

Holowinsky, A. W., and Schiff, J. A. (1970). Events surrounding the early development of *Euglena* chloroplasts. *Plant Physiol.* **45,** 339–347.

Holt, A. S. (1965). Nature, properties and distribution of chlorophylls. *In* "Chemistry and Biochemistry of Plant Pigments" (T. W. Goodwin, ed.), pp. 3–28. Academic Press, New York.

Holzer, H., and Holldorf, A. (1957). Isolierung von D-Glycerat-dehydrogenase, einige Eigenschaften des Enzyms und seine Verwendung zur enzymatisch-optischen Bestimmung von Hydroxypyruvat neben Pyruvat. *Biochem. Z.* **329,** 292–312.

Holzer, H., and Schröter, W. (1962). Zum Wirkungsmechanismus der Phosphoketolase. I. Oxydation verschiedener Substrate mit Ferricyanid Glykolsäure. *Biochim. Biophys. Acta* **65,** 271–288.

Homann, P. H., and Schmid, G. H. (1967). Photosynthetic reactions of chloroplasts with unusual structures. *Plant Physiol.* **42,** 1619–1632.

Hong, J.-S., Champion, A. B., and Rabinowitz, J. C. (1969). Concerning the source of "labile" sulfur in clostridial ferredoxin. *Eur. J. Biochem.* **8,** 307–313.

Hoober, K. J., Siekevitz, P., and Palade, G. E. (1969). Formation of chloroplast membranes in *Chlamydomonas reinhardi* y-l. *J. Biol. Chem.* **244**, 2621–2631.

Horton, A. A., and Hall, D. O. (1968). Determining the stoichiometry of photosynthetic phosphorylation. *Nature (London)* **218**, 386–388.

Humble, G. D., and Hsiao, T. C. (1969). Specific requirement of K for light-activated opening of stomata in epidermal strips. *Plant Physiol.* **44**, 230–234.

Hurd, R. G. (1968). Effects of CO_2-enrichment on the growth of young tomato plants in low light. *Ann. Bot. (London)* [N. S.] **32**, 531–542.

Imbamba, S. K.. and Moss, D. N. (1969). Effect of 2-chloro-4-ethylamino-6-isopropylamino-S-triazine (Atrazine) on physiological processes in leaves. *Agron. Abstr.* p. 26.

Imber, D., and Tal, M. (1970). Phenotypic reversion of flacca, a wilty mutant of tomato, by abscisic acid. *Science* **169**, 592–593.

Ipens, I. I., Stewart, D. W., Allen, L. H., Jr., and Lemon, E. R. (1967). Diffusive resistances at, and transpiration rates from leaves in situ within the vegetative canopy of a corn crop. *Plant Physiol.* **42**, 99–104.

Inoue, E. (1965). On the CO_2 concentration profiles within crop canopies. *J. Agr. Meteorol.* **20**, 137–140.

Ishizuka, Y. (1969). Engineering for higher yields. *In* "Physiological Aspects of Crop Yield," (J. D. Eastin, *et al.,* eds.) pp. 15–25. A.S.A. and C.S.S.A., Madison, Wisconsin.

Itoh, M., Izawa, S., and Shibata, K. (1963). Shrinkage of whole chloroplasts upon illumination. *Biochim. Biophys. Acta* **66**, 319–327.

Itoh, M., Yamashita, K., Nishi, T., Konishi, K., and Shibata, K. (1969). The site of manganese function in photosynthetic electron transport system. *Biochim. Biophys. Acta* **180**, 509–519.

Izawa, S., and Good, N. E. (1968). The stoichiometric relation of phosphorylation to electron transport in isolated chloroplasts. *Biochim. Biophys. Acta* **162**, 380–391.

Izhar, S., and Wallace, D. H. (1967). Studies on the physiological basis for yield differences. III. Genetic variation in photosynthetic efficiency of *Phaseolus vulgaris* L. *Crop Sci.* **7**, 457–460.

Jackson, W. A., and Volk, R. J. (1969). Oxygen uptake by illuminated maize leaves. *Nature (London)* **222**, 269–271.

Jackson, W. A., and Volk, R. J. (1970). Photorespiration. *Annu. Rev. Plant Physiol.* **21**, 385–432.

Jacobi, G. (1959). Über den Zusammenhang von Glykolsäure und lichtabhängiger Phosphorylierung. *Planta* **53**, 402–411.

Jacobson, B. S., Smith, B. N., Epstein, S., and Laties, G. G. (1970). The prevalence of carbon-13 in respiratory carbon dioxide as an indicator of the type of endogenous substrate. *J. Gen. Physiol.* **55**, 1–17.

Jagendorf, A. T. (1959). The relationship between electron transport and photophosphorylation in spinach chloroplasts. *Brookhaven Symp. Biol.* **11**, 236–258.

Jagendorf, A. T. (1967). Acid-base transitions and phosphorylation by chloroplasts. *Fed. Proc., Fed. Amer. Soc. Exp. Biol.* **26**, 1361–1369.

Jagendorf, A. T., and Neumann, J. (1965). Effect of uncouplers on the light-induced pH rise with spinach chloroplasts. *J. Biol. Chem.* **240**, 3210–3214.

Jagendorf, A. T., and Uribe, E. (1966). ATP formation by acid-base transition of spinach chloroplasts. *Proc. Nat. Acad. Sci. U. S.* **55**, 170–177.

Jakoby, W. B., Brummond, D. O., and Ochoa, S. (1956). Formation of 3-phosphoglyceric acid by carbon dioxide fixation with spinach leaf enzymes. *J. Biol. Chem.* **218**, 811–822.

James, A. T., and Nichols, B. W. (1966). Lipids of photosynthetic systems. *Nature (London)* **210**, 372–375.

James, W. O. (1953). "Plant Respiration." Oxford Univ. Press (Clarendon), London and New York.

Jeffers, D. L., and Shibles, R. M. (1969). Some effects of leaf area, solar radiation, air temperature, and variety on net photosynthesis in field-grown soybeans. *Crop Sci.* **9**, 762–764.

Jenkins, R., and Griffiths, D. E. (1970). Manganese in chloroplasts. *Biochem. J.* **116**, No. 4, 40P.

Jensen, R. G. (1969). Light control of CO_2 fixation in isolated spinach chloroplasts. *Abstr. Int. Bot. Congr., 11th, 1969,* p. 102.

Jensen, R. G., and Bassham, J. A. (1966). Photosynthesis by isolated chloroplasts. *Proc. Nat. Acad. Sci. U. S.* **56**, 1095–1101.

Jensen, R. G., and Bassham, J. A. (1968a). Photosynthesis by isolated chloroplasts. II. Effect of addition of cofactors and intermediate compounds. *Biochim. Biophys. Acta* **153**, 219–226.

Jensen, R. G., and Bassham, J. A. (1968b). Photosynthesis by isolated chloroplasts. III. Light activation of the carboxylation reaction. *Biochim. Biophys. Acta* **153**, 227–234.

Jensen, S. G. (1968). Photosynthesis, respiration and other physiological relationships in barley infected with barley yellow dwarf virus. *Phytopathology* **58**, 204–208.

Ji, T. H., Hess, J. L., and Benson, A. A. (1968). Studies on chloroplast membrane structure. I. Association of pigments with chloroplast lamellar protein. *Biochim. Biophys. Acta* **150**, 676–685.

Jimenez, E., Baldwin, R. L., Tolbert, N. E., and Wood, W. A. (1962). Distribution of C^{14} in sucrose from glycolate-C^{14} and serine-3-C^{14} metabolism. *Arch. Biochem. Biophys.* **98**, 172–175.

Johnson, H. S., and Hatch, M. D. (1968). Distribution of the C_4-dicarboxylic acid pathway of photosynthesis and its occurrence in dicotyledonous plants. *Phytochemistry* **7**, 375–380.

Joliffe, P. A., and Tregunna, E. B. (1968). Effect of temperature, CO_2 concentration, and light intensity on oxygen inhibition of photosynthesis in wheat leaves. *Plant Physiol.* **43**, 902–906.

Jones, M. B., and Mansfield, T. A. (1970). A circadean rhythm in the level of the carbon dioxide compensation point in *Bryophyllum* and *Coffea*. *J. Exp. Bot.* **21**, 159–163.

Jones, R., and Mansfield, T. A. (1970). Suppression of stomatal opening in leaves treated with abscisic acid. *J. Exp. Bot.* **21**, 714–719.

Kahn, A., and von Wettstein, D. (1961). Macromolecular physiology of plastids. II. Structure of isolated spinach chloroplasts. *J. Ultrastruct. Res.* **5**, 557–574.

Kalberer, P. P., Buchanan, B. B., and Arnon, D. I. (1967). Rates of photosynthesis by isolated chloroplasts. *Proc. Nat. Acad. Sci. U. S.* **57**, 1542–1549.

Kamen, M. D. (1963). "Primary Processes in Photosynthesis." Academic Press, New York.

Kandler, O., and Gibbs, M. (1956). Asymmetric distribution of C^{14} in the glucose phosphates formed during photosynthesis. *Plant Physiol.* **31**, 411–412.

Kaneda, T. (1969). Hydrocarbons in spinach: Two distinctive carbon ranges of aliphatic hydrocarbons. *Phytochemistry* **8**, 2039–2044.

Karlstam, B., and Albertsson, P. A. (1969). Demonstration of three classes of spinach chloroplasts by counter-current distribution. *FEBS Lett.* **5**, 360–363.

Kates, M. (1970). Plant phospholipids and glycolipids. *Advan. Lipid Res.* **8**, 225–265.

Katoh, S., and Takamiya, A. (1965). Restoration of NADP photoreducing activity of sonicated chloroplasts by plastocyanin. *Biochim. Biophys. Acta* **99**, 156–160.

Kawashima, N., and Wildman, S. G. (1970). Fraction I protein. *Annu. Rev. Plant Physiol.* **21**, 325–358.

Kearney, P. C., and Tolbert, N. E. (1962). Appearance of glycolate and related products of photosynthesis outside of chloroplasts. *Arch. Biochem. Biophys.* **98**, 164–171.

Kenten, R. H., and Mann, P. J. G. (1952). Hydrogen peroxide formation in oxidations catalysed by plant α-hydroxyacid oxidase. *Biochem. J.* **52**, 130–134.

Kiesselbach, T. A. (1949). The structure and reproduction of corn. *Neb., Agr. Exp. Sta., Res. Bull.* **161**.

King, R. W., and Evans, L. T. (1967). Photosynthesis in artificial communities of wheat, lucerne, and subterranean clover plants. *Aust. J. Biol. Sci.* **20**, 623–635.

King, R. W., Wardlaw, I. F., and Evans, L. T. (1967). Effect of assimilate utilization on photosynthetic rate in wheat. *Planta* **77**, 261–276.

Kirk, J. T. O. (1966). Nature and function of chloroplast DNA. *In* "The Biochemistry of Chloroplasts" (T. W. Goodwin, ed.), Vol. 1, pp. 319–340. Academic Press, New York.

Kirk, J. T. O., and Tilney-Bassett, R. A. E. (1967). "The Plastids." Freeman, San Francisco, California.

Kisaki, T., and Tolbert, N. E. (1969). Glycolate and glyoxylate metabolism by isolated peroxisomes or chloroplasts. *Plant Physiol.* **44**, 242–250.

Kiaski, T., and Tolbert, N. E. (1970). Glycine as a substrate for photorespiration. *Plant Cell Physiol.* **11**, 247–258.

Klein, A. O. (1964). Metabolism of threo-D_s-isocitric acid in detached leaves of *Bryophyllum calycinum*. *Plant Physiol.* **39**, 290–295.

Klein, A. O., and Vishniac, W. (1961). Activity and partial purification of chlorophyllase in aqueous systems. *J. Biol. Chem.* **236**, 2544–2547.

Knaff, D. B., and Arnon, D. I. (1969). Light-induced oxidation of a chloroplast b-type cytochrome at −189°C. *Proc. Nat. Acad. Sci. U. S.* **63**, 956–962.

Kohn, L. D., and Warren, W. A. (1970). The kinetic properties of spinach leaf glyoxylic acid reductase. *J. Biol. Chem.* **245**, 3831–3839.

Kohn, L. D., Warren, W. A., and Carroll, W. R. (1970). The structural properties of spinach leaf glyoxylic acid reductase. *J. Biol. Chem.* **245**, 3821–3830.

Kok, B. (1956). On the reversible absorption change at 705 mμ in photosynthetic organisms. *Biochim. Biophys. Acta* **22**, 399–401.

Kok, B. (1961). Partial purification and determination of oxidation reduction potential of the photosynthetic chlorophyll complex absorbing at 700 mμ. *Biochim. Biophys. Acta* **48**, 527–533.

Kok, B. (1969). Photosynthesis. *In* "Physiology of Plant Growth and Development" (M. B. Wilkens, ed.), pp. 335–379. McGraw-Hill, New York.

Kok, B., and Cheniae, G. M. (1966). Kinetics and intermediates of the oxygen evolution step in photosynthesis. *Curr. Top. Bioenerg.* **1**, 1–47.

Kok, B., and Cheniae, G. M. (1969). Kinetic and biochemical aspects of photosynthetic O_2 evolution. *Abstr. Int. Bot. Congr., 11th, 1969*, p. 114.

Kok, B., and Hoch, G. (1961). Spectral changes in photosynthesis. *In* "Light and Life" (W. D. McElroy and B. Glass, eds.), pp. 397–416. Johns Hopkins Press, Baltimore, Maryland.

Kolattukudy, P. E. (1965). Biosynthesis of wax in *Brassica oleracea*. *Biochemistry* **4**, 1844–1855.

Kolattukudy, P. E. (1966). Biosynthesis of wax in *Brassica oleracea*. Relation of fatty acids to wax. *Biochemistry* **5**, 2265–2275.

Kolattukudy, P. E. (1967). Mechanisms of synthesis of waxy esters in broccoli (*Brassica oleracea*). *Biochemistry* **6**, 2705–2717.

Kolattukudy, P. E. (1968). Biosynthesis of surface lipids. *Science* **159**, 1–8.

Kolattukudy, P. E. (1970). Biosynthesis of cuticular lipids. *Annu. Rev. Plant Physiol.* **21**, 163–192.

Kolesnikov, P. A. (1948). Oxidation of glycolic acid in green cells. *Dokl. Akad. Nauk SSSR* **60**, 1205–1207.

Kolesnikov, P. A., Petrochenko, E. I., and Zore, S. V. (1958). Enzymatic reduction of quinone by glycolic acid. *Dokl. Akad. Nauk SSSR* **123**, 729–732.

Kolesnikov, P. A., Petrochenko, E. I., and Zore, S. V. (1959). The relation between glycolic acid oxidase and polyphenol oxidase. *Fiziol. Rast.* **6**, 598–603.

Kortschak, H. P., Hartt, C. E., and Burr, G. O. (1965). Carbon dioxide fixation in sugarcane leaves. *Plant Physiol.* **40**, 209–213.

Kostychev, S. (1931). "Chemical Plant Physiology" (C. J. Lyon, transl.). McGraw-Hill (Blakiston), New York.

Kowallik, W., and Gaffron, H. (1967). Enhancement of respiration and fermentation in algae by blue light. *Nature (London)* **215**, 1038–1040.

Krall, A. R., and Bass, E. R. (1962). Oxygen dependency of *in vivo* photophosphorylation. *Nature (London)* **196**, 791–792.

Krasnovsky, A. A. (1969). Molecular arrangement of the pigment system in photosynthetic organisms: Action of light. *Abstr., IBP Meet. Moscow, 1969.*

Krause, G. H., and Bassham, J. A. (1969). Induction of respiratory metabolism in illuminated *Chlorella pyrenoidosa* and isolated spinach chloroplasts by the addition of vitamin K_5. *Biochim. Biophys. Acta* **172**, 553–565.

Kretovich, V. L., and Stepanovich, K. M. (1963). Synthesis of serine from hydroxypyruvic acid in a live plant. *Dokl. Akad. Nauk SSSR* **148**, 939–940.

Kriedemann, P. (1966). The photosynthetic activity of the wheat ear. *Ann. Bot. (London)* [N. S.] **30**, 349–363.

Krinsky, N. I. (1966). The role of carotenoid pigments as protective agents against photosensitized oxidations in chloroplasts. *In* "The Biochemistry of Chloroplasts" (T. W. Goodwin, ed.), Vol. 1, pp. 423–430. Academic Press, New York.

Krizek, D. T., Zimmerman, R. H., Klueter, H. H., and Bailey, W. A. (1969). Accelerated growth of birch and crabapple seedlings under CO_2 enriched atmospheres. *Plant Physiol.* **44**, 15S.

Krotkov, G. (1963). Effect of light on respiration. *In* "Photosynthetic Mechanisms in Green Plants," pp. 452–454. Natl. Acad. Sci.—Natl. Res. Council, Washington, D. C.

Krotkov, G., Runeckles, V. C., and Thimann, K. V. (1958). Effect of light on the CO_2 absorption and evolution by Kalanchoe, wheat and pea leaves. *Plant Physiol.* **33**, 289–292.

Kuiper, P. J. C. (1964). Dependence upon wavelength of stomatal movement in epidermal tissue of *Senecio odoris. Plant Physiol.* **39**, 952–955.

Kuiper, P. J. C. (1969). Effect of lipids on chloride and sodium transport in bean and cotton plants. *Plant Physiol.* **44**, 968–972.

Kung, S. D., and Williams, J. P. (1969). Chloroplast DNA from broad bean. *Biochim. Biophys. Acta* **195**, 434–445.

Kushida, H., Itoh, M., Izawa, S., and Shibata, K. (1964). Deformation of chloroplasts on illumination in intact spinach leaves. *Biochim. Biophys. Acta* **79**, 201–203.

Laetsch, W. M. (1970). Comparative ultrastructure of chloroplasts in plants with C-4 pathways of carbon fixation. *Plant Physiol.* **46S**, 22.

Lake, J. V. (1967). Respiration of leaves during photosynthesis. I. Estimates from an electrical analogue. *Aust. J. Biol. Sci.* **20**, 487–493.

Lam, T. H., and Shaw, M. (1970). Removal of phenolics from plant extracts by grinding with anion exchange resin. *Biochem. Biophys. Res. Commun.* **39**, 965–968.

Lascelles, J. (1965). The biosynthesis of chlorophyll. *In* "Biosynthetic Pathways in Higher Plants" (J. B. Pridham and T. Swain, eds.), pp. 163–177. Academic Press, New York.

Latzko, E., and Gibbs, M. (1968). Distribution and activity of enzymes of the reductive pentose phosphate cycle in spinach leaves and in chloroplasts isolated by different methods. *Z. Pflanzenphysiol.* **59**, 184–194.

Lee, R., and Gates, D. M. (1964). Diffusion resistance in leaves as related to their stomatal anatomy and micro-structure. *Amer. J. Bot.* **51**, 963–975.

Leech, R. M. (1966). Comparative biochemistry and comparative morphology of chloroplasts isolated by different methods. *In* "The Biochemistry of Chloroplasts" (T. W. Goodwin, ed.), Vol. 1, pp. 65–74. Academic Press, New York.

Leech, R. M. (1968). The chloroplast inside and outside the cell. *In* "Plant Cell Organelles" (J. B. Pridham, ed.), pp. 137–162. Academic Press, New York.

Lemon, E. (1967). Aerodynamic studies of CO_2 exchange between the atmosphere and the plant. *In* "Harvesting the Sun" (A. San Pietro, F. A. Greer, and T. J. Army, eds.), pp. 263–290. Academic Press, New York.

Lemon, E. (1969). Gaseous exchange in crop stands. *In* "Physiological Aspects of Crop Yield," (J. D. Eastin, *et al.,* eds.) pp. 117–137. A.S.A. and C.S.S.A., Madison, Wisconsin.

Lemon, E., Allen, L. H., Jr., and Müller, L. (1970). Carbon dioxide exchange of a tropical rain forest. Part II. *BioScience* **20**, 1054–1059.

Lemon, E. R., and Wright, J. L. (1969). Photosynthesis under field conditions. Assessing sources and sinks of carbon dioxide in a corn (*Zea mays*) crop using a momentum balance approach. *Agron. J.* **61**, 405–411.

Lemon, E. R., Wright, J. L., and Drake, G. M. (1969). Photosynthesis under field conditions. XB. Origins of short-time CO_2 fluctuations in a cornfield. *Agron. J.* **61**, 411–413.

Levine, R. P. (1969). The analysis of photosynthesis using mutant strains of algae and higher plants. *Annu. Rev. Plant Physiol.* **20**, 523–540.

Levitt, J. (1967). The mechanism of stomatal action. *Planta* **74**, 101–118.

Lichtenthaler, H. K., and Park, R. B. (1963). Chemical composition of chloroplast lamellae from spinach. *Nature (London)* **198**, 1070–1072.

Lichtenthaler, H. K., and Sprey, B. (1966). Über die osmiophilen globulären Lipideinschlüsse der Chloroplasten. *Z. Naturforsch. B* **21**, 690–697.

Lieberman, M., and Baker, J. E. (1965). Respiratory electron transport. *Annu. Rev. Plant Physiol.* **16**, 343–382.

Linacre, E. T. (1967). Further studies of the heat transfer from a leaf. *Plant Physiol.* **42**, 651–658.

Litz, R. E., and Kimmins, W. C. (1968). Plasmodesmata between guard cells and accessory cells. *Can. J. Bot.* **46**, 1603–1604.

Livne, A. (1964). Photosynthesis in healthy and rust-affected plants. *Plant Physiol.* **39**, 614–621.

Livne, A., and Racker, E. (1968). A new coupling factor for photophosphorylation. *Biochem. Biophys. Res. Commun.* **32**, 1045–1049.

Livne, A., and Racker, E. (1969). Partial resolution of the enzymes catalyzing photophosphorylation. *J. Biol. Chem.* **244**, 1339–1344.

Livne, A., and Vaadia, Y. (1965). Stimulation of transpiration rate in barley leaves by kinetin and gibberellic acid. *Physiol. Plant.* **18**, 658–664.

Loomis, R. S., and Williams, W. A. (1963). Maximum crop productivity: An estimate. *Crop Sci.* **3**, 67–72.

Loomis, R. S., and Williams, W. A. (1969). Productivity and the morphology of crop stands: Patterns with leaves. *In* "Physiological Aspects of Crop Yields," (J. D. Eastin, *et al.,* eds.) pp. 27–47. A.S.A. and C.S.S.A., Madison, Wisconsin.

Loomis, R. S., Williams, W. A., and Duncan, W. G. (1967). Community architecture and the productivity of terrestrial plant communities. *In* "Harvesting the Sun" (A. San Pietro,

F. A. Greer, and T. J. Army, eds.), pp. 291–308. Academic Press, New York.

Loomis, R. S., Williams, W. A., Duncan, W. G., Dovrat, A., and Nunez, A. F. (1968). Quantitative descriptions of foliage display and light absorption in field communities. *Crop Sci.* **8**, 352–356.

Loomis, W. D., and Battaile, J. (1966). Plant phenolic compounds and the isolation of plant enzymes. *Phytochemistry* **5**, 423–438.

Lord, J. M., and Merrett, M. J. (1968). Glycollate oxidase in *Chlorella pyrenoidosa. Biochim. Biophys. Acta* **159**, 543–544.

Lord, J. M., and Merrett, M. J. (1970). The pathway of glycollate utilization in *Chlorella pyrenoidosa. Biochem. J.* **117**, 929–937.

Ludwig, L. J., and Krotkov, G. (1967). The kinetics of labeling of the substrate for CO_2 evolution by sunflower leaves in the light. *Plant Physiol.* **42S**, 47.

Lundegårdh, H. (1927). Carbon dioxide evolution of soil and crop growth. *Soil Sci.* **23**, 417–453.

Lupton, F. G. H. (1969). Estimation of yield in wheat from measurements of photosynthesis and translocation in the field. *Ann. Appl. Biol.* **64**, 363–374.

Lyman, H., Epstein, H. T., and Schiff, J. A. (1961). Studies of chloroplast development in *Euglena. Biochim. Biophys. Acta* **50**, 301–309.

Lyttleton, J. W. (1962). Isolation of ribosomes from spinach chloroplasts. *Exp. Cell Res.* **26**, 312–317.

McCarty, R. E., and Racker, E. (1967). Effect of a coupling factor and its antiserum on photophosphorylation and hydrogen ion transport. *Brookhaven Symp. Biol.* **19**, 202–214.

McCarty, R. E., and Racker, E. (1968). Partial resolution of the enzymes catalyzing photophosphorylation. III. Activation of adenosine triphosphatase and ^{32}P-labeled orthophosphate-adenosine triphosphate exchange in chloroplasts. *J. Biol. Chem.* **243**, 129–137.

McCarty, R. E., Guillory, R. J., and Racker, E. (1965). Dio-9 inhibition of coupled electron transport and photophosphorylation in chloroplasts. *J. Biol. Chem.* **240**, 4822–4823.

MacLennan, D. H., Beevers, H., and Harley, J. L. (1963). "Compartmentation" of acids in plant tissues. *Biochem. J.* **89**, 316–327.

McNaughton, S. J., and Fullem, L. W. (1970). Photosynthesis and photorespiration in *Typha latifolia. Plant Physiol.* **45**, 703–707.

Mac Robbie, E. A. C. (1965). The nature of the coupling between light and active ion transport in *Nitella translucens. Biochim. Biophys. Acta* **94**, 64–73.

Majerus, P. W., Albers, A. W., and Vagelos, P. R. (1965). Acyl carrier protein. VII. The primary structure of the substrate-binding site. *J. Biol. Chem.* **240**, 4723–4726.

Mani, R. S., and Zalik, S. (1970). Physiochemical studies of bean and wheat chloroplast structural protein. *Biochim. Biophys. Acta* **200**, 132–137.

Mansfield, T. A. (1969). *In* "Physiological Aspects of Crop Yield," (J. D. Eastin, *et al.,* eds.) pp. 231–233. A.S.A. and C.S.S.A., Madison, Wisconsin.

Mapson, L. (1958). Metabolism of ascorbic acid in plants. Part I. Function. *Annu. Rev. Plant Physiol.* **9**, 119–150.

Marcus, A. (1959). Amino acid dependent exchange between pyrophosphate and adenosine triphosphate in spinach preparations. *J. Biol. Chem.* **234**, 1238–1240.

Marker, A. F. H., and Whittingham, C. P. (1967). The site of synthesis of sucrose in green plant cells. *J. Exp. Bot.* **18**, 732–739.

Marks, G. S. (1966). The biosynthesis of heme and chlorophyll. *Bot. Rev.* **32**, 56–94.

Marsh, H. V., Jr., Galmiche, J. M., and Gibbs, M. (1964). Respiration during photosynthesis. *Rec. Chem. Progr.* **25**, 259–271.

Marsh, H. V., Jr., Galmiche, J. M., and Gibbs, M. (1965). Effect of light on the citric acid cycle in *Scenedesmus. Plant Physiol.* **40**, 1013–1022.

Maruyama, H., Easterday, R. L., Chang, H. C., and Lane, M. D. (1966). The enzymatic carboxylation of phosphoenolpyruvate. I. Purification and properties of phosphoenolpyruvate carboxylase. *J. Biol. Chem.* **241**, 2405–2412.

Mateus Ventura, M. (1954). Action of enzymatic inhibitors on transpiration and the behavior of stomata. II. Action of sodium arsenite, 2,4-dinitrophenol, and Janus green on isolated leaves of *Stizolobium alterrimum. Rev. Brasil. Biol.* **14**, 153–161.

Mayer, A. M., and Friend, J. (1960). Localization and nature of phenolase in sugar-beet leaves. *J. Exp. Bot.* **11**, 141–150.

Mazelis, M. (1960). Formate oxidation by particulate preparations from higher plants. *Plant Physiol.* **35**, 386–391.

Mego, J. L., and Jagendorf, A. T. (1961). Effect of light on growth of Black Valentine bean plastids. *Biochim. Biophys. Acta* **53**, 237–254.

Meidner, H. (1962). The minimum intercellular space CO_2 concentration (Γ) of maize leaves and its influence on stomatal movements. *J. Exp. Bot.* **13**, 284–293.

Meidner, H. (1967). Further observations on the minimum intercellular space carbon-dioxide concentration (Γ) of maize leaves and the postulated roles of "photo-respiration" and glycollate metabolism. *J. Exp. Bot.* **18**, 177–185.

Meidner, H., and Mansfield, T. A. (1965). Stomatal responses to illumination. *Biol. Rev.* **40**, 483–509.

Meidner, H., and Mansfield, T. A. (1968). "Physiology of Stomata." McGraw-Hill, New York.

Menke, W. (1962). Structure and chemistry of plastids. *Annu. Rev. Plant Physiol.* **13**, 27–44.

Menke, W. (1966). The structure of the chloroplasts. *In* "The Biochemistry of Chloroplasts" (T. W. Goodwin, ed.), Vol. 1, pp. 3–22. Academic Press, New York.

Merrett, M. J., and Goulding, K. H. (1967). Glycollate formation during photoassimilation of acetate by *Chlorella. Planta* **75**, 275–278.

Mifflin, B. J., Marker, A. F. H., and Whittingham, C. P. (1966). The metabolism of glycine and glycollate by pea leaves in relation to photosynthesis. *Biochim. Biophys. Acta* **120**, 266–273.

Milthorpe, F. L., and Moorby, J. (1969). Vascular transport and its significance in plant growth. *Annu. Rev. Plant Physiol.* **20**, 117–138.

Mitchell, P. (1961). Coupling of phosphorylation to electron and hydrogen transfer by a chemiosmotic type of mechanism. *Nature (London)* **191**, 144–148.

Mittleheuser, C. J., and Van Steveninck, R. F. M. (1969). Stomatal closure and inhibition of transpiration induced by (RS)-abscisic acid. *Nature (London)* **221**, 281–282.

Miyachi, S., Kanai, R., and Benson, A. A. (1968). Aerobically bound CO_2 in *Chlorella* cells. *In* "Comparative Biochemistry and Biophysics of Photosynthesis" (K. Shibata *et al.*, eds.), pp. 246–252. Univ. of Tokyo Press, Tokyo.

Mollenhauer, H. H., Morré, D. J., and Kelley, A. G. (1966). The widespread occurrence of plant cytosomes resembling animal microbodies. *Protoplasma* **62**, 44–52.

Monteith, J. L. (1962). Measurement and interpretation of carbon dioxide fluxes in the field. *Neth. J. Agr. Sci.* **10**, 334–346.

Monteith, J. L. (1964). Evaporation and environment. *Symp. Soc. Exp. Biol.* **19**, 205–234.

Monteith, J. L. (1965). Light distribution and photosynthesis in field crops. *Ann. Bot. (London)* [N. S.] **29**, 17–38.

Monteith, J. L. (1969). Light interception and radiative exchange in crop stands. *In* "Physiological Aspects of Crop Yield," (J. D. Eastin, *et al.*, eds.) pp. 89–111. A.S.A. and C.S.S.A., Madison, Wisconsin.

Monteith, J. L., Sciecz, G., and Yabuki, K. (1964). Crop photosynthesis and the flux of carbon dioxide below the canopy. *J. Appl. Ecol.* **1**, 321–337.

Monteith, J. L., Szeicz, G., and Waggoner, P. E. (1965). The measurement and control of stomatal resistance in the field. *J. Appl. Ecol.* **2**, 345–355.

Moore, R. E., Springer-Lederer, H., Ottenhym, H. C. J., and Bassham, J. A. (1969). Photosynthesis by isolated chloroplasts. IV. Regulation by factors from leaf cells. *Biochim. Biophys. Acta* **180**, 368–376.

Mortimer, D. C. (1959). Some short-term effects of increased carbon dioxide concentration on photosynthetic assimilation in leaves. *Can. J. Bot.* **37**, 1191–1201.

Mortimer, D. C. (1965). Translocation of the products of photosynthesis in sugarbeet petioles. *Can. J. Bot.* **43**, 269–280.

Mortimer, D. C., and Terry, N. (1969). Measurement of translocation rate of sucrose in sugar beet. *Abstr. Int. Bot. Congr., 11th, 1969,* p. 152.

Moss, D. N. (1962a). The limiting carbon dioxide concentration for photosynthesis. *Nature* (*London*) **193**, 587.

Moss, D. N. (1962b). Photosynthesis and barrenness. *Crop Sci.* **2**, 366–367.

Moss, D. N. (1963). The effect of environment on gas exchange of leaves. *Conn., Agr. Exp. Sta., New Haven, Bull.* **664**, 86–101.

Moss, D. N. (1966). Respiration of leaves in light and darkness. *Crop Sci.* **6**, 351–354.

Moss, D. N. (1967). High activity of the glycolic acid oxidase system in tobacco leaves. *Plant Physiol.* **42**, 1463–1464.

Moss, D. N. (1968). Photorespiration and glycolate metabolism in tobacco leaves. *Crop Sci.* **8**, 71–76.

Moss, D. N., and Rasmussen, H. P. (1969). Cellular localization of CO_2 fixation and translocation of metabolites. *Plant Physiol.* **44**, 1063–1068.

Moss, D. N., Musgrave, R. B., and Lemon, E. R. (1961). Photosynthesis under field conditions. III. Some effects of light, carbon dioxide, temperature, and soil moisture on photosynthesis, respiration, and transpiration of corn. *Crop Sci.* **1**, 83–87.

Moss, D. N., Krenzer, E. G., Jr., and Brun, W. A. (1969). Carbon dioxide compensation points in related plant species. *Science* **164**, 187–188.

Moss, R. A., and Loomis, W. E. (1952). Absorption spectra of leaves. I. The visible spectrum. *Plant Physiol.* **27**, 370–391.

Mudd, J. B. (1967). Fat metabolism in plants. *Annu. Rev. Plant Physiol.* **18**, 229–252.

Mudd, J. B., and McManus, T. T. (1962). Metabolism of acetate by cell-free preparations from spinach leaves. *J. Biol. Chem.* **237**, 2057–2063.

Mudd, J. B., Van Vliet, H. H. D. M., and Van Deenen, L. L. M. (1969). Biosynthesis of galactolipids by enzyme preparations from spinach leaves. *J. Lipid Res.* **10**, 623–630.

Müllhofer, G., and Rose, I. A. (1965). The position of carbon–carbon bond cleavage in the ribulose diphosphate carboxydismutase reaction. *J. Biol. Chem.* **240**, 1341–1346.

Murata, Y. (1969). Physiological responses to nitrogen in plants. *In* "Physiological Aspects of Crop Yield," (J. D. Eastin, *et al.,* eds.) pp. 225–259. A.S.A. and C.S.S.A., Madison, Wisconsin.

Murata, Y. and Iyama, J. (1963). Studies on the photosynthesis of forage crops. II. Influence of air-temperature upon the photosynthesis of some forage and grain crops. *Proc. Crop Sci. Soc. Jap.* **31**, 315–322.

Nadler, K., and Granick, S. (1970). Controls on chlorophyll synthesis in barley. *Plant Physiol.* **46**, 240–246.

Nagai, J., and Bloch, K. (1966). Enzymatic desaturation of stearyl acyl carrier protein. *J. Biol. Chem.* **241**, 1925–1927.

Naylor, A. W., Rabson, R., and Tolbert, N. E. (1958). Aspartic-C^{14} acid metabolism in leaves, roots, and stems. *Physiol. Plant.* **11**, 537–547.

Neales, T. F., and Incoll, L. D. (1968). The control of leaf photosynthetic rate by the level of assimilate concentration in the leaf: A review of the hypothesis. *Bot. Rev.* **34,** 107–125.

Nelson, E. B., and Tolbert, N. E. (1969). The regulation of glycolate metabolism in *Chlamydomonas reinhardtii. Biochim. Biophys. Acta* **184,** 263–270.

Nelson, E. B., Cenedella, A., and Tolbert, N. E. (1969). Carbonic anhydrase levels in *Chlamydomonas. Phytochemistry* **8,** 2305–2306.

Neufeld, E. F., and Hall, C. W. (1964). Formation of galactolipids by chloroplasts. *Biochem. Biophys. Res. Commun.* **14,** 503–508.

Neumann, J., and Jagendorf, A. T. (1964). Dinitrophenol as an uncoupler of photosynthetic phosphorylation. *Biochem. Biophys. Res. Commun.* **16,** 562–567.

Neumann, J., and Jagendorf, A. (1965). Uncoupling photophosphorylation by detergents. *Biochim. Biophys. Acta* **109,** 382–389.

Nichols, B. W., and James, A. T. (1968). The function and metabolism of fatty acids and acyl lipids in chloroplasts. *In* "Plant Cell Organelles" (J. B. Pridham, ed.), pp. 163–197. Academic Press, New York.

Nishida, K. (1962). Studies on the re-assimilation of respiratory CO_2 in illuminated leaves. *Plant Cell Physiol.* **3,** 111–124.

Nishida, K. (1963). Studies on stomatal movement of crassulacean plants in relation to the acid metabolism. *Physiol. Plant.* **16,** 281–298.

Nobel, P. S. (1969). Light-induced changes in the ionic content of chloroplasts in *Pisum sativum. Biochim. Biophys. Acta* **172,** 134–143.

Noll, C. R., Jr., and Burris, R. H. (1954). Nature and distribution of glycolic acid oxidase in plants. *Plant Physiol.* **29,** 261–265.

Norman, J. M., and Tanner, C. B. (1969). Transient light measurements in plant canopies. *Agron. J.* **61,** 847–849.

Nössberger, J., and Thorne, G. N. (1965). Effect of removing florets or shading the ear of barley on production and distribution of dry matter. *Ann. Bot.* (*London*) [N. S.] **29,** 635–644.

Ongun, A., and Mudd, J. B. (1968). Biosynthesis of galactolipids in plants. *J. Biol. Chem.* **243,** 1558–1566.

Ongun, A., and Stocking, C. R. (1965a). Effect of light on the incorporation of serine into the carbohydrates of chloroplasts and nonchloroplast fractions of tobacco leaves. *Plant Physiol.* **40,** 819–824.

Ongun, A., and Stocking, C. R. (1965b). Effect of light and dark on the intracellular fate of photosynthetic products. *Plant Physiol.* **40,** 825–831.

Ongun, A., Thomson, W. W., and Mudd, J. B. (1968). Lipid composition of chloroplasts isolated by aqueous and nonaqueous techniques. *J. Lipid Res.* **9,** 409–415.

Öpik, H. (1968). Structure, function and developmental changes in mitochondria of higher plant cells. *In* "Plant Cell Organelles" (J. B. Pridham, ed.), pp. 47–88. Academic Press, New York.

Orth, G. M., and Cornwell, D. G. (1963). The isolation and composition of chloroplasts and etiolated plastids from corn seedlings. *Biochim. Biophys. Acta* **71,** 734–736.

Orth, G. M., Tolbert, N. E., and Jimenez, E. (1966). Rate of glycolate formation during photosynthesis at high pH. *Plant Physiol.* **41,** 143–147.

Osborne, T. B., and Wakeman, L. I. (1920). The proteins of green leaves. I. Spinach leaves. *J. Biol. Chem.* **42,** 1–26.

Osmond, C. B. (1969). β-Carboxylation photosynthesis and photorespiration in higher plants. *Biochim. Biophys. Acta* **172,** 144–149.

Osmond, C. B., and Avadhani, P. N. (1970). Inhibition of the β-carboxylation pathway of CO_2 fixation by bisulfite compounds. *Plant Physiol.* **45,** 228–230.

Osmond, C. B., Troughton, J. H., and Goodchild, D. J. (1969). Physiological biochemical and structural studies of photosynthesis and photorespiration in two species of *Atriplex*. *Z. Pflanzenphysiol.* **61,** 218–237.

Otsuki, Y., and Takebe, I. (1969). Isolation of intact mesophyll cells and their protoplasts from higher plants. *Plant Cell Physiol.* **10,** 917–921.

Ozbun, J. L., Volk, R. J., and Jackson, W. A. (1964). Effects of light and darkness on gaseous exchange of bean leaves. *Plant Physiol.* **39,** 523–527.

Packer, L. (1963). Structural changes correlated with photochemical phosphorylation in chloroplast membranes. *Biochim. Biophys. Acta* **75,** 12–22.

Pallas, J. E., Jr. (1965). Transpiration and stomatal opening with changes in carbon dioxide content of the air. *Science* **147,** 171–173.

Pallas, J. E., Jr. (1966). Mechanisms of guard cell action. *Quart. Rev. Biol.* **41,** 365–383.

Pallas, J. E., Jr., and Box, J. E., Jr. (1970). Explanation for the stomatal response of excised leaves to kinetin. *Nature (London)* **227,** 87–88.

Parenti, F., and Margulies, M. M. (1967). *In vitro* protein synthesis by plastids of *Phaseolus vulgaris*. I. Localization of activity in the chloroplasts of a chloroplast containing fraction from developing bean. *Plant Physiol.* **42,** 1179–1186.

Park, R., and Epstein, S. (1961). Metabolic fractionation of C^{13} and C^{12} in plants. *Plant Physiol.* **36,** 133–138.

Park, R. B., and Biggins, J. (1964). Quantasome: Size and composition. *Science* **144,** 1009–1011.

Parlange, J.-Y., and Waggoner, P. E. (1970). Stomatal dimensions and resistance to diffusion. *Plant Physiol.* **46,** 337–342.

Paulsen, J. M., and Lane, M. D. (1966). Spinach ribulose diphosphate carboxylase. I. Purification and properties of the enzyme. *Biochemistry* **5,** 2350–2357.

Pearce, R. B., and Lee, D. R. (1969). Photosynthetic and morphological adaptation of alfalfa leaves to light intensity at different stages of maturity. *Crop Sci.* **9,** 791–794.

Pearce, R. B., Brown, R. H., and Blaser, R. E. (1968). Photosynthesis of alfalfa leaves as influenced by age and environment. *Crop Sci.* **8,** 677–680.

Peaslee, D. E., and Moss, D. N. (1968). Stomatal conductivity in K-deficient leaves of maize. *Crop Sci.* **8,** 427–430.

Pendleton, J. W., Smith, G. E., Winter, S. R., and Johnston, T. J. (1968). Field investigations of the relationship of leaf angle in corn (*Zea mays*) to grain yield and apparent photosynthesis. *Agron. J.* **60,** 422–424.

Penman, H. L., and Schofield, R. K. (1951). Some physical aspects of assimilation and transpiration. *Symp. Soc. Exp. Biol.* **5,** 115–129.

Peterkofsky, A., and Racker, E. (1961). The reductive pentose phosphate cycle. III. Enzyme activities in cell-free extracts of photosynthetic organisms. *Plant Physiol.* **36,** 409–414.

Pierpoint, W. S. (1959). Mitochondrial preparations from the leaves of tobacco (*Nicotiana tabacum*). *Biochem. J.* **71,** 518–528.

Pierpoint, W. S. (1960). Mitochondrial preparations from the leaves of tobacco (*Nicotiana tabacum*). 2. Oxidative phosphorylation. *Biochem. J.* **75,** 504–510.

Pierpoint, W. S. (1962). Mitochondrial preparations from the leaves of tobacco (*Nicotiana tabacum*). 4. Separation of some components by density-gradient centrifuging. *Biochem. J.* **82,** 143–148.

Pierpoint, W. S. (1963). The distribution of succinate dehydrogenase and malate dehydrogenase among components of tobacco-leaf extracts. *Biochem. J.* **88,** 120–125.

Plamondon, J. E., and Bassham, J. A. (1966). Glycolic acid labeling during photosynthesis with $^{14}CO_2$ and tritiated water. *Plant Physiol.* **41,** 1272–1275.

Plaut, Z., and Gibbs, M. (1970). Glycolate formation in intact spinach chloroplasts. *Plant Physiol.* **45**, 470–474.

Porra, R. J., and Irving, E. A. (1970). Tetrapyrrole biosynthesis in plant and yeast organelles. *Biochem. J.* **116**, No. 4, 42P.

Porter, H. K. (1966). Leaves as collecting and distributing agents of carbon. *Aust. J. Sci.* **29**, 31–40.

Porter, J. W., and Anderson, D. G. (1967). Biosynthesis of carotenes. *Annu. Rev. Plant Physiol.* **18**, 197–228.

Poskuta, G., Nelson, C. D., and Krotkov, G. (1967). Effects of metabolic inhibitors on the rates of CO_2 evolution in light and in darkness by detached spruce twigs, wheat and soybean leaves. *Plant Physiol.* **42**, 1187–1190.

Possingham, J. V., and Spencer, D. (1962). Manganese as a functional component of chloroplasts. *Aust. J. Biol. Sci.* **15**, 58–68.

Preiss, J., and Kosuge, T. (1970). Regulation of enzyme activity in photosynthetic systems. *Annu. Rev. Plant Physiol.* **21**, 433–466.

Preiss, J., Ghosh, H. P., and Wittkop, J. (1967). Regulation of the biosynthesis of starch in spinach leaf chloroplasts. *In* "The Biochemistry of Chloroplasts" (T. W. Goodwin, ed.), Vol. 2, pp. 131–153. Academic Press, New York.

Pristupa, N. A. (1964). Redistribution ot radiactive assimilates in the leaf tissues of cereals. *Sov. Plant Physiol.* **11**, 31–36.

Pritchard, G. G., Griffin, W. J., and Whittingham, C. P. (1962). The effect of CO_2 concentration, light intensity and isonicotinyl hydrazide on the photosynthetic production of glycollic acid by *Chlorella*. *J. Exp. Bot.* **13**, 176–184.

Pritchard, G. G., Whittingham, C. P., and Griffin, W. J. (1963). The effect of isonicotinyl hydrazide on the photosynthetic incorporation of radioactive CO_2 into ethanol-soluble compounds of *Chlorella*. *J. Exp. Bot.* **14**, 281–289.

Pucher, G. W. (1942). The organic acids of the leaves of *Bryophyllum calicynum*. *J. Biol. Chem.* **145**, 511–523.

Puckridge, D. W. (1969). Photosynthesis of wheat under field conditions. II. Effect of defoliation on the carbon dioxide uptake of the community. *Aust. J. Agr. Res.* **20**, 623–634.

Pugh, E. L., and Wakil, S. J. (1965). Studies on the mechanism of fatty acid synthesis. XIV. The prosthetic group of acyl carrier protein and the mode of its attachment to the protein. *J. Biol. Chem.* **240**, 4727–4733.

Purdy, S. J., and Truter, E. V. (1963). Constitution of the surface lipid from the leaves of *Brassica oleracea* [var. *capitata* (*Winnigstadt*)]. III. Nonacosane and its derivatives. *Proc. Roy. Soc., Ser. B* **158**, 553–565.

Quayle, J. R., Fuller, R. C., Benson, A. A., and Calvin, M. (1954). Enzymatic carboxylation of ribulose diphosphate. *J. Amer. Chem. Soc.* **76**, 3610–3611.

Rabinowitch, E. I. (1945). "Photosynthesis and Related Processes," Vol. 1. Wiley (Interscience), New York.

Rabinowitch, E. I. (1951). "Photosynthesis and Related Processes," Vol. 2, Part 1. Wiley (Interscience), New York.

Rabinowitch, E. I. (1956). "Photosynthesis and Related Processes," Vol. 2, Part 2. Wiley (Interscience), New York.

Rabson, R., Tolbert, N. E., and Kearney, P. C. (1962). Formation of serine and glyceric acid by the glycolate pathway. *Arch. Biochem. Biophys.* **98**, 154–163.

Racker, E. (1957). The reductive pentose cycle. I. Phosphoribulokinase and ribulose diphosphate carboxylase. *Arch. Biochem. Biophys.* **69**, 300–310.

Racker, E. (1965). "Mechanisms in Bioenergetics." Academic Press, New York.

Racker, E. (1970). Function and structure of the inner membrane of mitochondria and chloroplasts. *In* "Membranes of Mitochondria and Chloroplasts" (E. Racker, ed.), pp. 127–171. Van Nostrand Reinhold, New York.

Ramirez, J. M., del Campo, F. F., and Arnon, D. I. (1968). Photosynthetic phosphorylation as energy source for protein synthesis and carbon dioxide assimilation by chloroplasts. *Proc. Nat. Acad. Sci. U. S.* **59**, 606–612.

Ranson, S. L. (1965). Plant acids. *In* "Biosynthetic Pathways in Higher Plants" (J. B. Pridham and T. Swain, eds.), pp. 179–198. Academic Press, New York.

Ranson, S. L., and Thomas, M. (1960). Crassulacean acid metabolism. *Annu. Rev. Plant Physiol.* **11**, 81–110.

Raschke, K. (1956). Über die physikalischen Beziehungen zwischen Wärmeübergangszahl, Strahlungsaustausch, Temperatur und Transpiration eines Blattes. *Planta* **48**, 200–238.

Raschke, K. (1958). Über den Einfluss der Diffusionswiderstünde auf die Transpiration und die Temperatur eines Blattes. *Flora (Jena)* **146**, 546–578.

Raschke, K. (1960). Heat transfer between the plant and the environment. *Annu. Rev. Plant Physiol.* **11**, 111–126.

Reed, M. L., and Graham, D. (1968). Control of photosynthetic carbon dioxide fixation during an induction phase in *Chlorella*. *Plant Physiol.* **43S**, 29.

Rehfeld, D. W., Randall, D. D., and Tolbert, N. E. (1970). Enzymes of the glycolate pathway in plants without CO_2-photorespiration. *Can. J. Bot.* **48**, 1219–1226.

Reifsnyder, W. E., and Furnival, G. M. (1970). Power-spectrum analysis of the energy contained in sunflecks. *Proc. Forest Microclimate Symp., 3rd. 1970,* pp. 117–118.

Rhoades, M. M. (1946). Plastid mutations. *Cold Spring Harbor Symp. Quant. Biol.* **11**, 202–207.

Richardson, K. E., and Tolbert, N. E. (1961a). Phosphoglycolic acid phosphatase. *J. Biol. Chem.* **236**, 1285–1290.

Richardson, K. E., and Tolbert, N. E. (1961b). Oxidation of glyoxylic acid to oxalic acid by glycolic acid oxidase. *J. Biol. Chem.* **236**, 1280–1284.

Robertson, R. N. (1968). "Protons, Electrons, Phosphorylation and Active Transport." Cambridge Univ. Press, London and New York.

Rosenberg, A. (1967). Galactosyl diglycerides: Their possible function in *Euglena* chloroplasts. *Science* **157**, 1191–1196.

Rosenberg, J. L. (1965). "Photosynthesis." Holt, New York.

Rosenberg, L. L., Capindale, J. B., and Whatley, F. R. (1958). Formation of oxalacetate and aspartate from phospho-enol-pyruvate in spinach leaf chloroplast extract. *Nature (London)* **181**, 632–633.

Ruben, S., Hassid, W. Z., and Kamen, M. D. (1939). Radioactive carbon in the study of photosynthesis. *J. Amer. Chem. Soc.* **61**, 661–663.

Ruben, S., Randall, M., Kamen, M. D., and Hyde, J. L. (1941). Heavy oxygen (O^{18}) as a tracer in the study of photosynthesis. *J. Amer. Chem. Soc.* **63**, 877–879.

Rutner, A. C. (1970). Estimation of the molecular weight of ribulose diphosphate carboxylase sub-units. *Biochem. Biophys. Res. Commun.* **39**, 923–929.

Rutner, A. C., and Lane, M. D. (1967). Nonidentical subunits of ribulose diphosphate carboxylase. *Biochem. Biophys. Res. Commun.* **28**, 531–537.

Saeki, T. (1963). Light relations in plant communities. *In* "Environmental Control of Plant Growth" (L. T. Evans, ed.), pp. 79–94. Academic Press, New York.

Samish, Y. and Koller, D. (1968). Estimation of photorespiration of green plants and of their mesophyll resistance to CO_2 uptake. *Ann. Bot.* (*London*) [N. S.] **32**, 687–694.

Sampson, J. (1961). A method of replicating dry or moist surfaces for examination by light microscopy. *Nature* (*London*) **191**, 932–933.

Sanderson, J. A., and Hulburt, E. O. (1955). Sunlight as a source of radiation. *Radiat. Biol.* **2**, 95–118.

Sane, P. V̇., and Park, R. B. (1970). Purification of photosystem I reaction centers from spinach stroma lamellae. *Biochem. Biophys. Res. Commun.* **41**, 206–210.

San Pietro, A. (1967). Electron transport in chloroplasts. *In* "Harvesting the Sun" (A. San Pietro, F. A. Greer, and T. J. Army, eds.), pp. 49–68. Academic Press, New York.

San Pietro, A., and Lang, H. M. (1956). Photosynthetic pyridine nucleotide reductase. I. Partial purification and properties of the enzyme from spinach. *J. Biol. Chem.* **231**, 211–229.

Santarius, K. A., and Heber, U. (1965). Changes in the intracellular levels of ATP, ADP, AMP and P_i and regulatory function of the adenylate system in leaf cells during photosynthesis. *Biochim. Biophys. Acta* **102**, 39–54.

Sastry, P. S., and Kates, M. (1964). Hydrolysis of monogalactosyl and digalactosyl diglycerides by specific enzymes in runner-bean leaves. *Biochemistry* **3**, 1280–1287.

Sastry, P. S., and Kates, M. (1966). Biosynthesis of lipids in plants. II. Incorporation of glycerophosphate-^{32}P into phosphatides by cell-free preparations from spinach leaves. *Can. J. Biochem.* **44**, 459–467.

Sawhney, B. L., and Zelitch, I. (1969). Direct determination of potassium ion accumulation in guard cells in relation to stomatal opening in light. *Plant Physiol.* **44**, 1350–1354.

Schmid, G. H. (1969). The effect of blue light on glycolate oxidase of tobacco. *Hoppe-Seyler's Z. Physiol. Chem.* **350**, 1035–1046.

Schmid, G. H., and Gaffron, H. (1967). Light metabolism and chloroplast structure in chlorophyll-deficient tobacco mutants. *J. Gen. Physiol.* **50**, 563–582.

Schmid, G. H., and Schwarze, P. (1969). Blue light enhanced respiration in a colorless *Chlorella* mutant. *Hoppe-Seyler's Z. Physiol. Chem.* **350**, 1513–1520.

Scholander, P. F., Hammel, H. T., Hemmingsen, E. A., and Bradstreet, E. D. (1964). Hydrostatic pressure and osmotic potential in leaves of mangroves and some other plants. *Proc. Nat. Acad. Sci. U. S.* **52**, 119–125.

Schou, L., Benson, A. A., Bassham, J. A., and Calvin, M. (1950). The path of carbon in photosynthesis. XI. The role of glycolic acid. *Physiol. Plant.* **3**, 487–495.

Semenenko, V. E. (1964). Characteristics of carbon dioxide gas exchange in the transition stages of photosynthesis upon changing from light to darkness, light induced evolution of carbon dioxide. *Sov. Plant Physiol.* **11**, 375–384.

Shaw, R. H., and Laing, D. R. (1966). Moisture stress and plant responses. *In* "Plant Environment and Efficient Water Use" (W. H. Pierre *et al.*, eds.), pp. 73–94. A.S.A. and C.S.S.A., Madison, Wisconsin.

Shetty, A. S., and Miller, G. W. (1969). Purification and general properties of δ-aminolaevulate dehydratase from *Nicotiana tabacum*. *Biochem. J.* **114**, 331–337.

Shibuya, I., Maruo, B., and Benson, A. A. (1965). Sulfolipid localization in lamellar lipoprotein. *Plant Physiol.* **40**, 1251–1256.

Shimshi, D. (1963). Effect of soil moisture and phenylmercuric acetate upon stomatal aperture, transpiration, and photosynthesis. *Plant Physiol.* **38**, 713–721.

Shin, M., Tagawa, K., and Arnon, D. I. (1963). Crystallization of ferredoxin-TPN reductase and its role in the photosynthetic apparatus of chloroplasts. *Biochem. Z* **338**, 84–96.

Sideris, C. P., Young, H. Y., and Chun, H. H. Q. (1948). Diurnal changes and growth rates as associated with ascorbic acid, titratable acidity, carbohydrate and nitrogenous fractions in the leaves of *Ananas comosus* (L.) Merr. *Plant Physiol.* **23**, 38–69.

Siegenthaler, P.-A., and Packer, L. (1965). Light dependent volume changes and reactions in chloroplasts. I. Action of alkenylsuccinic acids and phenylmercuric acetate and possible relation to mechanisms of stomatal control. *Plant Physiol.* **40**, 785–791.

Simoni, R. D., Criddle, R. S., and Stumpf, P. K. (1967). Fat metabolism in higher plants. XXXI. Purification and properties of plant and bacterial acyl carrier protein. *J. Biol. Chem.* **242**, 573–581.

Sirevåg, R., and Ormerod, J. G. (1970). Carbon dioxide-fixation in photosynthetic green sulfur bacteria. *Science* **169**, 186–188.

Sissakian, N. M. (1958). Enzymology of the plastids. *Advan. Enzymol.* **20**, 201–236.

Slack, C. R. (1968). The photoactivation of a phosphopyruvate synthase in leaves of *Amaranthus palmeri*. *Biochem. Biophys. Res. Commun.* **30**, 483–488.

Slack, C. R., Hatch, M. D., and Goodchild, D. J. (1969). Distribution of enzymes in mesophyll and parenchyma-sheath chloroplasts of maize leaves in relation to the C_4-dicarboxylic acid pathway of photosynthesis. *Biochem. J.* **114**, 489–498.

Slatyer, R. O. (1967). "Plant-Water Relationships." Academic Press, New York.

Slatyer, R. O. (1969). Physiological significance of internal water relations to crop yield. *In* "Physiological Aspects of Crop Yield," (J. D. Eastin, *et al.*, eds.) pp. 53–83. A.S.A. and C.S.S.A., Madison, Wisconsin.

Slatyer, R. O., and Bierhuizen, J. F. (1964). The influence of several transpiration suppressants on transpiration, photosynthesis, and water-use efficiency of cotton leaves. *Aust. J. Biol. Sci.* **17**, 131–146.

Smillie, R. M. (1956). Enzymic activities of sub-cellular particles from leaves. *Aust. J. Biol. Sci.* **9**, 81–91.

Smillie, R. M. (1969). Nucleic acid and protein metabolism in chloroplasts. *Abstr., IBP Meet.,* Moscow, 1969.

Smith, D., and Buchholtz, K. P. (1964). Modification of plant transpiration rate with chemicals. *Plant Physiol.* **39**, 572–578.

Smith, E. L. (1941). The chlorophyll-protein compound of the green leaf. *J. Gen. Physiol.* **24**, 565–582.

Smith, R. E., and Horowitz, B. A. (1969). Brown fat and thermogenesis. *Physiol. Rev.* **49**, 330–425.

Soderstrom, T. R. (1962). The isocitric acid content of crassulacean plants and a few succulent species from other families. *Amer. J. Bot.* **49**, 850–855.

Spencer, D. (1965). Protein synthesis by isolated spinach chloroplasts. *Arch. Biochem. Biophys.* **111**, 381–390.

Spencer, D., and Whitfeld, P. R. (1969). The characteristics of spinach chloroplast DNA polymerase. *Arch. Biochem. Biophys.* **132**, 477–488.

Springer-Lederer, H., El-Badry, A. M., Ottenheym, H. C. J., and Bassham, J. A. (1969). Inhibition of photosynthesis in isolated spinach chloroplasts by added fructose-1,6-diphosphatase. *Biochim. Biophys. Acta* **189**, 464–467.

Srivastava, S. K., and Krishnan, P. S. (1962). Oxalic acid oxidase in the leaves of *Bougainvillea spectabilis*. *Biochem. J.* **85**, 33–38.

Stafford, H. A., Magaldi, A., and Vennesland, B. (1954). The enzymatic reduction of hydroxypyruvic acid to D-glyceric acid in higher plants. *J. Biol. Chem.* **207**, 621–629.

Stålfelt, M. G. (1962). The effect of temperature on opening of the stomatal cells. *Physiol. Plant.* **15**, 772–779.

Stålfelt, M. G. (1966). The role of epidermal cells in the stomatal movements. *Physiol. Plant.* **19**, 241–256.

Stanhill, G. (1958). Effects of soil moisture on the yield and quality of turnips. II. Response at different growth stages. *J. Hort. Sci.* **33**, 264–274.

Steer, B. T., and Gibbs, M. (1969). Delta-aminolevulinic acid dehydrase in greening bean leaves. *Plant Physiol.* **44**, 781–783.

Stephens, G. R. (1969). Productivity of red pine. 1. Foliage distribution in tree crown and stand canopy. *Agr. Meteorol.* **6**, 275–282.

Stephens, G. R., and Waggoner, P. E. (1970). Carbon dioxide exchange of a tropical rain forest. Part. I. *BioScience* **20**, 1050–1053.

Stern, W. R., and Donald, C. M. (1962). Light relationships in grass-clover swards. *Aust. J. Agr. Res.* **13**, 599–614.

Stiller, M. (1962). The path of carbon in photosynthesis. *Annu. Rev. Plant Physiol.* **13**, 151–170.

Stocking, C. R. (1959). Chloroplast isolation in non-aqueous media. *Plant Physiol.* **34**, 56–61.

Stocking, C. R., and Larson, S. (1969). A chloroplast cytoplasmic shuttle and the reduction of extraplastid NAD. *Biochem. Biophys. Res. Commun.* **37**, 278–282.

Stocking, C. R., and Ongun, A. (1962). Intracellular distribution of some metallic elements in leaves. *Amer. J. Bot.* **49**, 284–289.

Stocking, C. R., Williams, G. R., and Ongun, A. (1963). Intracellular distribution of the early products of photosynthesis. *Biochem. Biophys. Res. Commun.* **10**, 416–421.

Stoy, V. (1963). The translocation of the C^{14}-labelled photosynthetic products from the leaf to the ear in wheat. *Physiol. Plant.* **16**, 851–866.

Stoy, V. (1965). Photosynthesis, respiration, and carbohydrate accumulation in spring wheat in relation to yield. *Physiol. Plant.* Suppl. IV, pp. 1–125.

Stumpf, P. K. (1969). Metabolism of fatty acids. *Annu. Rev. Biochem.* **38**, 159–212.

Stumpf, P. K., and James, A. T. (1963). Biosynthesis of long-chain fatty acids by lettuce chloroplast preparations. *Biochim. Biophys. Acta* **70**, 20–32.

Stutz, R. E., and Burris, R. H. (1951). Photosynthesis and metabolism of organic acids in higher plants. *Plant Physiol.* **26**, 226–243.

Sugiyama, T., Nakayama, N., and Akazawa, T. (1968). Activation of spinach leaf ribulose-1,5-diphosphate carboxylase activities by magnesium ions. *Biochem. Biophys. Res. Commun.* **30**, 118–123.

Syrett, P. J. (1966). The kinetics of isocitrate lyase formation in *Chlorella:* Evidence for the promotion of enzyme synthesis by photophosphorylation. *J. Exp. Bot.* **17**, 641–654.

Tagawa, K., and Arnon, D. I. (1962). Ferredoxins as electron carriers in photosynthesis and in the biological production and consumption of hydrogen gas. *Nature (London)* **195**, 537–543.

Tait, G. H. (1970). Glycine decarboxylase in *Rhodopseudomonas spheroides* and in rat liver mitochondria. *Biochem. J.* **118**, 819–830.

Takebe, I., Otsuki, Y., and Aoki, S. (1968). Isolation of tobacco mesophyll cells in intact and active state. *Plant Cell Physiol.* **9**, 115–124.

Tal. M. (1966). Abnormal stomatal behavior in wilty mutants of tomato. *Plant Physiol.* **41**, 1387–1391.

Tamiya, H. (1957). Mass culture of algae. *Annu. Rev. Plant Physiol.* **8**, 309–334.

Tanaka, A., Kawano, K., and Yamaguchi, J. (1966). Photosynthesis, respiration, and plant type of the tropical rice plant. *Int. Rice Res. Inst., Tech. Bull.* **7**.

Tanner, J. W. (1969). *In* "Physiological Aspects of Crop Yield," (J. D. Eastin, *et al.,* eds.) pp. 50–51. A.S.A. and C.S.S.A., Madison, Wisconsin.

Tanner, W., Loffler, M., and Kandler, O. (1969). Cyclic photophosphorylation *in vivo* and its relation to photosynthetic CO_2 fixation. *Plant Physiol.* **44**, 422–428.

Telfer, A., Cammack, R., and Evans, M. C. W. (1970). Hydrogen peroxide as the product of autoxidation of ferredoxin: Reduced either chemically or by illuminated chloroplasts. *FEBS Lett.* **10**, 21–24.

Thomas, M. D., and Hill, G. R. (1949). Photosynthesis under field conditions. *In* "Photosynthesis in Plants" (J. Franck, and W. E. Loomis, eds.), pp. 19–52. Iowa State Coll. Press, Ames, Iowa.

Thomas, M. D., Hendricks, R. H., and Hill, E. R. (1944). Apparent equilibrium between photosynthesis and respiration in an unrenewed atmosphere. *Plant Physiol.* **19**, 370–376.

Thompson, C. M., and Whittingham, C. P. (1967). Intracellular localization of phosphoglycollate phosphatase and glyoxylate reductase. *Biochim. Biophys. Acta* **143**, 642–644.

Thompson, C. M., and Whittingham, C. P. (1968). Glycollate metabolism in photosynthesizing tissue. *Biochim. Biophys. Acta* **153**, 260–269.

Thomson, W. W., and De Journett, R. (1970). Studies on the ultrastructure of the guard cells of *Opuntia*. *Amer. J. Bot.* **57**, 309–316.

Thornber, J. P. (1969). Chlorophyll-protein complexes and photochemical reaction centers. *Abstr. Int. Bot. Congr., 11th. 1969*, p. 218.

Thornber, J. P., Stewart, J. C., Hatton, M. W. C., and Bailey, J. L. (1967). Studies on the nature of chloroplast lamellae. II. Chemical composition and further physical properties of two chlorophyll–protein complexes. *Biochemistry* **6**, 2006–2014.

Thorne, G. N. (1965). Photosynthesis of ears and flag leaves of wheat and barley. *Ann. Bot. (London)* [*N.S.*] **29**, 317–329.

Thorne, G. N., and Evans, A. F. (1964). Influence of tops and roots on net assimilation rate of sugar-beet and spinach-beet and grafts between them. *Ann. Bot. (London)* [*N.S.*] **28**, 499–508.

Tobin, A. J. (1970). Carbonic anhydrase from parsley leaves. *J. Biol. Chem.* **245**, 2656–2666.

Togari, Y., Murata, Y., and Saeki, T. (1969a). Photosynthesis and utilization of solar energy. Level I experiments. Report II. JIBP/PP—Photosynthesis, Fac. Agr., Tokyo Univ.

Togari, Y., Murata, Y., and Saeki, T. (1969b). Photosynthesis and utilization of solar energy. Level I experiments. Report III. JIBP/PP—Photosynthesis, Fac. Agr., Tokyo Univ.

Tolbert, N. E. (1963). Glycolate pathway. In "Photosynthetic Mechanisms in Green Plants," pp. 648–662. Natl. Acad. Sci.—Natl. Res. Council, Washington, D. C.

Tolbert, N. E., and Burris, R. H. (1950). Light activation of the plant enzyme which oxidizes glycolic acid. *J. Biol. Chem.* **186**, 791–804.

Tolbert, N. E., and Cohan, M. S. (1953). Activation of glycolic acid oxidase in plants. *J. Biol. Chem.* **204**, 639–648.

Tolbert, N. E., and Yamazaki, R. K. (1969). Leaf peroxisomes and their relation to photorespiration and photosynthesis. *Ann. N. Y. Acad. Sci.* **168**, 325–341.

Tolbert, N. E., and Zill, L. P. (1956). Excretion of glycolic acid by algae during photosynthesis. *J. Biol. Chem.* **222**, 895–906.

Tolbert, N. E., Clagett, C. O., and Burris, R. H. (1949). Products of the oxidation of glycolic acid and l-lactic acid by enzymes from tobacco leaves. *J. Biol. Chem.* **181**, 905–914.

Tolbert, N. E., Oeser, A., Kisaki, T., Hageman, R. H., and Yamazaki, R. K. (1968). Peroxisomes from spinach leaves containing enzymes related to glycolate metabolism. *J. Biol. Chem.* **243**, 5179–5184.

Tolbert, N. E., Oeser, A., Yamazaki, R. K., Hageman, R. H., and Kisaki, T. (1969). A survey of plants for leaf peroxisomes. *Plant Physiol.* **44**, 135–147.

Tolbert, N. E., Yamazaki, R. K., and Oeser, A. (1970). Localization and properties of hydroxypyruvate and glyoxylate reductases in spinach leaf particles. *J. Biol. Chem.* **245**, 5129–5136.

Tregunna, B. (1966). Flavin mononucleotide control of glycolic acid oxidase and photorespiration in corn leaves. *Science* **151**, 1239–1241.

Tregunna, E. B., and Downton, J. (1967). Carbon dioxide compensation in members of the *Amaranthaceae* and some related families. *Can. J. Bot.* **45**, 2385–2387.

Tregunna, E. B., Krotkov, G., and Nelson, C. D. (1961). Evolution of carbon dioxide by tobacco leaves during the dark period following illumination with light of different intensities. *Can. J. Bot.* **39**, 1045–1059.

Tregunna, E. B., Krotkov, G., and Nelson, C. D. (1964). Further evidence on the effects of light on respiration during photosynthesis. *Can. J. Bot.* **42**, 989–997.

Tregunna, E. B., Krotkov, G., and Nelson, C. D. (1966). Effect of oxygen on the rate of photorespiration in detached tobacco leaves. *Physiol. Plant.* **19**, 723–733.

Tregunna, E. B., Smith, B. N., Berry, J. A., and Downton, W. J. S. (1970). Some methods for studying the photosynthetic taxonomy of the angiosperms. *Can. J. Bot.* **48**, 1209–1214.

Trip, P. (1969). Sugar transport in conducting elements of sugar beet leaves. *Plant Physiol.* **44**, 717–725.

Trip, P., and Gorham, P. R. (1968). Bidirectional translocation of sugars in sieve tubes of squash plants. *Plant Physiol.* **43**, 877–882.

Troughton, J. H. (1969). Plant water status and carbon dioxide exchange of cotton leaves. *Aust. J. Biol. Sci.* **22**, 289–302.

Trown, P. W. (1965). An improved method for the isolation of carboxy-dismutase. Probable identity with fraction I protein and the protein moiety of protochlorophyll holochrome. *Biochemistry* **4**, 908–918.

Turner, J. S., and Brittain, E. G. (1962). Oxygen as a factor in photosynthesis. *Biol. Rev.* **37**, 130–170.

Turner, N. C. (1969). Stomatal resistance to transpiration in three contrasting canopies. *Crop Sci.* **9**, 303–307.

Turner, N. C. (1970). Response of adaxial and abaxial stomata to light. *New Phytol.* **69**, 647–653.

Turner, N. C., and Graniti, A. (1969). Fusicoccin: A fungal toxin that opens stomata. *Nature (London)* **223**, 1070–1071.

Turner, N. C., and Incoll, L. D. (1971). The vertical distribution of photosynthesis in crops of tobacco and sorghum. *J. Appl. Ecol.* **8**, (in press).

Turner, N. C., and Parlange, J.-Y. (1970). Analysis of operation and calibration of a ventilated diffusion porometer. *Plant Physiol.* **46**, 175–177.

Turner, N. C., and Waggoner, P. E. (1968). Effects of changing stomatal width in a red pine forest on soil water content, leaf water potential, bole diameter and growth. *Plant Physiol.* **43**, 973–978.

Urbach, W., and Gimmler, H. (1968). Stimulation of glycolate excretion in algae by disalicylidenepropanediamine and hydroxypyridine-methanesulfonate. *Z. Naturforsch. B.* **23**, 1282–1283.

Uribe, E., and Jagendorf, A. T. (1968). Membrane permeability and internal volume as factors in ATP synthesis by spinach chloroplasts. *Arch. Biochem. Biophys.* **128**, 351–359.

Van der Veen, R. (1949). Induction phenomena in photosynthesis. *Physiol. Plant.* **2**, 217–234.

Van Poucke, M., Cerff, R., Barthe, F., and Mohr, H. (1970). Simultaneous induction of glycolate oxidase and glyoxylate reductase in white mustard seedlings by phytochrome. *Naturwissenschaften* **57**, 132–133.

Vennesland, B. (1963). Some flavin interactions with grana (seen in a different light). In "Photosynthetic Mechanisms of Green Plants," pp. 421–435. Natl. Acad. Sci.—Natl. Res. Council, Washington, D. C.

Vernon, L. P., Ke, B., and Shaw, E. R. (1967). Relationship of P700, electron spin resonance signal, and photochemical activity of a small chloroplast particle obtained by the action of Triton X-100. *Biochemistry* **6**, 2210–2220.

Vernon, L. P., Yamamoto, H. Y., and Ogawa, T. (1969). Partially purified photosynthetic reaction centers from plant tissues. *Proc. Nat. Acad. Sci. U. S.* **63**, 911–917.

Vickery, H. B. (1952). The behavior of isocitric acid in excised leaves of *Bryophyllum calicynum* during culture in alternating light and darkness. *Plant Physiol.* **27,** 9–17.

Vickery, H. B. (1954). The effect of temperature on the behavior of malic acid and starch in leaves of *Bryophyllum calicynum* cultured in darkness. *Plant Physiol.* **29,** 385–392.

Vickery, H. B. (1961). Chemical investigations of the tobacco plant. XI. Composition of the green leaf in relation to position in the stalk. *Conn., Agr. Exp. Sta., New Haven, Bull.* **640.**

Vickery, H. B. (1963). The metabolism of the organic acids of tobacco leaves. XIX. Effect of culture of excised leaves in solutions of potassium glutamate. *J. Biol. Chem.* **238,** 2453–2459.

Vickery, H. B., and Palmer, J. K. (1956). The metabolism of the organic acids of tobacco leaves. XI. Effect of culture of excised leaves in solutions of glycolate at pH 3 to pH 6. *J. Biol. Chem.* **221,** 79–92.

Vickery, H. B., and Wilson, D. G. (1958). Preparation of potassium dihydrogen L_s (+)-isocitrate from *Bryophyllum calicynum* leaves. *J. Biol. Chem.* **233,** 14–17.

Vickery, H. B., and Zelitch, I. (1960). The metabolism of the organic acids of tobacco leaves. XVII. Effect of culture of excised leaves on solutions of potassium pyruvate. *J. Biol. Chem.* **235,** 1871–1875.

Villemez, C. L., Swanson, A. L., and Hassid, W. Z. (1966). Properties of a polygalacturonic acid-synthesizing enzyme system from *Phaseolus aureus* seedlings. *Arch. Biochem. Biophys.* **116,** 446–452.

Virgin, H. I. (1956). Light-induced stomatal transpiration of etiolated wheat leaves as related to chlorophyll a content. *Physiol. Plant.* **9,** 482–493.

Virgin, H. I. (1964). Some effects of light on chloroplasts and plant protoplasm. *In* "Photophysiology" (A. C. Giese, ed.), Vol. 1, pp. 273–303. Academic Press, New York.

von Wettstein, D. (1966). On the physiology of chloroplast structures. *In* "The Biochemistry of Chloroplasts" (T. W. Goodwin, ed.), Vol. 1, pp. 19–22. Academic Press, New York.

von Wettstein, D. (1967). Chloroplast structure and genetics. *In* "Harvesting the Sun" (A. San Pietro, F. A. Greer, and T. J. Army, eds.), pp. 153–190. Academic Press, New York.

Vose, J. R., and Spencer, M. (1967). Energy sources for photosynthetic carbon dioxide fixation. *Biochem. Biophys. Res. Commun.* **29,** 532–537.

Voskresenskaya, N. P., Wiil, Y. A., Grishina, G. S., and Pärnik, T. R. (1970a). Effect of oxygen concentration and light intensity on the distribution of labelled carbon in photosynthesis products in bean plants. *Photosynthetica* **4,** 1–8.

Voskresenskaya, N. P., Grishina, G. S., Chmora, S. N., and Poyarkova, N. M. (1970b). The influence of red and blue light on the rate of photosynthesis and the CO_2 compensation point at various oxygen concentrations. *Can. J. Bot.* **48,** 1251–1257.

Wadsworth, R. M. (1960). The effect of artificial wind on the growth rate of plants in water culture. *Ann. Bot. (London)* [N. S.] **24,** 200–211.

Waggoner, P. E. (1965). Calibration of a porometer in terms of diffusive resistance. *Agr. Meteorol.* **2,** 317–329.

Waggoner, P. E. (1967). Moisture loss through the boundary layer. *In* "Biometeorology" (S. W. Tromp and W. H. Wiehe, eds.), Vol. 3, pp. 41–52. Swets & Zeitlinger, Amsterdam.

Waggoner, P. E. (1969a). Predicting the effect upon net photosynthesis of changes in leaf metabolism and physics. *Crop Sci.* **9,** 315–321.

Waggoner, P. E. (1969b). Environmental manipulation for higher yields. *In* "Physiological Aspects of Crop Yield." (J. D. Eastin, *et al.,* eds.) pp. 343–373. A.S.A. and C.S.S.A., Madison, Wisconsin.

Waggoner, P. E., and Bravdo, B.-A. (1967). Stomata and the hydrologic cycle. *Proc. Nat. Acad. Sci. U. S.* **57,** 1096–1102.

Waggoner, P. E., and Shaw, R. H. (1952). Temperature of potato and tomato leaves. *Plant Physiol.* **27**, 710–724.

Waggoner, P. E., and Simmonds, N. W. (1966). Stomata and transpiration of droopy potatoes. *Plant Physiol.* **41**, 1268–1271.

Waggoner, P. E., and Zelitch, I. (1965). Transpiration and the stomata of leaves. *Science* **150**, 1413–1420.

Waggoner, P. E., Moss, D. N., and Hesketh, J. D. (1963). Radiation in the plant environment and photosynthesis. *Agron. J.* **55**, 36–39.

Waggoner, P. E., Monteith, J. L., and Szeicz, G. (1964). Decreasing transpiration of field plants by chemical closure of stomata. *Nature (London)* **201**, 97–98.

Waldron, J. C., Glasziou, K. T., and Bull, T. A. (1967). The physiology of sugarcane. IX. Factors affecting photosynthesis and sugar storage. *Aust. J. Biol. Sci.* **20**, 1043–1052.

Waldron, J. D., Gouers, D. S., Chibnall, A. C., and Piper, S. H. (1961). Further observations on the paraffins and primary alcohols of plant waxes. *Biochem. J.* **78**, 435–442.

Walker, D. A. (1962). Pyruvate carboxylation and plant metabolism. *Biol. Rev.* **37**, 215–256.

Walker, D. A. (1964). Improved rates of carbon dioxide fixation by illuminated chloroplasts. *Biochem. J.* **92**, 22c–23c.

Walker, D. A. (1965). Correlation between photosynthetic activity and membrane integrity in isolated pea chloroplasts. *Plant Physiol.* **40**, 1157–1161.

Walker, D. A. (1966). Carboxylation in plants. *Endeavour* **25**, 21–26.

Walker, D. A. (1967). Photosynthetic activity of isolated pea chloroplasts. *In* "The Biochemistry of Chloroplasts" (T. W. Goodwin, ed.), Vol. 2, pp. 53–69. Academic Press, New York.

Walker, D. A., and Brown, J. M. A. (1957). Physiological studies on acid metabolism. 5. Effects of carbon dioxide concentration on phosphoenolpyruvic carboxylase activity. *Biochem. J.* **67**, 79–83.

Walker, D. A., and Hill, R. (1967). The relation of oxygen evolution to carbon assimilation with isolated chloroplasts. *Biochim. Biophys. Acta* **131**, 330–338.

Walker, D. A., and Zelitch, I. (1963). Some effects of metabolic inhibitors, temperature, and anaerobic conditions on stomatal movement. *Plant Physiol.* **38**, 390–396.

Walker, D. A., Cockburn, W., and Baldry, C. W. (1967). Photosynthetic oxygen evolution by isolated chloroplasts in the presence of carbon cycle intermediates. *Nature (London)* **216**, 597–599.

Wallihan, E. F. (1964). Modification and use of an electric hygrometer for estimating relative stomatal apertures. *Plant Physiol.* **39**, 86–90.

Wang, D., and Waygood, E. R. (1962). Carbon metabolism of C^{14}-labeled amino acids in wheat leaves. I. A pathway of glyoxylate-serine metabolism. *Plant Physiol.* **37**, 826–832.

Warburg, O., and Krippahl, G. (1960). Glykolsäurebildung in *Chlorella. Z. Naturforsch. B* **15**, 197–199.

Warburg, O., Krippahl, G., Jetschmann, K., and Lehmann, A. (1963). Chemie der photosynthese. *Z. Naturforsch. B.* **18**, 837–844.

Warburg, O., Krippahl, G., and Lehmann, A. (1969). Chlorophyll catalysis and Einstein's law of photochemical equivalence in photosynthesis. *Amer. J. Bot.* **56**, 961–971.

Wareing, P. F., Khalifa, M. M., and Treharne, K. J. (1968). Rate-limiting processes in photosynthesis at saturating light intensities. *Nature (London)* **220**, 453–457.

Warren Wilson, J. (1966). High net assimilation rates of sunflower plants in an arid climate. *Ann. Bot. (London)* [N. S.] **30**, 745–751.

Warren Wilson, J. (1967). Stand structure and light penetration. III. Sunlit foliage area. *J. Appl. Ecol.* **4**, 159–165.

Wassink, E. C. (1959). Efficiency of light energy conversion in plant growth. *Plant Physiol.* **34,** 356–361.

Wassink, E. C. (1963). Photosynthesis. *Comp. Biochem.* **5,** Part C, 347–492.

Watson, D. J. (1952). The physiological basis of variation in yield. *Advan. Agron.* **4,** 101–145.

Watson, D. J. (1958). The dependence of net assimilation rate on leaf-area index. *Ann. Bot.* (*London*) [N. S.] **22,** 37–54.

Watson, J. D. (1965). "Molecular Biology of the Gene." Benjamin, New York.

Waygood, E. R., Mache, R., and Tan, C. K. (1969). Carbon dioxide, the substrate for phosphoenolpyruvate carboxylase from leaves of maize. *Can. J. Bot.* **47,** 1455–1458.

Weissbach, A., Horecker, B. L., and Hurwitz, J. (1956). The enzymatic formation of phosphoglyceric acid from ribulose diphosphate and carbon dioxide. *J. Biol. Chem.* **218,** 795–810.

Wessels, J. S. C. (1963). Separation of the two photochemical systems of photosynthesis by digitonin fragmentation of spinach chloroplasts. *Proc. Roy. Soc., Ser. B* **157,** 345–355.

West, J., and Hill, R. (1967). Carbon dioxide and the reduction of indophenol and ferricyanide by chloroplasts. *Plant Physiol.* **42,** 819–826.

West, K. R., and Wiskich, J. T. (1969). The action of DIO-9 on photophosphorylation. *FEBS Lett.* **3,** 247.

Westlake, D. F. (1963). Comparisons of plant productivity. *Biol. Rev.* **38,** 385–425.

Whittingham, C. P., and Keys, A. J. (1969). Photosynthetic fixation in tobacco leaves. *Abstr. Int. Bot. Congr., 11th, 1969,* p. 237.

Whittingham, C. P., Hiller, R. G., and Bermingham, M. (1963). The production of glycollate during photosynthesis. *In* "Photosynthetic Mechanisms of Green Plants," pp. 675–683. Natl. Acad. Sci.—Natl. Res. Council, Washington, D. C.

Whittingham, C. P., Coombs, J., and Marker, A. F. H. (1967). The role of glycollate in photosynthetic carbon fixation. *In* "The Biochemistry of Chloroplasts" (T. W. Goodwin, ed.), Vol. 2, pp. 155–173. Academic Press, New York.

Widholm, J. M., and Ogren, W. L. (1969). Photorespiratory-induced senescence of plants under conditions of low carbon dioxide. *Proc. Nat. Acad. Sci. U. S.* **63,** 668–673.

Wildman, S. G., Cheo, C. C., and Bonner, J. (1949). The proteins of green leaves. III. Evidence of the formation of tobacco mosaic virus protein at the expense of a main protein component in tobacco leaf cytoplasm. *J. Biol. Chem.* **180,** 985–1001.

Williams, G. R., and Novelli, G. D. (1964). Stimulation of an *in vitro* amino acid incorporating system by illumination of dark-grown plants. *Biochem. Biophys. Res. Commun.* **17,** 23–27.

Willis, J. E., and Sallach, H. J. (1963). Serine biosynthesis from hydroxypyruvate in plants. *Phytochemistry* **2,** 23–28.

Wilson, A. T., and Calvin, M. (1955). The photosynthetic cycle. CO_2 dependent transients. *J. Amer. Chem. Soc.* **77,** 5948–5957.

Wilson, D. G., King, K. W., and Burris, R. H. (1954). Transamination reactions in plants. *J. Biol. Chem.* **208,** 863–874.

Winget, G. D., Izawa, S., and Good, N. E. (1965). The stoichiometry of photophosphorylation. *Biochem. Biophys. Res. Commun.* **21,** 438–443.

Winter, H., and Mortimer, D. C. (1967). Role of the root in the translocation of products of photosynthesis in sugarbeet, soybean, and pumpkin. *Can. J. Bot.* **45,** 1811–1822.

Wintermans, J. F. G. M., Helmsing, P. J., Polman, B. J. J., Van Gisbergen, J., and Collard, J. (1969). Galactolipid transformations and photochemical activities of spinach chloroplasts. *Biochim. Biophys. Acta* **189,** 95–105.

Wishnick, M., and Lane, M. D. (1969). Inhibition of ribulose diphosphate carboxylase by cyanide. Inactive ternary complex of enzyme, ribulose diphosphate, and cyanide. *J. Biol. Chem.* **244**, 55–59.

Wishnick, M., Lane, M. D., Scrutton, M. C., and Mildvan, A. S. (1969). The presence of tightly bound copper in ribulose diphosphate carboxylase from spinach. *J. Biol. Chem.* **244**, 5761–5763.

Wittwer, S. H., and Robb, W. (1964). Carbon dioxide enrichment of greenhouse atmospheres for food crop production. *Econ. Bot.* **18**, 34–56.

Wolf, F. T. (1969). Plants with high rates of photosynthesis. *Biologist* **51**, 147–155.

Wood, H. G., and Utter, M. F. (1965). The role of CO_2 fixation in metabolism. *Essays Biochem.* **1**, 1–27.

Woodward, R. B. *et al.,* (1960). The total synthesis of chlorophyll. *J. Amer. Chem. Soc.* **82**, 3800–3802.

Yagi, T., and Benson, A. A. (1962). Plant sulfolipid. V. Lysosulfolipid formation. *Biochim. Biophys. Acta* **57**, 601–603.

Yamamoto, Y., and Beevers, H. (1960). Malate synthetase in higher plants *Plant Physiol.* **35**, 102–108.

Yamashita, K., Itoh, M., and Shibata, K. (1969). Activation by manganese of photochemical oxygen evolution and $NADP^+$ photoreduction in chloroplasts. *Biochim. Biophys. Acta* **189**, 133–135.

Yamashita, T., and Butler, W. L. (1968). Photoreduction and photophosphorylation with Tris-washed chloroplasts. *Plant Physiol.* **43**, 1978–1986.

Yemm, E. W., and Bidwell, R. G. S. (1969). Carbon dioxide exchange in leaves. I. Discrimination between $^{14}CO_2$ and $^{12}CO_2$ in photosynthesis. *Plant Physiol.* **44**, 1328–1334.

Yocum, C. F., and San Pietro, A. (1969). Ferredoxin-reducing substance (FRS) from spinach. *Biochem. Biophys. Res. Commun.* **36**, 614–620.

Yocum, C. F., and San Pietro, A. (1970). Ferredoxin-reducing substance (FRS) from spinach. II. Separation and assay. *Arch. Biochem. Biophys.* **140**, 152–157.

Yocum, C. S., Allen, L. H., and Lemon, E. R. (1964). Photosynthesis under field conditions. VI. Solar radiation balance and photosynthetic efficiency. *Agron. J.* **56**, 249–253.

Zak, E. G. (1965). Effect of molecular oxygen on the formation of amino acids in photosynthesizing *Chlorella* under various conditions of illumination. *Fiziol. Rast.* **12**, 263–267.

Zelawski, W. (1967). A contribution to the question of the CO_2- evolution during photosynthesis in dependence on light intensity. *Bull. Acad. Pol. Sci., Ser. Sci. Biol.* **15**, 565–570.

Zelitch, I. (1953). Oxidation and reduction of glycolic and glyoxylic acids in plants. II. Glyoxylic acid reductase. *J. Biol. Chem.* **201**, 719–726.

Zelitch, I. (1955). The isolation and action of crystalline glyoxylic acid reductase from tobacco leaves. *J. Biol. Chem.* **216**, 553–575.

Zelitch, I. (1957). α-Hydroxysulfonates as inhibitors of the enzymatic oxidation of glycolic and lactic acids. *J. Biol. Chem.* **224**, 251–260.

Zelitch, I. (1958). The role of glycolic acid oxidase in the respiration of leaves. *J. Biol. Chem.* **233**, 1299–1303.

Zelitch, I. (1959). The relationship of glycolic acid to respiration and photosynthesis in tobacco leaves. *J. Biol. Chem.* **234**, 3077–3081.

Zelitch, I. (1961). Biochemical control of stomatal opening in leaves. *Proc. Nat. Acad. Sci. U. S.* **47**, 1423–1433.

Zelitch, I. (1963). The control and mechanisms of stomatal movement. *Conn., Agr. Exp. Sta., New Haven, Bull.* **664**, 18–42.

Zelitch, I. (1964a). Reduction of transpiration of leaves through stomatal closure induced by alkenylsuccinic acid. *Science* **143**, 692–693.

Zelitch, I. (1964b). Organic acids and respiration in photosynthetic tissues. *Annu. Rev. Plant Physiol.* **15**, 121–142.

Zelitch, I. (1965a). The relation of glycolic acid synthesis to the primary photosynthetic carboxylation reaction in leaves. *J. Biol. Chem.* **240**, 1869–1876.

Zelitch, I. (1965b). Environmental and biochemical control of stomatal movement in leaves. *Biol. Rev.* **40**, 463–482.

Zelitch, I. (1966). Increased rate of net photosynthetic carbon dioxide uptake caused by the inhibition of glycolate oxidase. *Plant Physiol.* **41**, 1623–1631.

Zelitch, I. (1967). Control of leaf stomata: Their role in transpiration and photosynthesis. *Amer. Sci.* **55**, 472–486.

Zelitch, I. (1968). Investigations on photorespiration with a sensitive ^{14}C-assay. *Plant Physiol.* **43**, 1829–1837.

Zelitch, I. (1969a). Stomatal control. *Annu. Rev. Plant Physiol.* **20**, 329–350.

Zelitch, I. (1969b). Mechanisms of carbon fixation and associated physiological responses. *In* "Physiological Aspects of Crop Yield," (J. D. Eastin, *et al.*, eds.) pp. 207–226. A.S.A. and C.S.S.A., Madison, Wisconsin.

Zelitch, I., and Barber, G. A. (1960). Oxidative phosphorylation and glycolate oxidation by particles from spinach leaves. *Plant Physiol.* **35**, 205–209.

Zelitch, I., and Day, P. R. (1968a). Glycolate oxidase activity in algae. *Plant Physiol.* **43**, 289–291.

Zelitch, I., and Day, P. R. (1968b). Variation in photorespiration. The effect of genetic differences in photorespiration on net photosynthesis in tobacco. *Plant Physiol.* **43**, 1838–1844.

Zelitch, I., and Gotto, A. M. (1962). Properties of a new glyoxylate reductase from leaves. *Biochem. J.* **84**, 541–546.

Zelitch, I., and Ochoa, S. (1953). Oxidation and reduction of glycolic and glyoxylic acids in plants. I. Glycolic acid oxidase. *J. Biol. Chem.* **201**, 707–718.

Zelitch, I., and Waggoner, P. E. (1962a). Effect of chemical control of stomata on transpiration and photosynthesis. *Proc. Nat. Acad. Sci. U. S.* **48**, 1101–1108.

Zelitch, I., and Waggoner, P. E. (1962b). Effect of chemical control of stomata on transpiration of intact plants. *Proc. Nat. Acad. Sci. U. S.* **48**, 1297–1299.

Zelitch, I., and Walker, D. A. (1964). The role of glycolic acid metabolism in opening of leaf stomata. *Plant Physiol.* **39**, 856–862.

Zelitch, I., and Zucker, M. (1958). Changes in oxidative enzyme activity during the curing of Connecticut shade tobacco. *Plant Physiol.* **33**, 151–155.

Zucker, M. (1963). Experimental morphology of stomata. *Conn., Agr. Exp. Sta., New Haven, Bull.* **664**, 1–17.

Zucker, M., and Stinson, H. T., Jr. (1962). Chloroplasts as the major protein-bearing structures in *Oenothera* leaves (evening primrose). *Arch. Biochem. Biophys.* **96**, 637–644.

Zweig, G., and Avron, M. (1965). On the oxidation-reduction potential of the photoproduced reductant of isolated chloroplasts. *Biochem. Biophys. Res. Commun.* **19**, 397–400.

Subject Index

A

Acetabularia
  nucleic acids in enucleated cells of, 44, 46
  photorespiration in, 190
Alfalfa
  diffusive resistance, 237
  productivity of, 270, 292
  rates of CO_2 assimilation in, 278, 280, 285
  sulfolipase of, 56
Algae
  cyclic photophosphorylation in hydrogen adapted, 77
  ferredoxin from, 75
  glycolic acid metabolism in, 189–190
  manganese-deficient, 73
  peroxisomes in, 16
  productivity of, 270, 275
Amaranthus
  $^{14}CO_2$ labeling in, 109

glycolic acid metabolism, 188
phosphoenolpyruvate carboxylase, 108
photorespiration in leaves of, 164
photosynthetic quotient in leaves, 128–129
rates of CO_2 assimilation in, 244, 254
Atriplex
  $^{14}CO_2$ labeling in, 110
  glycolic acid metabolism, 188, 198
  O_2 concentration on CO_2 assimilation in, 144
  phosphoenolpyruvate carboxylase, 108
  photorespiration in leaves of, 154, 156, 164
  rates of CO_2 assimilation in, 254

B

Barley
  ^{14}C distribution after photosynthesis in $^{14}CO_2$, 94, 139